AF777903

In Quest of Mineral Wealth

This formal ceremonial headdress, believed to have been worn by a Middle Sicán nobleman (ca. A.D. 1000), was buried at the bottom of an 11-meter-deep shafttomb at the north base of the Huaca Loro pyramid in the Batán Grande region, northern Peru.

It was assembled from a gold mask, biconical forehead ornament, cylindrical crown, parabolic head ornament, and 90 gold feathers excavated from this tomb. Together they measure ca. 100 cm x 60 cm and are estimated to have originally weighed about 2 kg. Neutron activation analysis of rub samples indicate that the components were made from 10-15 karat gold sheets. The headdress as a whole amply displays unparalleled mastery of Sicán gold sheetmetal working.

Photo by Yutaka Yoshii.

In Quest of Mineral Wealth

Aboriginal and Colonial Mining and Metallurgy in Spanish America

edited by

Alan K. Craig
and Robert C. West

GEOSCIENCE AND MAN, vol. 33

Geoscience Publications
Department of Geography and Anthropology
Louisiana State University
Baton Rouge, LA 70893-6010

The Geoscience and Man series of symposia, monographs, and collections of papers in geography, anthropology and geology is published and distributed by Geoscience Publications, Department of Geography and Anthropology, Louisiana State University. A list of works from Geoscience Publications can be found at the back of this volume.

Series Co-Editors: Kent Mathewson and Gregory Stone

Geoscience Publications Committee: Chair, Jay D. Edwards, associate professor of anthropology; ex officio, Carville Earle, chair of the Department of Geography and Anthropology; Nina S. Lam, professor of geography; Kent Mathewson, associate professor of geography; M. Jill Brody, associate professor of anthropology; Anne Mosher, assistant professor of geography; Gregory Stone, assistant professor of geography; Joyce Jackson, assistant professor of anthropology; Mary Lee Eggart, research associate, art director; and Esther Shaffer, managing editor.

© Copyright 1994 Department of Geography and Anthropology, Louisiana State University, Baton Rouge, LA 70893-6010. All rights reserved.

Library of Congress Cataloging-in-Publication Data

In quest of mineral wealth: aboriginal and colonial mining and metallurgy in Spanish America / edited by Alan K. Craig and Robert C. West
p. cm.—(Geoscience and man : v.33)
Includes bibliographical references.
ISBN 0-938909-57-6
1. Mines and mineral resources—Latin America—History—To 1830. 2. Metallurgy—Latin America—History—To 1830. 3. Human geography—Latin America. I. Craig, Alan K. II. West, Robert Cooper, 1913- . III. Series: Geoscience and man : v. 33.
TN27.5.A23 1994
669'.098—dc20 94-19560

COVER DESIGN: Mary Lee Eggart. Artwork: adapted from an illustration depicting a means of mixing mercury with comminuted silver ore in patio process, Guanajuato, Mexico, early 19th century. Reproduced from the cover article of *El Movimiento Histórico en México*, Sept. 1, 1940, no. 5.

Contents

Introduction

Both editors of this volume of Geoscience and Man have had long-term interests in mining and its effects on the physical and cultural landscape of Spanish America. Our combined experience includes both commercial and scholarly efforts, and it was easy to agree in 1990 to organize a symposium on Latin American mining as part of the 47th Congress of Americanists held in New Orleans the following year. Publishing the results has been more challenging than either of us would have predicted, yet more satisfying than we might have expected. This is largely because, in an era of facile talk about the need for inter-disciplinary research, we have been fortunate to actually find it. Our appeal for international collaboration has been met by the participants in many ways that have made our tasks more rewarding than they otherwise might have been. We sincerely thank all of the participants, as well as the editorial committee of Geoscience Publications for allowing us the opportunity to explore a subject that has profoundly impacted the political economy of the Americas and, indeed, much of the world.

Scholarly leadership in the field of mining has never been in the hands of geographers. Since the great polymath Alexander von Humboldt visited much of Spanish America, it has been historians, and lately economic historians, who have made the most numerous and significant contributions. We are pleased, however, to support a vigorous group of young archaeologists whose discoveries of New World prehistoric metallurgical technology show that it reached unexpected levels of organization and sophistication prior to the Conquest. Moreover, our invitations were also accepted by a new wave of Spanish scholars determined to clarify the transfers of technology between Old World and New.

We are also pleased to acknowledge the existence of a thriving group of Mexican scholars whose interest in mining also was coincidentally formalized in 1990. Since then they have made great progress in a series of three inter-American meetings sponsored by the Instituto Nacional de Antropología e Historia, Mexico City, under the direction of Dra. Inés Herrera.

In Quest of Mineral Wealth: Aboriginal and Colonial Mining and Metallurgy in Spanish America, edited by Alan K. Craig and Robert C. West, 1994. Geoscience and Man, vol. 33, pp. 1-4. Dept. of Geography & Anthropology, Louisiana State University, Baton Rouge, LA 70893-6010, U.S.A.

The 19 chapters in this volume deal mainly with the extraction and processing of the precious metals, gold and silver, although one essay considers the mining of mercury in Spain; another, the aboriginal exploitation of turquoise in Mexico; and still another describes early extraction of petroleum for the manufacture of tar in coastal Ecuador. Most of the essays stress ore processing and the metallurgical technologies; less material is presented on methods of underground mining and the problem of mine drainage.

Those in Part 1 examine prehistoric aboriginal mining and metallurgy in Spanish America; those dealing with the exploitation of metals during the colonial period have been placed in Part 2.

Based on review of recent literature, a brief article by geographer Robert West on aboriginal metallurgy and metalworking in Spanish America opens Part 1. This is followed by archaeologist Phil Weigand's essay on prehistoric exploitation of turquoise in Mexico and southwestern United States, which reveals some of the best evidence of aboriginal subsurface mining techniques as well as the significance of the trading system based on this mineral among the early civilizations of Mesoamerica. As one of the leading authorities on the development of pre-Spanish metallurgy in South America, Izumi Shimada outlines recent archaeological finds in Peru, which indicate the surprising sophistication in metalworking that was invented by the aboriginal smiths. Archaeologist Gray Graffam and colleagues present evidence of early prehistoric copper mining in the hyperarid Atacama Desert of northern Chile. To conclude Part 1, historian Susan Ramírez briefly analyses the late Inkan socioeconomic and political control of mining and metallurgy in northern Peru, as seen by early Spanish colonists who recorded their observations in documents currently housed in various Hispanic archives.

Part 2 opens with articles on colonial mining of silver in Mexico, where exploitation of the metal dates from the early sixteenth century. Historian Otis Young reviews subsurface mining and ore processing used by Spaniards, concluding that the ridicule later heaped on their supposed primitive practices by Europeans was unwarranted. The earliest silver production by Spaniards in central Mexico (1531-1555) with the opening of the mines of Taxco and Sultepec, where German practices were first introduced, is reviewed by Robert West, followed by an essay by engineer-historian Ramón Sánchez Flores on the various tools and installations used in the amalgamation of silver ores that was perfected in 1555 by the Spaniard Bartolomé de Medina — a process that revolutionized silver production in Colonial Spanish America.

In his essay Alvaro Sánchez-Crispín deals with the geography of mining settlement in Mexico from the sixteenth through the eighteenth centuries, indicating the advance northward from central into northern New

Spain; he emphasizes the thesis that the discovery of silver mines encouraged the development of surrounding farms and livestock ranches to supply the mining population. By contrast, in the following article historian Salvador Alvarez shows that in central Chihuahua, northern New Spain, farms and ranches were established prior to the development of mining, a settlement process thought to be necessary to ensure the continuation of mining activity under the control of rich *hacendados*.

The three essays that follow consider, in part, the transfer of mining techniques between Europe and Spanish America. Like Otis Young, previously mentioned, historians Julio Sánchez Gómez and Guillermo Mira Delli-Zotti, both of Spain, discount the perception that late eighteenth-century engineers from Europe held in regard to the primitiveness of mining practices in the New World. But they point out the lack of administrative expertise in the colonial mining industry, as compared to that in Europe. In his essay the Spanish scholar José Luis Peset traces the transfer of mining administrative practices from Europe to New Spain in the late eighteenth century with the establishment of the School of Mines in Mexico City, staffed with European professors and under the guidance of the Basque chemist Fausto de Elhuyar. Although different in subject matter, the article by economic historian Rafael Dobado González on the mercury mines of Almadén in Spain emphasizes the importance of Spanish mercury production for the silver output in Mexico, where amalgamation was employed in practically all mines, especially those of Zacatecas, Guanajuato and Pachuca.

Anthropologist-historian María Clemencia Ramírez de Jara offers the only study on gold mining in New Granada (Colombia), the main producer of that precious metal in Spanish America. She argues that the mining frontier was closely associated with the Church missionary frontier in southwestern Colombia, and fluctuated according to the degree of Indian hostility.

Four essays present various aspects of colonial mining and ore processing in the Viceroyalty of Peru, which vied with New Spain as the major producer of silver in Spanish America. Geologist-geographer Alan Craig describes sixteenth-century developments in ore processing at Potosí with emphasis on the use of hydraulic stamp mills in relation to water supply, the disastrous floods that caused closure of many mills at the close of the century, and the gradual decline of silver production thereafter. Closely related is historian Peter Bakewell's account of the first contracts made for the construction of hydraulic stamp mills in Potosí after the introduction of the amalgamation process in the 1570s; using detailed specifications found in documents of the period, he was able to reconstruct drawings depicting the operation of the mills. The little-known silver mines of Huantajaya in the Atacama Desert are portrayed by historian

Kendall Brown and Alan Craig, who show that, despite adverse environmental conditions, production was substantial, especially during the last half of the eighteenth century. The attempt of the Spanish crown to modernize mining technology and administration in Peru during the late eighteenth century by dispatching a team of German specialists is examined by historian John Fisher; this episode parallels similar attempts in Mexico, described by José Luis Peset, previously mentioned.

Finally, archaeologist Karen Stothert's article on colonial and post-colonial exploitation of shallow oil deposits in coastal Ecuador for the manufacture of tar (abbreviated by the editors to conform to the parameters of this volume) describes one of the earliest commercial uses of the mineral petroleum in the New World.

In sum, among the various aspects involving the quest for minerals in Spanish America during the pre-Hispanic and colonial periods, the following facts stand out: one is the sophisticated metallurgical techniques developed by the aboriginal metalsmiths; another, the emphasis that they and the elite placed on the sacredness of mines and the religious significance of copper, silver, gold and their alloys. Moreover, aboriginal knowledge of mining and metallurgy proved to be essential to Spaniards in their quest for gold and silver during the early colonial period, when Indian laborers were used to prospect for ore deposits and to work in the smelters. But the crowning Spanish-European contribution was the perfection of the amalgamation process to reduce low-grade silver ores in Mexico and Peru, which led to the great wealth that Spain derived from its American colonies and which for nearly 350 years (1550s-1890s) was used to produce much of the world's silver.

Alan K. Craig
Boca Raton, Florida

Robert C. West
Baton Rouge, Louisiana

PART 1: ABORIGINAL

Aboriginal Metallurgy and Metalworking in Spanish America: A Brief Overview

Robert C. West, Louisiana State University
Baton Rouge, Louisiana, U.S.A.

Abstract

Prior to the Spanish Conquest, knowledge of metallurgy and metalworking had evolved and spread over much of the area occupied by high Indian civilizations in the New World. Copper, gold, and silver and their alloys were the main metals involved. These were fashioned mainly as ornaments used in religious ceremony and for the enhancement of elite cultural status; the manufacture of metal tools and weapons was secondary and evolved relatively late. Archaeological investigation has indicated that mining practices appear to have been rudimentary, whereas metallurgical techniques were relatively advanced and complex. Sophisticated metalworking first developed in Andean South America, from which metallurgical knowledge spread northward as far as Mexico.
Key words: pre-Hispanic mining, pre-Hispanic metallurgy, Spanish America

At the time of the Spanish Conquest, three main areas in which metallurgy and metalworking were well developed included (1) the Peruvian Center, (2) the Colombian-Lower Central American Center, and (3) the Mexican Center, as shown in figure 1.[1]

The Peruvian Center

Aboriginal metallurgy was first developed in the Peruvian Center, which encompassed most of the central Andean highlands and adjacent Pacific coast, as well as the Altiplano of Alto Perú (Bolivia) with extensions southward along the Atacama desert of northern Chile and the mountain valleys of northwestern Argentina. Recent archaeological finds suggest that metallurgy first arose in the central Andean area ca. 1500 B.C. (Burger 1992, 98, 127). Practices involved the use of gold, obtained from placer deposits, and either native or oxidized copper carbonate ores mined from

In Quest of Mineral Wealth: Aboriginal and Colonial Mining and Metallurgy in Spanish America, edited by Alan K. Craig and Robert C. West, 1994. Geoscience and Man, vol. 33, pp. 5-20. Dept. of Geography & Anthropology, Louisiana State University, Baton Rouge, LA 70893-6010, U.S.A.

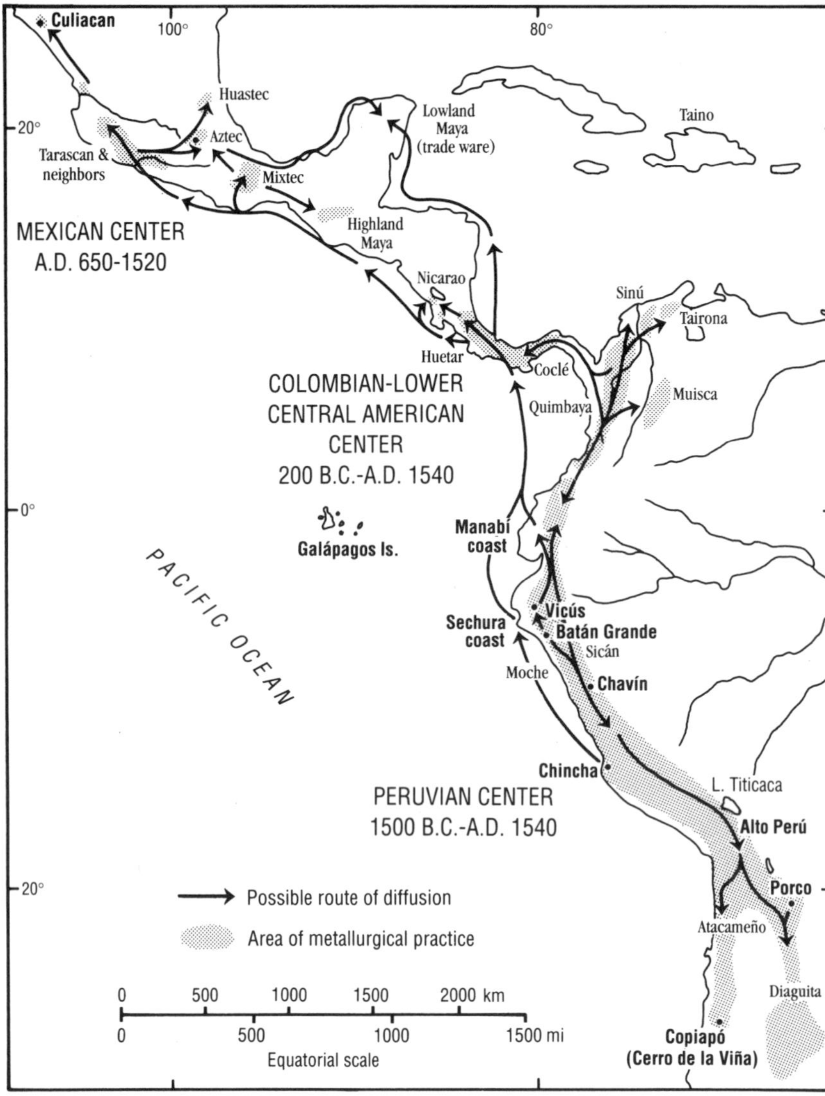

Fig. 1. Pre-Hispanic metallurgy in South and Middle America.

surficial deposits and easily smelted. Through hammering and annealing, the native and refined metals were made into small thin sheets. These were ornamented with designs in relief or used to fashion objects in the round, the sheets being joined by welding or by crimping the edges. These processes became the major metalworking techniques of the Peruvian

Center (Lechtman 1980), although various forms of casting were developed in certain areas. By the beginning of the Christian era, metalsmiths along the northern Peruvian coast, especially within the Moche culture area, perfected techniques of gilding and silvering objects made of copper-silver-gold (*tumbaga*) and copper-gold-silver alloys so that the surface resembled the sacred precious metals of gold and silver, processes termed "surface metallurgy" by some investigators (Lechtman 1984a).[2]

During the period A.D. 1000-1476 (Late Intermediate), alloys of copper-arsenic (in coastal Peru) and copper-tin (in the Bolivian highlands) were smelted to form true bronzes. At the site of Batán Grande in the coastal desert of northern Peru, archaeologists in 1982-83 found evidence of smelting copper-arsenic ores from which the aboriginal metalsmiths produced an arsenical bronze that was cast in molds of fired clay. First developed around A.D. 900, this and other metallurgical techniques involving the use of copper ores persisted in this area for perhaps 600 years until the Spanish Conquest, making it one of the major aboriginal mining and metallurgical centers in the New World (Shimada et al. 1982; Shimada and Merkel 1991; Shimada, this volume; Lechtman, 1988, 348). In the altiplano of southern Peru and western Bolivia, where copper and tin ores are abundant, true bronze, made from copper-tin alloys had become common by A.D. 1000 and was used for casting both ornaments and implements. Later, among the Inka it was the preferred alloy for the manufacture of tools and weapons that were spread throughout the Empire, giving the commoners access to new domestic utensils and improved weapons for the Inkan armies. Gold and silver, however, remained as sacred metals for the Inkan dynasty, priests, and nobility (Bennett 1946a, 28; Lechtman 1980, 1984b).

In terms of present archaeological evidence it is difficult to determine the place or places of origin of metallurgy and the diffusion of techniques within the Peruvian Center. Excavations of Initial Period sites (1800-1000 B.C.) have revealed tiny sheets of hammered gold foil; and expansion of the Chavín religious cult over much of the northern Central Andes during the Early Horizon Period (1000-200 B.C.) introduced the use of large objects of hammered gold pieces embossed with Chavín-style motifs, examples of which have been found at various sites (Burger 1992, 201-202). The famous Mochica gold and copper-gold alloy pieces fashioned in the round from hammered sheet metal may have stemmed from practices developed earlier. Later such techniques may have been introduced into the highlands of southern Ecuador where graves contained spectacular gold and gilded copper objects (Collier 1946, 777; Murra 1946, 799). In northwestern Argentina the metallurgical practices of the Diaguita people, including the manufacture of copper-tin bronze, appear to have had their origin in the Altiplano of Bolivia (Marqués Miranda 1946, 646; González 1979, 187-188).

The Colombian-Lower Central American Center

In contrast to the mountainous highlands and desert coastal milieu of metallurgy in the Peruvian Center, that in the Columbian-Lower Central American Center arose in the tropical lowland valleys between the northern Andean ranges. It extended from northwestern Ecuador to northern Colombia and eventually into Panama and Costa Rica. Whereas the Peruvian metalsmiths worked mainly with copper and its alloys, the Colombians and Central Americans concentrated on gold and a gold-copper alloy known as *tumbaga* (Bennett 1946b; Root 1961). The Colombians perfected the lost-wax (*cire-perdue*) process of casting finely fashioned figurines, masks, breastplates and other ornaments, all of which served as objects within the various chiefdoms and were accumulated by the political and religious elite to be used in ceremonies and as offerings to deities; such objects were also buried with the elite on their death.

Gold was obtained mainly from placer deposits, abundant in and along streams tributary to the Cauca and lower Magdalena rivers. Indians also extracted substantial amounts of precious metals from lode deposits (gold and silver ores), where archaeological evidence of pre-Hispanic mining has been found at the Cerro de Buriticá in Antioquia and at Mariquita near the Magdalena Valley (Restrepo 1888, 8; West 1952, 54). Copper deposits in Colombia are few, but thin ingots of refined copper, called *naipes*, may have been traded in from northern Peru (A. Craig, pers. comm. 1993). Gold alloyed with various proportions of silver and copper produced the *tumbaga* that was the prime metal used by the Colombian smiths.[3]

As in Peru, in Colombia the earliest known metal objects were formed by hammering and possibly by annealing gold to produce thin sheets that were decorated by embossing (repoussé) (Root 1961, 243). Possibly by A.D. 100 hollow casting of gold and *tumbaga* objects by means of the lost-wax process was perfected by the Quimbaya, who lived along the middle Cauca Valley and who are thought to have been the leading goldsmiths of Colombia (Root 1961, 254). They also achieved the techniques of gilding objects made of gold-rich *tumbaga* so that the surface resembled pure gold in color (*mise-en-couleur*). A similar technique was common among the Mochica in northern Peru, who habitually gilded objects made of gold-copper alloys. The period A.D. 400-700 is considered by some to have been the "classical period in Colombia metallurgy," when the Quimbaya were producing their best goldwork (Root 1961, 255). Thereafter similar techniques appear to have diffused into the Caribbean coastal region where the Sinú carried on the Quimbaya metallurgical tradition. Even the Muisca and Tairona in the northern Andean highlands adopted similar metalcrafts, importing native gold from the Cauca Valley and lower Magdalena River area through trade. Colombian metallurgical techniques probably spread into Panama and Costa Rica by trade overland and via rivers through the present Colombia-Panama border (Bray 1984, 305).

Early Spanish accounts indicate a wealth of gold objects among the Coclé and other aboriginal groups in Panama and Costa Rica. Undoubtedly many of these objects had been traded in from the Sinú, but others carry local metalworking styles that incorporate regional designs adopted from motifs on local pottery. (Bray 1984, 326; Cooke 1984, 292). Such evidence suggests that natives were using practices such as casting and gilding to produce the large quantity of metal artifacts that Spaniards looted from burials in central Panama.[4]

Emergence of goldworking in central Panama dates from A.D. 200-300 (Graham 1993, 12) and in Costa Rica at least by A.D. 700, if not earlier (Stone and Balser 1967). Much later, Central American techniques of goldsmithing had reached into southwestern Nicaragua, where the Nicarao, Nahuat-speaking immigrants from Mexico, were famed for their exquisite goldcraft, as reported by early Spanish observers (Stanislawski 1983, 7).

Chronological dates that have been suggested for presence of metallurgical practices in certain areas of Colombia and lower Central America must be highly speculative. Except for the Caribbean lowlands of Colombia there have been few stratigraphic excavations undertaken. One of the earliest carbon 14 (^{14}C) dates based on charcoal taken from a site associated with gold artifacts in the Quimbaya area corresponds to A.D. 110. Probably all of the gold and *tumbaga* objects contained in private collections and in the famous Museo del Oro in Bogotá have been purchased from *guaqueros* (professional grave looters), who are often unsure of the provenience of items, much less the stratigraphic information needed to establish a chronological framework of reference (Dussán de Reichel 1979, 50).

The Mexican Center

In the Mexican Center, metallurgy and metalworking developed much later than in the two areas previously discussed, West Mexico being the earliest locus of the craft in Mesoamerica. According to Dorothy Hosler, who lately has been the prime investigator of metallurgy in Mesoamerica, A.D. 650 is "the earliest secure date for a metal object found in West Mexico." (Hosler and Stresser-Pean 1992, 1219, n. 5; see also Hosler, 1988a, 1988b, 1988c, 1990).

As in Peru, metallurgy and metalworking in West Mexico, especially among the Tarascans and their neighbors, was based mainly on copper and its alloys. Although some utilitarian implements, such as needles and fishhooks were made, most metal objects were considered to be sacred, to be used for adornment in religious ceremonies and to enhance the social and political status of the elite. (Pollard 1987; Hosler 1988a).

Deposits of native copper and copper ores were abundant in western Mexico. Gold was placered from deposits along streams flowing from highland to the coast in Colima and Michoacán, and silver ores may have been mined from small deposits in the interior. The most frequent copper objects fashioned were small bells cast with the lost-wax method. These were used by priests to imitate sounds associated with certain deities while performing religious rites. According to Hosler, during an initial period of metallurgy that lasted ca. A.D. 650 to 1200 or 1300, the west Mexican metalsmiths worked almost exclusively with native copper and easily smelted oxidized copper ores. But from 1300 to the Spanish Conquest in 1521 they worked with alloys of copper-gold and copper-silver to produce objects such as small discs, bracelets, diadems, and masks that were gilded or silvered by hammering and annealing to form a gold or silver surface that represented the sacred colors, much as did the Peruvian and Colombian metalsmiths in South America. Such objects were worn by priests and members of the elite during religious ceremonies, as described in the *Relación de Michoacán* written just after the Spanish Conquest in the early 1520s (Craine and Reindorp 1970). Accumulated for generations, these religious items formed the large Tarascan "treasures" that Spaniards looted from storerooms and graves in the Lake Pátzcuaro area of Michoacán during the Conquest.[5] Because of the surface color of the objects, the Spaniards probably thought them to be of pure gold or silver, but when melted down and assayed the resulting bullion carried low percentages of the precious metals, much to the chagrin of the looters, who labeled their take "*plata baja*" or "*oro de baja ley*." Few if any objects of pure gold or silver from West Mexico exist today in private or museum collections and only small pieces of hammered or cast gold have been found in tombs untouched by looters (Borbolla 1944).

The metalsmiths of West Mexico also alloyed copper with arsenic ores and copper with tin to produce bronze. Both ceremonial and utilitarian objects were made of bronze, which was stronger and more durable than copper or copper-gold and silver alloys. Bells and other sacred pieces cast with copper-tin alloy produced a bronze with a golden color, while those of arsenic bronze gave a silvery hue (Hosler 1988a; Hosler and Stresser-Pean 1992). Bronze objects of practical use included needles, fishhooks, tweezers, axeheads, awls, and (possibly) agricultural coa blades.

The relatively late date of the appearance of metals in a peripheral area of Mesoamerica and the similarity of the techniques employed by the native metalsmiths to those developed in South America have led many scholars to suggest that metal objects and metallurgical ideas were introduced into West Mexico from Peru and Ecuador by traders using watercraft capable of long-distance voyages along the Pacific coast of South and Central America and southwestern Mexico (Arsandaux and Rivet 1921,

275; Willey 1955, 584-585; Coe 1962, 132; Meighan 1969, 14; Mountjoy 1969, 26). Moreover, the conspicuous absence of early use of metals in northern Central America and southeastern Mexico strengthens the possibility of diffusion from South America by sea rather than by land. Striking similarities occur between other material traits of coastal Ecuador and parts of West Mexico, but are absent in intervening areas. For example, Kelly (1970) refers to the presence of stirrup-spout pots in archaeological sites in Colima, a distinctive pottery style found also in Ecuador and northern Peru, where it was the hallmark of Moche culture (100 B.C.-A.D. 800) and which probably originated in the Chavín period (800-400 B.C.). Meighan (1969) considers the shafttombs found in Colima and Jalisco of western Mexico and their similarity with those of Ecuador. If such material traits were involved in the cultural transfers which occurred via maritime trade routes, that activity may have been going on long before the introduction of metallurgy from South America into western Mexico (Paulsen 1977). And if such trading voyages had been in progress for more than a millennium before the Spaniards arrived, they probably occurred intermittently with decades, if not centuries, between trips. A Spanish document of 1525 reporting on conditions in Mexico after the Conquest mentions that natives at Zacatula, a port on the southwest coast, tell that their forefathers would relate how strange Indians in large canoes laden with "exquisite" things would come in by sea from the south to trade, often staying several months (West 1961). Such trading voyages probably involved low-bulk goods like metals or precious stones to exchange for exotics like the spiny oyster (*Spondylus*) which occurs along the Pacific coast of Central America and southwestern Mexico and was highly prized in South America for its religious meaning.

In his studies on aboriginal watercraft along the Pacific coast of South America, Clinton Edwards (1960, 1965a, 1965b, 1969) has established the existence of large pre-Spanish canoes and rafts capable of extensive coastwise and offshore voyaging. At least two coastal Indian groups were noted for their expertise in navigation, one along the Manabí coast of Ecuador, the other around the Bay of Sechura in northern Peru (fig. 1). Both had developed (1) the sailing raft constructed of balsa logs and guided by means of centerboards and (2) the large dugout canoe also equipped with sail. With such craft an active coastwise trade system evolved, probably at an early date. Sailing rafts voyaged as far south as central Peru (Hosler 1988a, 848) and large canoes as far north as western Mexico. Through this trading system the diffusion of metallurgical knowledge involving hammered sheet metals and copper alloys from Peru, the lost-wax method of casting from Colombia, and the practice of gilding from both areas to western Mexico may well have taken place. Although large balsa sailing rafts were capable of oceanic voyages, canoe navigators probably remained in sight of coasts, where food and fresh water were available, where sandy beaches afforded easy landing sites, and where occasional coastal lagoons provided safe passage free from storm waves.

During the Post Classic period (after A.D. 900) metallurgical knowledge had spread from western Mexico in several directions into parts of northern and central Mesoamerica. Copper objects, apparently smelted from local ores, have been found in archaeological sites along the west coast at Culiacán and Guasave, Sinaloa (Kelly 1945; Ekholm 1942); and by A.D. 1100 copper trade ware from Mexico first appeared in Hohokam culture in the American Southwest (Pendergast 1962, 535; Haury 1976, 278). Just when and how the Culhua-México (Aztec) people in the Valley of Mexico acquired metallurgy remains unexplained, but trade items of copper alloys (e.g. bells) may have come from Michoacán in western Mexico and knowledge of casting gold objects from the Mixtec of Oaxaca, who were the foremost goldsmiths of Mesoamerica. The Mixtec's knowledge of lost-wax casting of gold and its alloys may have been introduced by sea traders from Costa Rica or Panama (Caso 1965, 915-916). Recently archaeologists discovered sites in Huastec Indian country, northeastern Mexico, that contained metal artifacts, mainly bells cast of copper alloys, including copper-arsenic and copper-tin bronzes (Hosler and Stresser-Pean, 1992). Dated as Post Classic (13th century A.D.), many of these items may have been acquired through trade with the Valley of Mexico, but some of the bells appear to have been cast locally.

Lacking deposits of metal ores in their limestone geological environment, the lowland Maya of the Yucatan Peninsula never developed full-fledged metallurgical crafts. They appear to have relied on metal trade goods from Costa Rica and Panama or from central Mexico, a development that occurred in Late Classic and Post Classic periods (A.D. 800-1540). The best repository of such goods is the "sacred cenote," a limestone sink, in the ceremonial center of Chichén-Itzá, into which gold and copper artifacts (among other sacred objects such as jade and copal incense) were thrown as "sacrifices" to deities. Among the metal found are pieces of hammered gold sheets and cast figurines of *tumbaga* thought to have been carried in by long-distance sea traders from Costa Rica along the Caribbean coast. Other objects consist of cast copper bells, probably from Central Mexico or Oaxaca. Perhaps the closest activity to the development of a metalworking craft was the use of imported sheet gold on which the Maya embossed by repoussé their own art designs (Coggins and Shane 1984; Bray 1977).

The highland Maya of Guatemala and western Honduras are said to have practiced techniques such as casting copper bells and objects of *tumbaga* alloy during the Late Classical period through knowledge introduced from Central Mexico or from southern Central America (Borhegyi 1965;

Woodbury 1965). Much of the metallic material found in archaeological context in those areas, however, may prove to be trade goods rather than locally made products.

Mining Techniques

Far more investigations have been undertaken on ancient metallurgy in the Americas than on mining techniques, and this lack of interest is more apparent for South and Central America than for Mexico. Spanish chroniclers gave only cursory descriptions of the way Indians extracted ores, citing such statements as "shallow holes or caves" (e.g. Sahagún [ca. 1570] 1938, 3:(chapt. 2) 29, 51).

In Mexico, archaeological investigation of aboriginal mining techniques has dealt more with the extraction of non-metallic substances, such as cinnabar and turquoise, than with ores such as copper. Adolph Langenscheidt's work (1970, 1982, 1985) on the cinnabar mines in the Sierra Gorda of Querétaro has revealed the presence of sizable adits, shafts and galleries dug with hafted hammerstones and probable use of fire-setting to spall sections of hard rock.[6] According to ^{14}C analysis of charcoal from the site, such activity occurred as early as A.D. 1, in Late Formative, a millennium before the introduction of metallurgy into central Mexico. Phil Weigand (1968, and this volume) reported on similar evidence of relatively sophisticated mining activity, but dated later as Classic (A.D. 200-900), for the turquoise mines of Chalchihuites in Zacatecas. Pedro Hendrichs (1940, 315-16, 326; 1943-44:I, 194 ff.) has dealt with the pre-Spanish copper mines (mainly in western Guerrero state), noting the small adits dug into hillsides in order to follow oxidized veins of copper (malachite, azurite, cuprite) with the use of hafted hammerstones and possibly wooden wedges.

In Colombia, European miners reported seeing traces of old Indian excavations at lodes containing gold and silver ores in various parts of the northern Andes, as at Buriticá and Mariquita. Shallow vertical or inclined shafts a meter wide and no more than five or seven meters (three *estados*) deep were sunk along outcropping veins. Excavations were limited to the upper weathered and oxidized zone and were made with wooden dibbles and crude hammerstones (White 1884, 245; West 1952, 54).

In northern Chile occasional accidental finds of a desiccated body of a native miner killed by collapse of a tunnel roof has given some evidence of aboriginal mining techniques. The body of the famous "copper man" found in a collapsed tunnel at Chuquicamata was surrounded by various tools such as a hafted hammerstone, wooden pry sticks and a slate-bladed shovel with wooden handle. Although originally thought to be of the sixteenth century, the unfortunate miner was apparently excavating atacamite, a hydrous copper chlorate prized for its ease of reduction with indigenous smelting techniques (Bird 1979). Alan Craig (Brown and Craig, this volume) also reports on a similar find at the mine of Huantajaya, also in arid northern Chile.

Pyrotechny

One of the most difficult problems in the development of metallurgy in the ancient New World was the attainment of sufficiently high temperatures in the reduction process to smelt metals and their ores. In the Peruvian Center the knowledge of eutectic temperatures probably was learned through experimentation. For example, the melting point of copper alone is 1,083°C and that of silver 960°C. But the copper-silver alloy has a melting point of 779°C., a temperature in reach of the early metalsmiths using primitive furnaces fueled with charcoal (Root 1949, 212-213). The attainment of eutectic temperatures in this way may have encouraged the practice of alloying in both the Old and New Worlds in ancient times, and in the Americas the idea must have diffused northward from Peru and Colombia into Mexico with other aspects of metallurgical practices.

In the ancient Old World, metalsmiths attained high temperatures in small furnaces with the aid of hand-operated bellows to supply a blast of air and increased amount of oxygen into a mixture of ore and burning charcoal. Not until its introduction by Europeans, however, was this instrument known in the New World. In its stead the Indians of Peru developed the blow tube made of a hollowed stem of cane through which a worker would blow to produce an air blast directed toward the burning coals within a clay furnace (fig. 2b). In this way a sufficiently high temperature was attained to smelt ore. A ceramic nozzle, or *tuyère,* was placed on the furnace end of the tube to protect it from the coals. Scores of broken nozzles have been found by archaeologists excavating the important metallurgical site of Batán Grande in northern Peru (Shimada et al. 1982; Shimada and Merkel 1991).

A simpler form of the blow tube was used by Mexican metalsmiths, but the idea of an air blast to raise temperatures in their small brazier-like clay furnaces likely was introduced into western Mexico from Peru or Colombia (fig. 2a). When Spaniards took over the aboriginal copper industry of Michoacán in 1533 Indians were employed as both miners and smeltermen, the latter using the blow tube in operation of their small furnaces; these were later replaced by larger Spanish furnaces equipped with sheepskin bellows (Barrett 1987, 15, 26, 64).[7]

On the altiplano of southern Peru and western Bolivia ingenious smelting furnaces called *huayras* were perfected by local metalsmiths probably long before the Inka period (fig. 2c). The sides of each furnace were perforated by small openings through which wind could pass to create a blast sufficiently strong to raise temperatures for smelting ores. The furnaces were placed on mountain slopes to take advantage of strong ka-

Fig. 2. Types of aboriginal furnaces used to smelt ores in pre-Hispanic Spanish America. Artist's conception of a) Use of blow tubes (*cañutos, sopletes*) in pre-Hispanic Mexico: i, from Codex Mendoza; ii, from Codex Tlótzin; iii, from Florentine Codex. b) use of blow tubes at Batán Grande, pre-Conquest northern Peru, from illustration in Shimada and Merkel 1991. c) a *huayra*, from photograph in Peele 1893. Artwork by Mary Lee Eggart.

tabatic air flow which occurs downslope at night (Craig, this volume). Apparently use of the *huayra* furnace did not expand much beyond the altiplano in pre-Hispanic times, although remains that have been interpreted as *huayras* have been noted as far south as northwestern Argentina (Márques Miranda 1946, 646).

Old and New World Metallurgy Compared

Metallurgy in the Old World began in Southwestern Asia ca. 7300 B.C., in the New World in Peru ca. 1500 B.C., a time differential of nearly six millennia.[8] Despite that, the development of processing metals and their ores followed a similar sequence in both areas, beginning with hammering and annealing of native gold or copper, followed by reduction by smelting and casting with development of alloys, including the bronzes. Only the use of iron and the technique of cupelation, both Old World metallurgical traits, were not developed in ancient America (Patterson 1971, 286).

A great difference, however, existed between Old and New Worlds in terms of the value system associated with metals and their refinement (Easby 1965, Patterson 1971, Lechtman 1984b, Hosler 1988a). In the Old, the utilitarian value of metals was emphasized, employing copper, bronze and iron for tools and weapons; in the New, metallurgy was fundamentally a religious activity, wherein gold, copper and its alloys were used mainly for ornaments, worn and manipulated by priests in religious ceremonies and for the enhancement of political power and prestige by the elite.

Endnotes

1. Metalworking was practiced by North American Indians of eastern United States as early as 4200 B.C., based on exploitation of large deposits of native copper in the Lake Superior region. The use of native copper, however, rather than copper ores, did not involve knowledge or use of metallurgy, but only the working of the pure metal by hammering to make various objects such as ornaments and small tools that were traded throughout a large area of what is now eastern and central United States (Patterson 1971, 287, 298).

2. Lechtman (1984a, 1988, 370-373) describes two techniques used by early Peruvian smiths in "surface metallurgy": (a) Depletion gilding and silvering, performed by hammering the alloy, bringing copper to the surface where it is dissolved or "pickled" off by application of acidic plant juice, thus enriching the gold or silver surface; and (b) Electrochemical replacement plating, done by placing objects made of the alloy in a solution of mixed corrosive minerals, which dissolves the surface copper, leaving a thin gold-enriched film. Both procedures were tested successfully in the laboratory with modern equipment.

3. Recent neutron activation analyses reveal a wide range of alloy components between gold-silver-copper that comprise what is now termed *tumbaga*. These now require special terminology based on some standard range of percentage of the dominant metal. *Tumbaga, sensu strictu*, should be reserved for artifacts that correspond closely to the composition of the Malayan objects from which the term originated.

4. Because of lack of archaeological evidence (e.g. remains of furnaces, slag, etc.) in Panamanian and Costa Rican sites, some writers (e.g. Sauer 1969, 276-277; Helms 1979, 146, 153) discounted the assumption of others that metallurgy was practiced in those areas, thus expressing their opinion that most or all metal artifacts found in archaeological context were actually traded in from Colombia. Both Bray and Cooke, however, vigorously rejected such statements for reasons stated.

5. The *Relación de Michoacán* lists the caches of "gold" and "silver" (actually copper-gold and copper-silver alloys) stored in houses of the Tarascan elite in their capital, Tzintzuntzan, and on islands within Lake Pátzcuaro. In his own house the Tarascan chief, or *calzonci*, kept 40 chests, 20 filled with gold objects, 20 with silver; on the island of Apuputo, 20 chests of "fine silver" fashioned as round plaques, or "shields," 200 in each chest; on the island of Xanecho [Janicho], 8 chests of round silver "shields" plus 400 *curindas*, or silver bars, all dedicated to the moon goddess; and on the island of Pecandan, 4 chests of fine silver, each con-

taining 100 round silver shields and 20 of gold. More treasure was stored on the islands of Huranden and Utuyo (Craine and Reindorp 1970, 78, 82). Apparently the Spaniards looted all of these caches upon their arrival in Michoacán in 1521 and carried the booty to the Valley of Mexico to be melted down, cast in bars and assayed.

6. Wittich (1922, 146) and Zaragoza (1993) found archaeological evidence of prehistoric mining of cinnabar as far north as the Guadalcázar area in central San Luis Potosí state. Although in the seventeenth century Spaniards first exploited the silver ores around Guadalcázar, in the late 19th century the area was one of Mexico's leading producers of mercury (Wittich 1922, 167).

7. According to Pedro Hendrichs (1943-44, I: 210), Indian housewives in western Guerrero state use small blow tubes called *sopletes* rather than the more common fire fan to encourage flames in their crude kitchen stoves.

8. The idea espoused especially by the German anthropologist Robert Heine-Geldern (1954) that metallurgical knowledge was first introduced into the New World from eastern Asia via trans-Pacific contact has now been discounted (Easby 1965, 97 n. 10; Patterson 1971, 309).

References

Arsandaux, H. and Rivet, P. 1921. Contribution à l'étude de la métallurgie mexicaine. *Journal de la Société des Américanistes de Paris* 13:261-280.

Barrett, E. 1987. *The Mexican colonial copper industry*. Albuquerque: University of New Mexico Press.

Bennett, W. C. 1946a. The Andean highlands: An introduction. In *Handbook of South American Indians*, ed. J. Steward, 2:1-60. Washington, D.C.: Smithsonian Institution, Bureau of American Ethnology, Bull. 143.

——. 1946b. The archeology of Colombia. In *Handbook of South American Indians*, ed. J. Steward, vol. 2, 823-864. Bureau of American Ethnology Bulletin 143. Washington, DC: Smithsonian Institution.

Bird, J. 1979. The "Copper Man": A Prehistoric miner and his tools from northern Chile. In *Pre-Columbian metallurgy of South America*, ed. E. Benson, 105-131. Washington, D.C.: Dumbarton Oaks Research Library and Collections.

Borbolla, Rubín de la, D. F. 1944. Orfebrería tarasca. *Cuadernos Americanos* 3(3): 127-138

Borhegyi, S. F. de. 1965. Archaeological synthesis of the Guatemalan highlands. In *Handbook of Middle American Indians*, ed. G. Willey, 2:3-58. Austin: University of Texas Press.

Bray, W. 1977. Maya metalwork and its external connections. In *Social processes in Maya prehistory*, ed. N. Hammond, 365-403. New York: Academic Press.

——. 1984. Across the Darien Gap. In *The Archaeology of lower Central America*, eds. F. W. Lange and D. Stone, 305-338. Albuquerque: University of New Mexico Press.

Burger, R. L. 1992. *Chavín and the origins of Andean civilization*. London: Thames and Hudson.

Caso, A. 1965. Lapidary work, goldwork and copperwork from Oaxaca. In *Handbook of Middle American Indians*, ed. G. Willey, 3:896-930. Austin: University of Texas Press.

Coe, M. D. 1962. *Mexico*. New York: Frederick A. Praeger.

Coggins, C. C. and Shane, O. C. 1984. *Cenote of sacrifice. Maya treasures from the sacred well at Chichén-Itzá*. Austin: University of Texas Press.

Collier, D. 1946. The archeology of Ecuador. In *Handbook of South American Indians*, ed. J. Steward, 2:767-784. Washington, D.C.: Smithsonian Institution, Bureau of American Ethnology, Bull. 143.

Cooke, R. 1984. Archaeological research in central and eastern Panama. In *Archaeology of lower Central America*, eds. F. W. Lange and D. Stone, 263-302. Albuquerque: University of New Mexico Press.

Craine, E. R. and Reindorp, R. C. 1970. *The chronicles of Michoacán.* Norman: University of Oklahoma Press.

Dussán de Reichel, A. 1979. Some observations on the prehistoric goldwork of Colombia. In *Pre-Columbian metallurgy of South America,* ed. E. Benson, 41-52. Washington, D.C.: Dumbarton Oaks Research Library and Collections.

Easby, D. T., Jr. 1965. Pre-Hispanic metallurgy and metalworking in the New World. American Philosophical Society *Proceedings* 109: 89-98.

Edwards, C. R. 1960. Sailing rafts of Sechura: History and problems of origin. *Southwestern Journal of Anthropology* 16: 368-391.

——. 1965a. *Aboriginal watercraft in the Pacific coast of South America.* Ibero-Americana, no. 47. Berkeley and Los Angeles: University of California Press.

——. 1965b. Aboriginal sail in the New World. *Southwestern Journal of Anthropology* 21:351-358.

——. 1969. Possibilities of pre-Columbian maritime contacts among New World civilizations. In *Pre-Columbian contacts within nuclear America,* eds. J. Kelley and C. Riley. Mesoamerican Studies no. 4, 3-10. Carbondale, IL: University of Southern Illinois Press.

Ekholm, G. 1942. *Excavations at Guasave, Sinaloa, Mexico.* Anthropological Papers 38, pt. 2. New York: American Museum of Natural History.

González, A. R. 1979. Pre-Columbian metallurgy of northwest Argentina: Historical development and cultural process. In *Pre-Columbian metallurgy of South America,* ed. E. Benson, 133-202. Washington, D.C.: Dumbarton Oaks Research Library and Collections.

Graham, M. 1993. Displacing the center: Constructing prehistory in Central America. In *Reinterpreting prehistory in Central America,* ed. M. Graham, 1-38. Niwot, CO: University Press of Colorado.

Haury, E. 1976. *The Hohokam. Desert farmers and craftsmen.* Tucson: University of Arizona Press.

Heine-Geldern, R. 1954. Die asiatische Herkunft der südamerikanischen Metalltechnik. *Paideuma* 5:347-423.

Helms, M. W. 1979. *Ancient Panama: Chiefs in search of power.* Austin and London: University of Texas Press.

Hendrichs, P. 1940. Datos sobre la técnica minera prehispánica. *México Antiguo* 5:148-160, 179-194, 311-238.

——. 1943-44. *Por Tierras Ignotas. Viajes y observaciones en la región del Río de las Balsas.* 2 vols. Mexico City: Editorial Cultura.

Hosler, D. 1988a. Ancient west Mexican metallurgy: South and Central American origins and west Mexican transformations. *American Anthropologist* 90(4):832-855.

——. 1988b. Ancient west Mexican metallurgy: A technological chronology. *Journal of Field Archaeology* 15(2):191-217.

——. 1988c. The metallurgy of ancient West Mexico. In *The beginning of the use of metals and alloys,* ed. R. Maddin, 328-343. Cambridge, MA: MIT Press.

——. 1990. The development of ancient Mesoamerican metallurgy. *Journal of Metals* 42(5):44-46.

Hosler, D. and Stresser-Pean, G. 1992. The Huastec region: A second locus for the production of bronze alloys in ancient Mesoamerica. *Science* 257:1215-1220.

Kelly, I. 1945. *Excavations at Culiacán, Sinaloa.* Ibero-Americana no. 25. Berkeley and Los Angeles: University of California Press.

——. 1970. Vasijas de Colima con boca de estribo. *Boletín del Instituto Nacional de Antropología e Historia* (Mexico) 42:26-30.

Langenscheidt, A. 1970. *Minería prehispánica en la Sierra de Querétaro.* Mexico City: Secretaría del Patrimonio Nacional.

——. 1985. Bosquejo de la minería prehispánica de México. *Quipu* 2(1): 37-58. (Mexico.)

Langenscheidt, A., and Tang Lay, Carlos. 1982. La minería prehispánica en la Sierra Gorda. In *Mining and mining techniques in ancient Mesoamerica*, eds. P. C. Weigand and G. Gwynne. Special Issue *Anthropology* 6(1&2): 135-148.

Lechtman, H. 1980. The central Andes: Metallurgy without iron. In *The coming of the age of iron*, eds. T. Wertime and J. Muhly, 267-334. New Haven: Yale University Press.

——. 1984a. Pre-Columbian surface metallurgy. *Scientific American* 250(6):56-63.

——. 1984b. Andean value system and the development of prehistoric metallurgy. *Technology and Culture* 25(1):1-36.

——. 1988. Traditions and styles in central Andean metallurgy. In *The beginning of the use of metals and alloys*, ed. R. Maddin, 344-378. Cambridge, MA: MIT Press.

Márques Miranda, F. 1946. The Diaguita of Argentina. In *Handbook of South American Indians*, ed. J. Steward, 2:637-654. Washington, D.C.: Smithsonian Institution, Bureau of American Ethnology, Bull. 143.

Meighan, C. 1969. Cultural similarities between western Mexico and Andean regions. In *Pre-Columbian contacts within nuclear America*, eds. J. Kelley and C. Riley, Mesoamerican Studies no. 4, 11-25. Carbondale, IL: Southern Illinois University Press.

Mountjoy, J. 1969. On the origin of west Mexican metallurgy. In *Pre-Columbian contacts within nuclear America*, eds. J. Kelley and C. Riley, Mesoamerican Studies no. 4, 26-42. Carbondale, IL: Southern Illinois University Press.

Murra, J. 1946. The historic tribes of Ecuador. In *Handbook of South American Indians*, ed. J. Steward, 2:785-821. Washington, D.C.: Smithsonian Institution, Bureau of American Ethnology, Bull. 143.

Patterson, C. 1971. Native copper, silver and gold accessible to early metallurgists. *American Antiquity* 36:286-321.

Paulsen, A. 1977. Patterns of maritime trade between south coastal Ecuador and western Mesoamerica, 1500 B.C. - A.D. 600. In *The sea in the pre-Columbian world*. ed. E. Benson, 141-166. Washington, D.C.: Dumbarton Oaks research Library and Collections.

Peele, R., Jr. 1893. A primitive smelting furnace. *The School of Mines Quarterly* 15:8-10.

Pollard, H. 1987. The political economy of prehispanic Tarascan metallurgy. *American Antiquity* 52:741-752.

Restrepo, V. 1888. *Estudio sobre las minas de oro y plata de Colombia*, 2nd. ed. Bogotá: Imprenta de Silvestre y Compañía.

Root, W. 1949. Metallurgy. In *Handbook of South American Indians*, ed. J. Steward, 5:205-225. Washington, D.C.: Smithsonian Institution, Bureau of American Ethnology, Bull. 143.

——. 1961. Pre-Columbian metalwork of Colombia and its neighbors. In *Essays of pre-Columbian art and archaeology*, ed. S. Lothrop, 242-257. Cambridge, MA: Harvard University Press.

Sahagún, Fr. B. de. [ca. 1570] 1938. *Historia general de las cosas de Nueva España*. 5 vols. Mexico City: Editorial Robredo.

Sauer, C. 1969. *The early Spanish main*. Berkeley: University of California Press.

Shimada, I., Epstein, S., and Craig, A. 1982. Batán Grande: A prehistoric metallurgical center in Peru. *Science* 216:952-959.

Shimada, I. and Merkel, J. 1991. Copper alloy metallurgy in ancient Peru. *Scientific American* 265:80-86.

Stanislawski, D. 1983. *The transformation of Nicaragua, 1519-1548*. Ibero-Americana no. 54. Los Angeles and Berkeley: University of California Press.

Stone, D. and Balser, C. 1967. *Aboriginal metalwork in lower Central America*. San José, Costa Rica: Editorial Antonio Lehmann.

Weigand, P. 1968. The mines and mining techniques of the Chalchihuites Culture. *American Antiquity* 33:45-61.

West, R. 1952. *Colonial placer mining in Colombia*. Louisiana State University Studies, Social Science Series 2. Baton Rouge, LA: Louisiana State University.

——. 1961. Aboriginal sea navigation between Middle and South America. *American Anthropologist* 63(1):133-135.

White, R. B. 1884. Notes on the aboriginal races of the northwestern provinces of South America. *Journal of the Anthropological Institute of Great Britain and Ireland* 13:240-256.

Willey, G. 1955. Prehistoric civilizations of nuclear America. *American Anthropologist* 57:571-593.

Wittich, E. 1922. La geología de la región minera de Guadalcázar. *Memorias de la Sociedad Científica "Antonio Alzate."* 40:145-178.

Woodbury, R. 1965. Artifacts of the Guatemalan highlands: In *Handbook of Middle American Indians*, ed. G. Willey, 2:163-179. Austin: University of Texas Press.

Zaragoza Ocaña, D. M. 1993. Guadalcázar, San Luis Potosí: una zona minera prehispánica. [Paper presented at III Reunión de Historiadores de la Minería Latinoamericana, Taxco, Mexico, November 22-26, 1993.]

Observations on Ancient Mining within the Northwestern Regions of the Mesoamerican Civilization, with Emphasis on Turquoise

Phil C. Weigand, Centro de Estudios Antropológicos
Colegio de Michoacán, Zamora, Mexico

Abstract

It has long been recognized that mining and long distance trade were important and crucial economic activities for the colonial societies of northern New Spain. Less known is the fact that these activities had hoary pre-Hispanic antecedents. The focus of this study is to document aspects of the pre-Hispanic trade structure for turquoise. We believe that the evolution of the turquoise trade played a crucial role in the cultural intensification of the social systems along Mesoamerica's northern frontier, including those located in the southwestern portions of the United States. In this sense, turquoise procurement and trade is the pre-Hispanic structural equivalent to the colonial period exploitation of silver. Special attention is given to the Chalchihuites region. The vast mining system there dates to the Mesoamerican Classic period, and probably sets the stage for the expansion of the mineralogical exploration of zones farther north. The roles that Chaco Canyon and Paquimé (Casas Grandes) played in turquoise procurement and trade will be examined in addition.
Key words: turquoise, pre-Hispanic trade, Southwest U.S., Mesoamerica.

Introduction

Mineral procurement is an enduring theme in both northwestern Mexico and the southwestern United States. Mining and mining settlements have given this great area much of its distinctive sociohistorical and cultural character. The popular mind forcefully associates mining with these signatures: the vivid images of field rushes; unstable social situations; constant warfare against nonsedentary neighbors of the miners; transitory populations; quick flowerings and equally rapid declines of settlements and trading posts; and hardship. But much of what we regard as colonial

In Quest of Mineral Wealth: Aboriginal and Colonial Mining and Metallurgy in Spanish America, edited by Alan K. Craig and Robert C. West, 1994. Geoscience and Man, vol. 33, pp. 21-35. Dept. of Geography & Anthropology, Louisiana State University, Baton Rouge, LA 70893-6010, U.S.A.

or modern characteristics of the mining activities have in reality far deeper roots. It is our contention that mineral exploration, exploitation, and procurement were in fact among the organizing postulates, in a structural sense, in the formation of the ancient Mesoamerican trade structure, and that this activity was important, in select areas such as the northwest, at many or all levels of economic organization. Many minerals were targeted for exploitation in the north: malachite, azurite, native copper, chrysocolla, cuprite, jasper, agate, agatized petrified wood, chert, flint, obsidian, and salt, as well as many more. The theme of rare resource exploitation of an arid land region by, and in service to, a neighboring civilization should not surprise us: this is one of world history's most perdurable socioeconomic and political themes. I argue that the relationships between the northwestern sections of pre-1848 Mexico and ancient Mesoamerica were no exception. But, from even the longest list of minerals procured from the aforementioned area, one—chemical turquoise—stands out as an item of major importance.

As Thompson (1978, 145) observed:

> It is interesting to note that *xiuitl*, the Mexican [Nahuatl] name for the year, also meant turquoise and, by extension, rain, both because of its color, which is that of the Tlalocs, and because both turquoise and rain were precious things.

As López Austin et al. (1991) point out, this association probably is over 2,000 years old, and thus is an ideological corollary and organizing principle, among others, for most, if not all, Mesoamerican religious patterns. The association was more complex than Thompson suggests. Turquoise was also linked to the very concept of time, which in turn was symbolized in political power. Turquoise was the most important precious stone in the royal diadems; it had become a metaphor for holy speech and wisdom. Turquoise became an extravagantly valuable possession and status marker, laden with so much symbolism that it is difficult for us to comprehend. Turquoise appears at almost every explanatory and symbolic juncture within the Mesoamerican ideological system. By the late Post Classic period, it had even outstripped its competitor gem, jade, in consumption, popularity, and importance. When Cortéz landed in Veracruz in 1519, the rich gifts presented to him a short time afterwards by Moctezuma II, the emperor of the Culhua Mexica, were not just intended to be the usual magnificent tokens of esteem from one ruler to another. Rather they were religious offerings of a devout emperor to one of his gods, for Moctezuma apparently was convinced that Cortéz was in fact Quetzalcoatl. Sahagún (1989) describes the magnificent gifts presented to Cortéz:

> First was the array of Quetzalcoatl: a serpent mask of turquoise; a quetzal feather headband and a mirror for the small of the back...like a turquoise shield, of turquoise mosaic—encrusted with turquoise, glued with turquoise...they bore in their arms the array of the gods...they adorned the Captain himself; they put on him the turquoise mosaic serpent mask; with it went the quetzal feather headband.

No series of societies, ever before or afterwards, have valued turquoise as much as those of ancient Mesoamerica.

Since turquoise, as a mineral, does not occur close to the ancient centers of consumption within Mesoamerica, its procurement required a special effort. Garman Harbottle and I (Weigand, Harbottle and Sayre 1977; Weigand and Harbottle 1992; Harbottle and Weigand 1992) have estimated that over a million individual pieces of turquoise have been recovered through excavations, or are in museum and private collections, from Mesoamerican sites of all time periods. This vast quantity assuredly is only a small percentage of that which actually circulated throughout Mesoamerica during antiquity. This quantity can only be explained as the result of a systematized effort at acquisition. The impact that this procurement effort had on societies near the natural outcrops must have been considerable.

Early Procurement and Mineralogical Activity in the North

The earliest societies which displayed a major interest in mining along the northern marches of the emerging civilization date to the Formative period. Langenscheidt (Langenscheidt and Lay 1982) has demonstrated a large-scale mining complex, beginning in the Formative and targeting cinnabar, for the Sierra de Querétaro. It is estimated that between 2,000 and 3,000 individual mines existed within the complex, though, of course, not all of these belong to the earliest periods of development. The deepest mines are ca. 60 m, and the longest tunnels ca. 200 m. Thus, while there are a large number of mines, in reality they are all rather small when compared to those found in the Chalchihuites region of Zacatecas.

The mining complex near Chalchihuites is monumental in every sense of that word (Weigand 1968, 1982). It is the largest and most monumental such complex yet reported for ancient Mesoamerica. There are about 800 individual mines from which millions of tons of spoil were taken. The landscape around these mines is still degraded, such was the degree and scale of earth and rock excavation. The largest mine yet measured has interior chambering and tunnels that stretch for over 3 km (P. Jiménez pers. comm.). While most are smaller than this, these mines are nonetheless impressive even by today's standards. The Chalchihuites mining complex probably began during the Canutillo phase (ca. A.D. 200-500), but it reached its peak development during the Alta Vista and Vesuvio phases (ca. A.D. 500-800). The few ^{14}C dates that exist from the mine interiors fall within this latter time period.

Mining in the deposits around Chalchihuites was multipurpose: a variety of blue/green stones, especially malachite and azurite (some of which looks remarkably like chemical turquoise); cinnabar; limonite; hematite; a weathered chert/flint (a highly carveable, white stone); iron pyrite; and possibly native copper were all recovered from these excavations.

The engineering of the mines was sophisticated. Entrances were usually of the adit variety, though many shafts also were dug. The natural caliche cap helped support the overburden, and the deposits that were exploited almost always are beneath this cap. The deposits that were mined consist of semi-sorted, semi-consolidated (by the actions of a caliche cement mixed with hematite), alluvial gravel beds. These gravel beds are very ancient.

While the caliche cap provided some margin of safety for the Chalchihuites miners, they were obviously not satisfied with that. The chambers and tunnels are all engineered with arched supports in the natural deposit. In the great chambers that often mark the first rooms beyond the entryways, large columns of the natural deposit were left in place to offer roof support. Support columns are found throughout the most elaborate mines, every place where the need was apparent. Air holes were frequently dug at regular intervals to service the longest tunnels. They are circular and about one meter in diameter. These air holes probably doubled as emergency exits, given that, despite the engineering safety precautions, the mines were and are dangerous.

The rooms within the mines still show the activities of the miners as they left them: hammerstone impact scars, pick marks, loosened gravel blocks, both still within the walls of the mine and on the floor of the rooms; the blackened, sooty walls and ceilings from the *ocote* (pine) splints used for lighting; broken hammerstones and picks discarded within the mines; the broken vine handles for the hammerstones and picks; broken water gourds; various types of threads and twines; and so on. Indeed, the broken handles and pine splints number in the thousands in some mines.

The tunnels and rooms were often spacious. The miners could easily stand up and move around within them. Access ways were kept very clean from debris. Much of what was mined was not removed to the surface; a great deal of preliminary sorting apparently was accomplished inside. Perhaps as much as 25 percent of the rock generated by mining was left within, neatly stacked in piles in exhausted rooms and chambers, and sometimes held in place by terrace retaining walls. In general, the mines were cleaner and safer than those of the colonial period mines that we have inspected from the same region. Despite this observation, mining therewithin was arduous and danger-ridden. Roof collapses were frequent, though the miners often cleaned them or by-passed them to continue their work further in the interior. While dry, the heat levels were, and are, very high. The limited air circulation must have been nightmarish: high levels of dust generated by the actual mining and smoke from the pine splints were the worst contaminants.

The Chalchihuites culture must have been well organized in order to have fielded such a demanding mining operation. Surface indications of this high level of social organization abound: small, but elegant pyramids, elaborate courts and halls of columns, an observatory, skull racks, elite burials accompanied by human sacrifice, and so on are all in evidence, especially at the largest ceremonial center of the zone at Alta Vista (Kelley 1971, Kelley and Kelley 1971, Kelley n.d.). In addition, there appears to have been a settlement hierarchy that operated to exploit the rather limited agricultural resources of this semi-arid region as well as to maximize access and control over the best mining outcrops (Weigand 1982).

It is important to understand the Chalchihuites cultural phenomenon because it was obviously they who gave the major impetus for the systematic acquisition of chemical turquoise within the Mesoamerican trade structure during the Classic period. Aside from their own mining operations, the peoples of the Chalchihuites zone were busy acquiring chemical turquoise from other regions, as a commodity for trade/exchange beyond their own lands, as well as for elite consumption within their own culture. Some of this turquoise was from fairly nearby deposits in Zacatecas, San Luis Potosí, and Coahuila. While these deposits were near the frontiers of actual civilization, they were not as extensively exploited as the higher-quality deposits found further north. Neutron activation analysis (NAA) (Weigand and Harbottle 1992, Harbottle and Weigand 1992) has shown the presence of both these fairly accessible turquoises as well as those from the Cerrillos area of New Mexico in the artifact collections of the Chalchihuites culture, beginning about A.D. 500 when the first chemical turquoises appear.

Turquoises appeared within Mesoamerica prior to the development of the Chalchihuites culture. For example: Rosa María Reina Robles (1989) recovered turquoises in status burials in Guerrero which date from ca. 600 B.C.; we have seen turquoise recovered from shaft tombs of the El Arenal and Ahualulco phases (300 B.C.-A.D. 200, and A.D. 200-400, respectively) in Jalisco. Obviously, turquoise was beginning to circulate during the Formative period. But the first large collections of turquoise in Mesoamerica apparently belong to the Classic period. Apparently, it was the experienced miners and traders of the Chalchihuites zone that introduced this mineral in quantity to the Mesoamerican trade structure. The trading and/or collaborating partner of the Chalchihuites culture closest to the higher-grade, northern deposits was the Hohokam. Indeed, during the Gila Butte phase of the Hohokam (ca. A.D. 500), we see both the presence of turquoise in their artifact inventory and Chalchihuites-like motifs on their ceramics, plus the presence of high status objects made from pseudo-cloisonné (Haury 1976, Kelley 1966). During the Classic period, turquoise is found with growing frequency from Quintana Roo, where the recent excavations of Felipe Rodríguez (pers. comm.) dated these finds to A.D. 600, to the Great Kiva network among the Anasazi (Windes 1992); in other words, from one end of the civilization to the other.

Thus, the growth in demand for turquoise within the Mesoamerican civilization is a chronological corollary of the earliest phases of sociocultural intensification of the Hohokam and Anasazi within the southwestern United States (Weigand 1992) (fig. 1). This demand, for the most prestigious gem stone that the Mesoamericans defined, cannot be viewed in isolation. The evolution of the sociocultural systems found in northwestern sections of pre-1848 Mexico must be contextualized on this score. To do otherwise would be like analyzing the contemporary political and economic order in Saudi Arabia without reference to their petroleum deposits

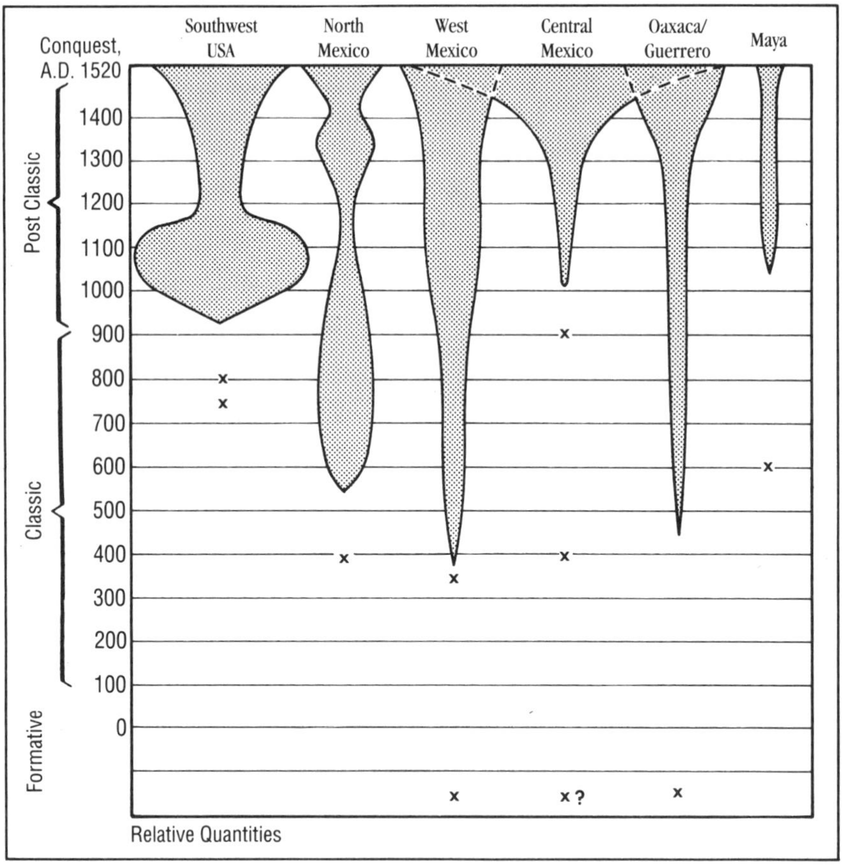

Fig. 1. Presence of turquoise artifacts through time throughout various zones of ancient Mesoamerica.

and without mentioning the world oil market. It is possible to do that, but the result is not credible.

These statements are not meant to be reductionistic, as there are many other reasons why the sociocultural systems of northwestern sectors of Mesoamerica underwent intensification. But given the extremely high status and value that turquoise had within Mesoamerica, it is completely logical to insist that the ancient sociocultural systems of northwest Mexico and the southwest United States, where the higher quality deposits for turquoise exist, must be contextualized by consideration of this demand profile, just as we do the silver mining societies of the same areas during the colonial period.

Turquoise Procurement, Trade and Mining

How many turquoise deposits were explored during the first phases of interest by the Hohokam and Chalchihuites cultures has not yet been assessed, though it seems reasonable to posit two areas within New Mexico (Cerrillos, Azure), one in Arizona (Courtland/Gleason), one in California (Halloron), and at least one each in Zacatecas and Coahuila. The next phase of turquoise procurement represents a considerable amplification of both demand and production. By ca. A.D. 900, a comprehensive wave of reorganization had been concluded within most sectors of the Mesoamerican civilization (cf. Weaver 1992). Accompanying the collapse of many of the Classic period cultures, such as the Teuchitlán Tradition in the Occidente, the Petén Maya, Teotihuacán, and Monte Albán, was a vast enlargement of the boundaries of the sociocultural influence of the civilization. At first glance, this appears to be a paradox: the component core entities within the civilization were collapsing while exterior influences appear to have enjoyed a quantum expansion. To help conceptualize the apparent paradox, we must distinguish between "trade networks" and the "trade structure." The former represents the particularistic arrangements that trading/exchange partners have during particular points or brief periods of time. Networks collapse when one or several of the partners stop participating, for whatever reasons. A "trade structure," however, represents the perdurable flow of goods over the *longue durée*, over millennia, unaffected except briefly by the rise and fall of specific partners. In addition, these structures are characteristically expansive over long periods.

In the southwestern United States, the presence and significance of Mesoamerican traits has been hotly contested. At points, the argument seems based in an unconscious nationalism concerning the "Mesoamerican connection." This is all the more discouraging because most of the proponents involved in denying a Mesoamerican presence, or relegating it to unimportance, have little or no preparation in the Spanish language or the archaeology and ethnohistory of Mesoamerica. By the early Postclassic period, however, these influences are certainly present if an objective stance is utilized in assessing them. I will not repeat the argument (see Weigand 1992), but rather just list the characteristics and traits that appear to indicate, at least to me, a growing Mesoamerican participation within the northwest sectors of pre-1848 Mexico. At Chaco Canyon, for example, we see large rectangular platforms in front of Pueblo Bonito, the employment of rubble core-veneer masonry as an engineering technology in architecture, an array of elite and status goods, which include parrot feathers and copper bells. Given the population implosion at Chaco Canyon, the road system, which served in part to provision the Canyon, and the appearance of a ward/barrio-like urban configuration, it is completely fair to ask whether or not outside influences could have played a role within the revolution that the San Juan Anasazi underwent just after A.D. 900 (cf. Doyel 1992, Lekson 1986, Crown and Judge 1991). Similar arguments can be made for the Hohokam regional system (cf. Crown and Judge 1991; Wilcox 1991).

The distribution of turquoise during Chaco's heyday is worth special note: in the northwest sectors of pre-1848 Mexico, turquoise was quite rare outside of the canyon per se. At Chaco Canyon, the estimates of turquoise recovered from excavation vary from 200,000 to 500,000 pieces—an incredible amount, by whichever count, and one which does not even begin to consider how much may have been exported. The highly uneven distributional profile for turquoise within the this area strongly suggests that Chaco had acquired, for all practical purposes, a working monopoly on turquoise procurement and redistribution. Of the roughly 200 pieces analyzed by NAA, the Cerrillos mines are strongly suggested as one of the primary sources. Add to this observation two others: 1) at Chaco, the technology of beveled edges for micromosaic manufacture was introduced, along with the standardization of tessarae blanks (a Mesoamerican unit measurement), and 2) the occurrence of Cerrillos turquoises at Guasave (Sinaloa), Ixtlán del Río (Nayarit), and Zacoalco and Las Cuevas (Jalisco) in levels coeval with Chaco Canyon. All of this cannot be coincidence, and is most logically explained by reference to a single trade structure which encompasses northwest Mexico, the southwestern United States, and the rest of Mesoamerica (fig. 2).

After Chaco Canyon collapsed, turquoise became popular within the northwestern sectors of pre-1848 Mexico in a general fashion. In other words, for the first time, turquoise became popular in a general sense within the zones of procurement. At the same time, its popularity within the corelands of the Mesoamerican civilization was soaring. The vast majority of the spectacular turquoise mosaic artifacts (cf. Carmichael 1970; Harbottle and Weigand 1992) are from the late Postclassic. Demand had risen geometrically. Into this sociological situation, the rise of Paquimé

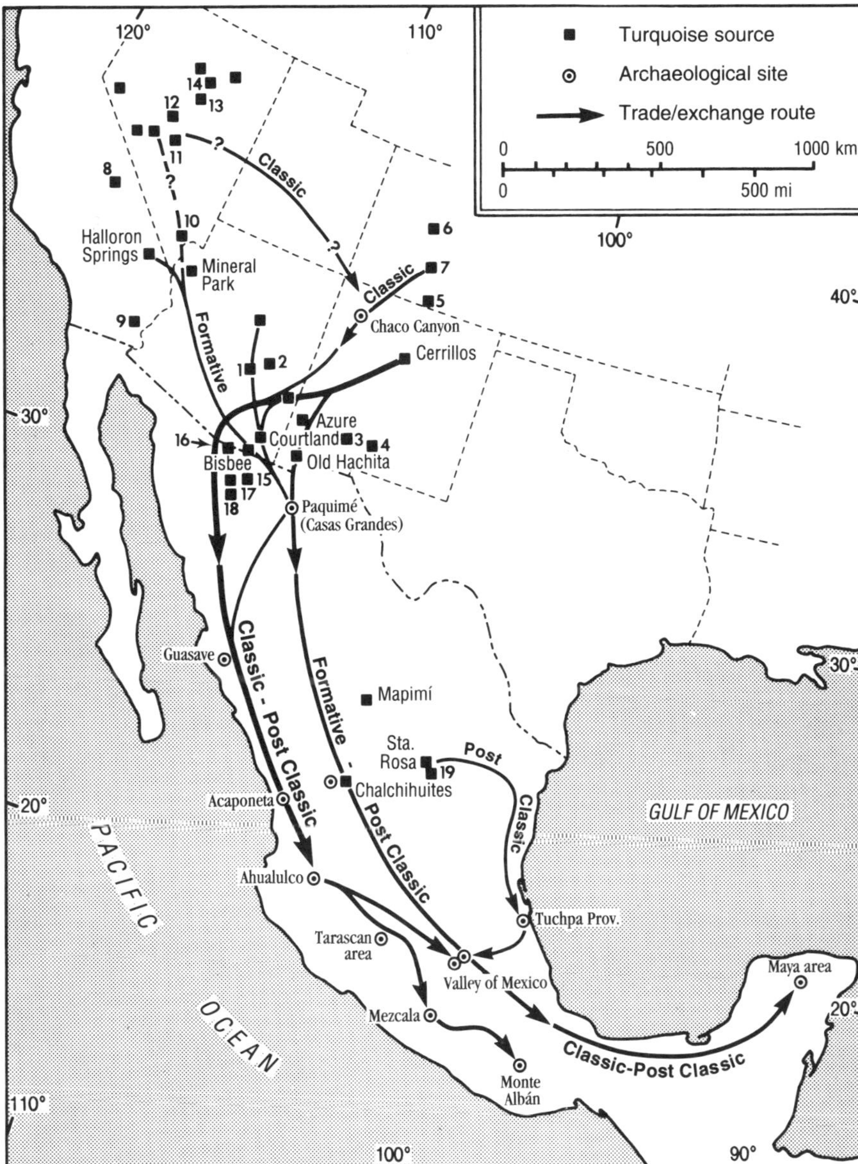

Fig. 2. Suggested trade/exchange routes for turquoise throughout pre-Hispanic Mesoamerica. Names of numbered turquoise sources: 1. Sleeping Beauty, 2. Canyon Creek, 3. White Signal, 4. Jarilla Mountains, 5. La Jarra, 6. Leadville, 7. Villa Grove, 8. Last Chance, 9. Quartz Mountains, 10. Crescent Peak, 11. Warm Springs, 12. New Pass Range, 12. Grass Valley, 14. Crescent Valley, 15, Cananeita, l6. Campo Frío, 17. Arroyo Cuitaca, 18. El Verde, and 19. Beta Pérez.

(Casas Grandes) needs to be contextualized. Since Paquimé is located south of the United States/Mexico border, there has been less dispute about its Mesoamerican character. Di Peso's original chronology (1974) has been corrected by recent work by J. Dean (n.d.), placing its heyday firmly after Chaco's demise, and within the period of geometrically increased demand in the north as well as within the core of the civilization.

Neutron activation analysis of the turquoises from Paquimé suggests the response of mineralogically-oriented societies to the changed circumstances: increased demand resulted in the opening of new, and more remote, sources. An analogy: when the price of oil reached the new and dramatic levels of the past twenty years, new fields which had been up to that point unprofitable because of remoteness and logistics were opened up — the north shore of Alaska and the North Sea. When the demand for turquoise rose dramatically, new sources in the Californias, Nevada, New Mexico, Arizona, Colorado, Sonora, and Chihuahua were opened up. A single piece of agatized turquoise may be from the Mapimí region of Chihuahua, as this is the only known source for this variety of the mineral. The old and perdurable source zones, such as the Cerrillos area, remained in production, as well. A number of these new sources, plus all of the older ones, are represented in the collections analyzed from Paquimé. Apparently, Paquimé either had no desire, or lacked the ability, to actually promote the variety of monopoly that Chaco had attempted for turquoise procurement and distribution. No one source ever again dominated the artifact inventory as completely as had the Chaco/Cerrillos pattern, except, of course, in strictly local configurations.

Paquimé's role as a rare resource procurer, with special reference to turquoise, has been questioned ever since Di Peso first proposed this specialized function for it. It is true that the collections of turquoise at Paquimé are nothing of the order of those found at Chaco Canyon, but then Chaco had neither precedent or antecedent in this sector of Mesoamerica. Only around 5,000 pieces of turquoise have been recovered at Paquimé. However, the argument that minimizes the importance of the turquoises found at Paquimé makes absolutely no allowance for the quantity that must have been exported or redistributed from the center to other sites, both local and distant. What is certain is that the first great field rush on the resource appears to be a chronological corollary of the rise and heyday of Paquimé. While this could be coincidence, it is more logical to assume that there is a structural relationship between the two occurrences (fig. 2).

Paquimé's geographical disposition, plus the trade materials of obvious Mesoamerican manufacture (especially from the Occidente) allowed Di Peso (1974) to posit an important role for the center within the Mesoamerican trade structure. The architectural considerations add weight to his observation. While it is true that most of the exotic Mesoamerican ceremonial buildings are outside the habitation area, this is also true for Mesoamerican sites within the metropolitan zones of the civilization: ceremonial buildings are specialized, and hence concentrated in ceremonial areas, just as at Paquimé. The "peripheral" location of the Mesoamerican ceremonial buildings cannot be used as an argument against the Mesoamerican character of the center. One must also consider the presence of breeding pens for parrots and the very strong possibility of copper metallurgy at the site. These are two very important indicators for a specialized dedication to selected aspects of the overall Mesoamerican technological base. Paquimé is strategically placed for easy access to a wide variety of malachite, azurite, and native copper sources, in addition to turquoise outcrops. Its prosperity and ceremonial adornment must be, in part, related to the wealth it was able to generate from its specializations and ability to attract rare resources.

The Turquoise Mines

The turquoise outcrop areas, where we have found indications of ancient mining, follow (fig. 2).

- Arizona: Sleeping Beauty, Bisbee, Mineral Park, Canyon Creek, Courtland/Gleason.
- New Mexico: Azure/Tyrone, White Signal, Old Hachita, Jarilla Mountains, Cerrillos.
- Colorado: La Jarra, Leadville, Villa Grove.
- California: Halloron Springs, Last Chance, Quartz Mountains.
- Nevada: Crescent Peak, Warm Springs, New Pass Range, Grass Valley, Crescent Valley.
- Sonora: Cananeita, Campo Frío, Arroyo Cuitaca, Los Campito, El Verde.
- Chihuahua: Mapimí.
- Coahuila: Beta Pérez.
- Zacatecas: Santa Rosa, Saucedo de Mulatos.
- San Luís Potosí: Salinas.

Most evidences for mining are represented by the occurrences of hammerstones and/or superficial quarries. Chambered mines for turquoise are rather rare, and mostly concentrated in the southwestern United States in areas where turquoise quality is the highest and the deposits are extensive. It should be noted, in addition, that our survey is ongoing and far from complete.

Aside from superficial quarries and shallow adits (the former frequent in the Nevada sources and the latter present at Canyon Creek, as examples), the best preserved examples of simple chambers come from the

Old Hachita area (fig. 3). Amid the numerous shallow pits, we found two shaft and chamber mines profiled by a wide contemporary trench excavated to rip open the turquoise deposit for easier access. The veins of turquoise here, as in most outcrops, are semi-discontinuous. At the La Jara outcrop, very narrow and constricted tunnels are visible. These are, for all practical purposes, crevice accesses following the interface of two different rock strata.

Clearly, excavations in hard rock for these stringers was arduous work. The scars of hammerstones and picks indicate that the mining technology was roughly the same for these mines as for the ones in the Chalchihuites area, even though the actual labor employed in mining was considerably more time consuming and difficult. No evidence of fire-cracked rock has been located to date.

Great open pits exist only in two localities: the Mt. Chalchihuitl pit at Cerrillos, New Mexico, and the Toltec mine, at Halloron Springs, California. Both open pits have been altered in significant fashions by historic and

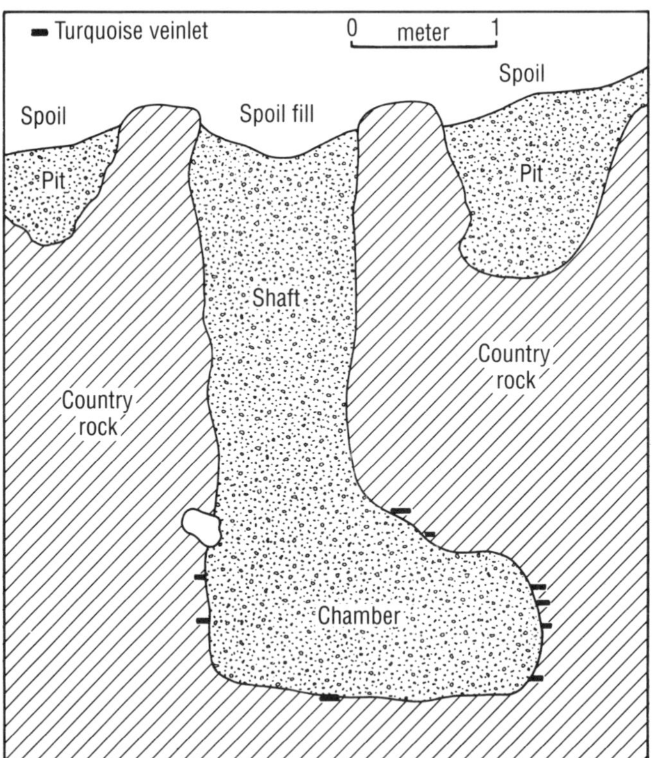

Fig. 3. Profile of sectioned pit-shaft complex located at Old Hachita, New Mexico.

contemporary mining, but both appear to be ancient in at least their basic configurations. The Mt. Chalchihuitl pit was close to the large Anasazi pueblo of San Marcos, and in a general area of farmland and settled populations. Exploitation of this pit was restricted during historic times, though the restrictions were often contested. Mathien and Warren (1985) have offered a good descriptive introduction to turquoise mining both at the pit and within the district. Thousands of tons of spoil have been removed from this open pit over the centuries. The last several excavations within the pit are crevices, and one of these involved ambitious underground tunneling.

The Toltec pit is located within one of the more extreme sectors of the Mojave Desert. There were no sedentary populations close by, nor agricultural resources for some distance. Even water is scarce. The chalky matrix preserves the scars of excavations quite well, but we did not see conclusive evidence for hammerstones or stone picks. As at Mt. Chalchihuitl, thousands of tons of spoil have been removed from this pit.

However, nowhere in the turquoise mining tradition in the northwestern sectors of pre-1848 Mexico would the careful architectural attention be given to develop large underground chambers and spacious tunnels as that given in the Chalchihuites zone. Hard rock vs. semi-consolidated alluvial gravel are certainly a major part of the explanation. The other part probably is the different levels of sociological commitment to mining. The Chalchihuites sociocultural system (defined as the system in and around Chalchihuites per se, which was dedicated to mining) was specialized in mining and mineral procurement in a fashion not commonly seen for other zones within Mesoamerica. They were obviously producing for export, and they were probably a colonial development (at least in response, if not actually in formation) to begin with. No society that mined turquoise appears to have been that committed and specialized.

Conclusions

We have observed that mining and miners played an important, indeed formative, role in the opening up and development of the colonial order in the northwest. One of Di Peso's many insights into the culture history of the northwestern sectors of pre-1848 Mexico was to recognize that this pattern may have had far older roots. His arguments lost force because of the difficulties of his chronology for Paquimé, and his insistence on a mercantile, or "*pochteca*," model for the interrelationship between the northwest and the core zones of the civilization. The emotional reaction of many archaeologists in the southwestern United States, who apparently felt that "their" cultures were losing something if somehow affiliated with Mesoamerica, meant that the basic contribution of his argument, which stressed systematic contacts between the northwest and the core regions, was ignored. The similar arguments forwarded by Kelley suffered more or less the same fate. By viewing mining as an indicator, rather than as an explanator, of the macroeconomic connective tissue between the northwest and the corelands of Mesoamerica, we can examine the zones of turquoise outcrops from the perspective of rare resource procurement in arid lands, and thus use economic theory to help explain these apparent interrelationships. The concepts of "trade structure" and "layered economies," promoted by Braudel (1982), for example, might help us avoid the emotionalism evident in assessing the sociocultural nature of the ancient societies along the contemporary Mexico/United States divide.

Acknowledgment

This work would have been impossible except for the laboratory and field aid of Dr. Garman Harbottle and Acelia García de Weigand, respectively. Dr. J. Charles Kelley and Dr. Charles Di Peso were very important to me in sponsoring and supporting the early phases of this research. The continuing intellectual support for our Ancient Mining Project by Dr. Kelley, Dra. Beatriz Braniff, and Dr. David Wilcox is gratefully acknowledged. All of us who study turquoise have an enormous debt to Joseph Pogue, whose classic 1915 study is always at hand.

References

Braudel, F. 1982. *Wheels of commerce: Civilization and capitalism...*, vol. 2. New York: Harper and Row.

Carmichael, E. 1970. *Turquoise mosaics from Mexico.* London: Trustees of the British Museum.

Crown, P., and Judge, W. J., eds. 1991. *Chaco and Hohokam: Prehistoric regional systems in the American Southwest.* Santa Fe, New Mexico: School of American Research Advanced Seminar Series.

Dean, J. n.d. Paper presented to the Amerind Foundation Conference in honor of Charles Di Peso. Dragoon, Arizona.

Di Peso, C. 1974-76. *Casas Grandes: A fallen trading center of the Gran Chichimeca.* 8 vols. Dragoon, Arizona: Amerind Foundation, and Flagstaff, Arizona: Northland Press.

Doyel, D., ed. 1992. *Anasazi regional organization and the Chaco system.* Maxwell Museum of Anthropology, Anthropological Papers #5. Albuquerque, New Mexico: University of New Mexico Press.

Harbottle, G., and Weigand, P. C. 1992. Turquoise in pre-Columbian America. *Scientific American* 266(2): 78-85.

Haury, E. 1976. *The Hohokam: Desert farmers and craftsmen.* Tucson, Arizona: University of Arizona Press.

Kelley, J. C. 1966. Mesoamerica and the southwestern United States. In *Handbook of Middle American Indians*, v. 4, eds. G. Ekholm and G. Willey, 95-110. Austin, TX: University of Texas Press.

——. 1971. Archaeology of the northern frontier: Zacatecas and Durango. In *Handbook of Middle American Indians*, v. 11, eds. G. Ehkolm and I. Bernal, 768-801. Austin, TX: University of Texas Press.

——. n.d. The ceremonial center in Chalchihuites culture. Ms.

Kelley, J. C., and Kelley, E. A. 1971. The decorated wares. Part 1 of *An introduction to the ceramics of the Chalchihuites culture of Zacatecas and Durango, Mexico.* Southern Illinois University Museum Mesoamerican Studies #5. Carbondale, Illinois.

Langenscheidt, A., and Lay, C. T. 1982. La Minería Prehispánica en la Sierra Gorda. In *Mining and mining techniques in ancient Mesoamerica*, eds. P.C. Weigand and G. Gywnne. Special issue: *Anthropology* 6(1&2): 135-148.

Lekson, S. 1986. *Great Pueblo architecture of Chaco Canyon.* Albuquerque, New Mexico: University of New Mexico Press.

López Austin, A., Lujan, L. L., and Sugiyama, S. 1991. The temple of Quetzalcoatl at Teotihuacan: Its possible ideological significance. *Ancient Mesoamerica* 2:93-105.

Mathien, F. J., and Warren, A. H. 1985. Prehistoric and historic turquoise mining in the Cerrillos district. In *Southwest culture history: Collected papers in honor of Albert Schroeder*, ed. Charles Lange, 93-127. Archaeological Society of New Mexico.

Reina Robles, R. M. 1989. El Preclásico en el Estado de Guerrero: Descubrimientos recientes. In *El Preclásico o Formativo, avances y perspectivas*, ed. Martha Carmona Macías, 65-80. Mexico City: Museo Nacional de Antropología/Instituto Nacional de Antropología e Historia.

Sahagún, B. de. 1989. *Historia general de las cosas de Nueva España.* Mexico City: Consejo de la Cultura y las Artes.

Thompson, J. E. S. 1978. An introduction. In *Maya hieroglyphic writing*, 3rd. ed. Norman, Oklahoma: University of Oklahoma Press.

Weaver, M. 1992. *The Aztecs, Mayas and their predecessors*, 3rd edition. New York: Academic Press.

Weigand, P. C. 1968. The mines and mining techniques of the Chalchihuites culture. *American Antiquity* 33:45-61.

——. 1982. Mining and mineral trade in Prehispanic Zacatecas. In *Mining and mining techniques in Ancient Mesoamerica*, eds. P. C. Weigand and G. Gywnne. Special issue: *Anthropology*, 6(1&2): 87-134.

——. 1992. The macroeconomic role of turquoise within the Chaco Canyon system. In *Anasazi regional organization and the Chaco Canyon system*, ed. D. Doyel, 169-173. Maxwell Museum of Anthropology, Anthropological Papers #5. Albuquerque, New Mexico: University of New Mexico Press.

Weigand, P. C., and Harbottle, G. 1992. The role of turquoise in the ancient Mesoamerican trade structure. In *The American Southwest and Mesoamerica: Systems of prehistoric exchange*, eds. J.E. Ericson and T.G. Baugh, 159-177. New York: Plenum Press.

Weigand, P. C., Harbottle, G., and Sayre, E. 1977. Turquoise sources and source analysis: Mesoamerica and the southwestern U.S.A. In *Exchange systems in prehistory*, eds. J. E. Ericson and T. X. Earle, 15-34. New York: Academic Press.

Wilcox, D. 1991. Hohokam social complexity. In *Chaco and Hohokam: Prehistoric regional systems in the American Southwest*, eds. P. Crown and W. J. Judge, 253-275. School of American Research Advanced Seminar Series, Santa Fe, New Mexico.

Windes, T. 1992. Blue notes: The Chacoan turquoise industry in the San Juan Basin. In *Anasazi regional organization and the Chaco system*, ed. D. Doyel, 159-168. Maxwell Museum of Anthropology, Anthropological Papers #5. Albuquerque, New Mexico: University of New Mexico Press.

Pre-Hispanic Metallurgy and Mining in the Andes: Recent Advances and Future Tasks

Izumi Shimada, Southern Illinois University, Carbondale, Illinois, U.S.A.

Abstract

A critical review of fieldwork and laboratory analyses conducted over the past 10 years defines current understanding of pre-Hispanic Andean metallurgy and mining, and elucidates future research tasks. While there have been various notable field discoveries, particularly on the north and central coasts of Peru, as well a valuable documentation of collections of metal objects, several problems in basic conception and organization of relevant research linger. Archaeometric analysis is becoming ever more sophisticated, but is still too dependent upon on skewed and/or unprovenienced samples. Field discoveries usually derive from fieldwork focused on issues unrelated to metallurgy and mining. In general, our current view of Andean metallurgy, including the behavioral and cultural significance of metal objects and scale of production, is too heavily based on isolated finds or examination of finished products from funerary or ritual contexts. We need comprehensive fieldwork focused on metallurgy, ideally centered at production centers. Examination of mines and their ores should form an integral part of such research, mirroring their functional interdependence with metallurgical sites. Pre-Hispanic mines identified thus far have been typically small and shallow and associated with small gangue and scant talus. Only through an interdisciplinary team approach with a sustained regional focus will we gain an integrated picture of metallurgical production and its products. It is also urged here that archaeologists include colonial sites and tackle worthwhile issues involving cross-fertilization of native and imported technologies and adaptation of the latter to the Andean environment.

Key words: Andes, Peru, pre-Hispanic mining, metallurgy.

Introduction

Grave looting, tribute collection, and the relentless search for silver and gold mines during the colonial era together with artifact collections established during the nineteenth and twentieth centuries have long indicated that the key loci of pre-Hispanic metallurgy and mining in the New World

In Quest of Mineral Wealth: Aboriginal and Colonial Mining and Metallurgy in Spanish America, edited by Alan K. Craig and Robert C. West, 1994. Geoscience and Man, vol. 33, pp. 37-73. Dept. of Geography & Anthropology, Louisiana State University, Baton Rouge, LA 70893-6010, U.S.A.

were situated within the Andean countries of western South America. Scientific studies that began around the turn of this century, in fact, have shown that Andean metallurgy clearly stood out in regard to technical innovations and sophistication, the extent and intensity of production, and the symbolic and economic importance of the products.

Today, at least three major loci of pre-Hispanic metallurgical developments have been recognized. A long succession of pre-Hispanic cultures on the north coast of Peru (table 1; fig. 1), including the Cupisnique,

Table 1: Succession of major pre-Hispanic cultures of the North Coast of Peru

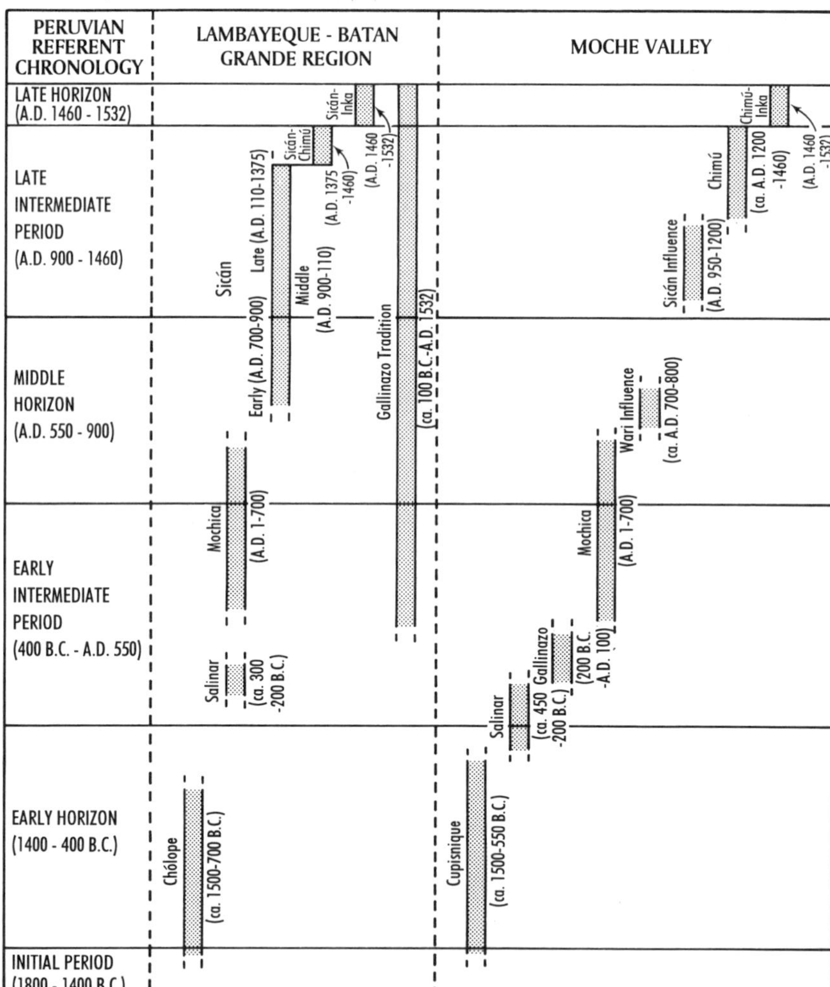

Fig. 1. Western South America showing archaeological sites and cultures mentioned in the text. (Batán Grande region rivers are shown in fig. 4.)

Mochica (or Moche), and Sicán (or Lambayeque —often misidentified as Chimú), excelled in sheetmetal working of copper, gold, and silver, as well as their varied alloys starting in the first millennium B.C. (Lechtman 1980, 1984a, 1988; Shimada 1994). Depletion-gilded *tumbaga* sheets[1] were widely used to emulate gold sheets during the Mochica and Sicán eras (Lechtman 1984a, b; Shimada and Griffin 1994; in press). Starting around A.D. 900, the large-scale production of arsenical copper[2] led to the permanent replacement of copper as the utilitarian metal of the north coast (Lechtman 1979a, 1981; Nordenskiöld 1921; Shimada 1985; Shimada and Merkel 1991). Chiripa, Tiwanaku (or Tiahuanaco) and other cultures in the extensive *altiplano* (high plateau) region covering far southern Peru and neighboring western Bolivia, northern Chile, and northwestern Argentina (fig. 1) are known to have developed another prominent metallurgical tradition focusing on copper and copper-tin bronze (González 1979, 1992a, b; Lechtman 1979a, 1980, 1988; Mathewson 1915; Nordenskiöld 1921). In this region, gold and gold alloys apparently never attained the level of prominence found among the pre-Inkaic north coast cultures, either qualitatively or quantitatively.

Diverse late pre-Hispanic cultures of central and northern Colombia (Muisca, Quimbaya, Sinú, and Tairona; fig. 1) appear to have had a different technological focus, emphasizing lost-wax casting[3] of gold alloys (Bray 1978; Plazas and Falchetti 1985, 1986, 1992). Though the artifacts are typically small personal ornaments, the complexity of form thus achieved is impressive.

The above pre-Hispanic developments were closely related to the reliable supply of minerals (including native metals) and their intrinsic qualities. For example, north coast and *altiplano* cultures developed two distinct bronzes, arsenic-copper and copper-tin, respectively, based on locally available ores (Merkel and Shimada 1988; Merkel et al. 1994; Shimada and Merkel 1991; cf. Lechtman 1976, 1979a, 1991). Large-scale copper and tin mining continues to this day in the *altiplano* and adjacent slopes, while the presence of arsenic in the former has discouraged modern exploitation of the local copper deposits.

The zeal with which colonists searched for gold and silver sources throughout the Andes has been extensively documented (Berthelot 1978, 1986). Even rudimentary colonial descriptions and drawings of indigenous *plateros* (gold- and silversmiths) and their techniques have survived. Colombian gold working seems to have been based on locally abundant placer gold (Plazas and Falchetti 1985, 1986, 1992). At the same time, the sources of gold for the prominent pre-Hispanic north coast gold-working tradition have remained an enigma (see below; also Shimada and Montenegro in press).

The above is a broad-stroke, preliminary characterization that awaits a great deal of elaboration and verification. Most likely, it is overly simplistic and underestimates the creative technical and technological intercourse among the loci identified thus far. For example, the recent documentation of copper-tin bronze in Colombia (Scott 1980), a copper alloy with high bismuth content at Machu Picchu (Gordon and Rutledge 1984), and copper-arsenic and copper-arsenic-tin bronze in northwest Argentina (Fester 1962; González 1979) adds wrinkles to the above characterization.

There has been little sustained research clearly focused on pre-Hispanic metallurgy and mining with well integrated field and laboratory facets. Thus, often new field finds are the coincidental or tangential products of research dealing with issues other than metallurgy or mining. Such research usually lacks necessary specialists and/or sampling schemes. Even when a battery of advanced analytical techniques is brought to bear for detailed metallographic and compositional characterization of metal objects, the range and depth of behavioral and cultural inferences that can be gained are limited as the objects are frequently unprovenienced or derived from funerary settings. Though the symbolic importance of Andean metal objects has been emphasized by some researchers, the overwhelming portion of analyzed objects derives from graves heavily imbued with symbolism. The sample of metal objects to be analyzed needs to be more representative of the total range of contexts in which they were used before we can properly define their cultural value. In general, laboratory analyses without benefit of updated field data have led to various misconceptions and erroneous conclusions.

In spite of these problems, over the past decade we have seen many important advances in our understanding of pre-Hispanic Andean metallurgy and mining. The number of archaeological publications reporting metallurgical activities has increased as archaeologists have become more aware of their diagnostic material evidence and the cultural importance of metal and its production. Heather Lechtman's research in Andean archaeometallurgy has been instrumental in bringing about this heightened awareness. This paper discusses those advances, particularly in Peru and its immediate vicinity, and, based on my experiences over the past 15 years as an archaeologist interested in this topic, proposes a research approach and directions to redress the above problems. It is intended to complement the fine overviews of Andean archaeometallurgical research and the state of our understanding up to the mid-1980s offered by Bray (1985) and Lechtman (1980,1988).

Pre-Hispanic Metallurgy

The last 10 years have seen systematic documentation of extant collections and field discoveries that have filled major gaps in the basic temporal-spatial parameters of pre-Hispanic metallurgical and mining activities. Under the aegis of the Kommission für Allgemeine und Vergleichende Archäologie of the Deutschen Archäologischen Instituts, Bonn, E. F. Mayer (1986, 1992) has published major compilations of pre-Hispanic metal arms and tools in Argentina, Chile, and Ecuador. A companion volume in this series is an exhaustive compilation of mostly cast bronze plaques from northwest Argentina and the surrounding region by A. R. González (1992a). He discusses the symbolism and cultural processes associated with their diverse forms and evolution over a some 2000-year span (up to early colonial era; also see González 1979; 1992b). These three volumes will undoubtedly remain as the reference works for many years to come. Though not as detailed, the Banco de la República of Colombia under the guidance of C. Plazas and A. M. Falchetti (1985, 1986, 1992) has maintained a similar interest in the basic documentation and diffusion of knowledge concerning Colombian goldwork.

Field discoveries have been more numerous but mostly coincidental to other research interests. We will examine some of more notable examples from south to north, oldest to the most recent. Sites and issues relating to the *altiplano* metallurgical tradition have received little attention in the past relative to the north coast of Peru. This situation is likely to change with the recent surge in archaeological fieldwork there. The initiation of fieldwork at the site of Ramaditas in northern Chile (Graffam et al., this volume) holds much promise to clarify what should be a long sequence of copper and copper alloy metallurgy in this important region. As part of the current popularity of the archaeology of Inka Empire in northwest Argentina, the late pre-Hispanic exploitation of regional mineral resources and associated metallurgical production has come to receive more attention (D'Altroy 1993; González 1983). However, as seen in the aforementioned works by González and Mayer, there remains much to be elucidated about the pre-Inka metallurgy of northwest Argentina. Clarification of developments in western Bolivia will also be pivotal in advancing our understanding of the *altiplano* metallurgical tradition, which appears to have a complex character with copper-arsenic, copper-tin, and bismuth bronzes (Gordon and Rutledge 1984; 1987). In spite of a few tantalizing reports on early copper metallurgy (see Lechtman 1979a for summary), detailed, follow-up studies have not been published.

We still do not know when copper metallurgy began in Peru but it is likely to date back to the second millennium B.C. Important in this regard is the recent discovery of fragments of thin copper foil in the construction fill of the "west wing" and on the superimposed floors atop the central mound of the U-shaped pyramid complex of Mina Perdida in the Lurín Valley just south of Lima (R. Burger 1994, pers. comm.). The fragments have been dated to the late Initial Period based on associated ceramics and radiocarbon dates (2870±90 BP and 2960±90 BP [I-16,762 and 15,557]). Analyses by Robert Gordon indicate that the foil was produced by the cold hammering of native copper. Evidence of annealing was found on one sample.

This discovery is not surprising given indications of some differentiation in offerings among contemporaneous graves at the site and previous reports of copper mining in the adjacent highland region of central Peru and inferred coast-highland trade in copper by 1200-650 B.C. (Browman 1975; MacNeish et al 1975; also pers. comm. from Browman cited by Lechtman 1980, 281). Ongoing excavation of the necropolis at Tablada de Lurín (Makowski and Cornejo 1993; Makowski, pers. comm. 1993) has yielded some small copper, gilded copper, and *tumbaga* personal ornaments. Copper and gilded copper objects recovered earlier from the same cemetery once thought much older, are now dated to the transition between the Early Horizon and the Early Intermediate Period, ca. 200 B.C. to A.D. 100 (95 percent confidence level), roughly coeval with Mochica style objects from the Vicús region some 900 km to the northwest (fig. 1). The upper reaches of the central coast and contiguous highland region of Peru are rich in copper, silver, and other ores and it would not be surprising to learn that the region supported another important center of pre-Hispanic metallurgy. We await better archaeological documentation of the beginnings of copper mining and metallurgy in the central highlands. The importance of metals in the late prehistory of this area is clearly indicated by the relative abundance of "copper and silver" artifacts recovered in excavations of Wanka II and III (Late Intermediate Period and Late Horizon) burials and households in the upper Mantaro Valley (Earle et al. 1987, 98). Inka domination (Wanka III) apparently was accompanied by a marked increase in the availability of copper (copper-tin bronze) to commoners. A concurrent decline in the number of silver objects is believed to reflect the Inka's attempt to control prestige items. Since local production of silver is inferred from the presence of small lumps of lead (i.e., silver as an impurity of the lead ore), the levels of silver production before and after the arrival of the Inka remain to be clarified.

The custom of placing copper discs in the mouths of the deceased was persistent and widespread on the northern coast of Peru (Donnan and Mackey 1978). At Puémape in the lower Jequetepeque Valley, a Cupisnique burial was found with a gilded copper disc in the mouth (Elera and Pinilla 1990). Two radiocarbon assays of associated samples suggest that this burial may well date between ca. 1500 and 1300 B.C. (corrected dates at 68 percent confidence level; J. Pinilla, pers. comm. 1993), a surprisingly early date for gilded copper. A later Cupisnique burial (estimated 800-700 B.C.) excavated at Mound II, Huaca La Merced in the La Leche Valley farther north was found with a gold ring and copper nose ring (Maguiña

1992). An Idol Phase burial at Kuntur Wasi (prior to "Chavín influence") with two copper discs has an associated radiocarbon date (uncorrected) 2860 ± 60 B.P. (910 ± 60 B.C; University of Tokyo Ancient Andean Civilization Research Project 1992). It seems likely that copper-based metallurgy was established in northern and/or central Peru at least by the latter part of the second millennium B.C.

The beginning of metal use in the Andes is commonly assumed to have been the direct working of gold with stone hammers (Lechtman 1980, 275). However, as seen above this is no longer certain. The earliest known "gold worker's tool kit" and gold foil comes from the site of Waywaka (Grossman 1972) some 200 km. west of Cuzco and has been dated to ca. 1500 B.C., but improved understanding of the associated ceramics now suggests a date closer to 1000 B.C.

Reflecting the intensity of archaeological fieldwork and grave looting, most of the recent discoveries relevant to early gold working in Peru come from the Lambayeque and adjacent valleys on the north coast. For a long time, due to the paucity of archaeological attention, a cache of Late Cupisnique[4] (400-200 B.C. ?) style gold crowns and other sheet ornaments recovered in Chongoyape in the upper Lambayeque Valley (Lothrop 1941, 1951) seemed to be an isolated example. We now know that this is not the case. Excavated burials at the Late Cupisnique ceremonial-residential site of Morro de Etén (mouth of the Lambayeque Valley) contained various small personal ornaments fashioned out of sheet gold (Elera 1986, 1992). A gold crown with an elaborate cut-out design looted from a sand dune west of Cerro Ventarrón in the same valley is technically and stylistically quite similar to a group of Cupisnique style crowns and pectorals found in the tombs at Kuntur Wasi in the upper Jequetepeque Valley to the south (University of Tokyo Ancient Andean Civilization Research Project 1992). The latter has five associated radiocarbon dates (uncorrected) ranging from 2410 ± 50 to 2710 ± 80 BP (460 ± 50 to 760 ± 80 B.C.). Numerous "Cupisnique" gold objects have been attributed to nearby Cerro Corbacho in the mid-Zaña Valley (Alva 1992), though many are of questionable authenticity on stylistic and technical grounds. Another cache of sheet gold ornaments bearing repoussé designs in a local variant of Cupisnique iconography was found at Loma de Macanche in the upper Piura Valley to the north (J. Centurion pers. comm. 1992). In all these cases, gold objects were not found with copper objects and the relationship between the presumably coterminous sheet gold working and copper-based metallurgy remains unclear. From the beginning of its use, gold may have been reserved for elite burials and rituals.

Other recent findings from the Lambayeque region, notably the results of the excavation of the Mochica elite tombs at Sipán (ca. A.D. 1-350; Alva 1988, 1990; Alva and Donnan 1993) and the ongoing Sicán Archaeological Project centered in Batán Grande (see below), have reinforced the view that the region was one of the primary metallurgical centers of the New World throughout much of its prehistory.

Numerous objects of gold, silver, copper, and their alloys from the Sipán tombs (fig. 2) clearly justify the fame the Mochica metallurgy had already gained for its sophistication and innovative quality (Donnan 1992; Jones 1979; Lechtman 1980, 1988; Shimada 1994). However, the broader sociopolitical and economic contexts of the Sipán tombs remain to be clarified (see Shimada 1994). Along with advances in the conservation of excavated metal objects (Alva et al. 1989), we also await the results of their technical analysis and what they may tell us of the metals employed as well as the techniques and organization of their production. The gilded copper objects from Sipán would serve as a good test case for the electrochemical gilding process proposed for the Mochica gilded copper pieces from the Vicús region some 200 km to the northwest (Lechtman 1979b; Lechtman et al. 1982) that are technically and stylistically quite similar. Donnan (1990, 32-33) argues that these objects belonged to the Mochica "royalty" in each valley and suggests that the tradition of interring the de-

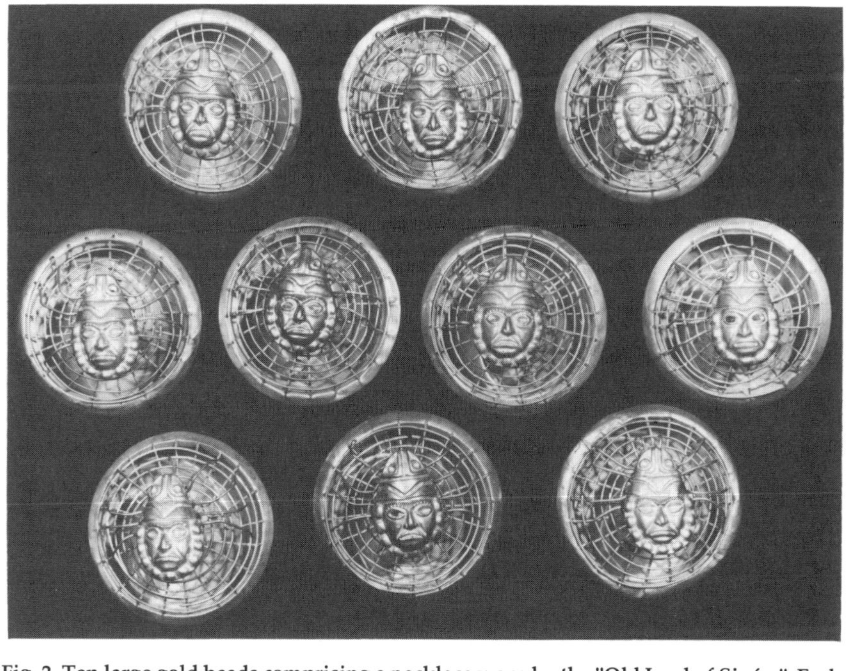

Fig. 2. Ten large gold beads comprising a necklace worn by the "Old Lord of Sipán." Each bead depicts a spider on a web. The body of the spider is a man's head. Diameter c a. 8.3 cm. Photo taken by Yutaka Yoshii with permission of Walter Alva.

ceased elite with his accumulated wealth provided a continuous demand for sumptuary goods and thus "substantial numbers of skilled craftsmen" were maintained by the inferred royal courts. There are at least two sites in the nearby Zaña Valley that show evidence of Mochica copper production (Shimada 1994; also Lechtman 1976), but, to date, we cannot specify where the precious metal objects found at Sipán were produced. Similarly, no gold workshops or royal palaces have been identified in and around Vicús, though some remains of copper working have been found nearby (Kaulicke 1991b, 413-414; Shimada 1988).

The technical and technological questions surrounding Sipán and Vicús metal objects are essentially the same and may be best seen from a macro-regional perspective that extends northward to the La Tolita-Tumaco region straddling the Ecuador-Colombia border (fig.1). The site of La Tolita with numerous *tolas* or earthen mounds on an island at the mouth of Santiago River where placer mining of gold still continues has long been the scene of intense grave looting for gold and other precious metal ornaments. The classic study by P. Bergsøe 1937, 1938) of tools, unfinished objects, and other workshop remains, as well as finished ornaments showed the diversity and sophistication of metalworking techniques used, including gilding, granulation, and various chemical (metallurgical) joinings. His claim of wire-drawing is questionable, but platinum working has since then been confirmed (Scott and Bray 1980; Bray 1988; also see Scott and Doehne 1990). Recent archaeological fieldwork at and around the site of La Tolita suggests that it served as a major regional metalworking center during the first several centuries A.D. with its beginnings dating back perhaps as early as the fifth century B.C. (Tolstoy and DeBoer 1989; Valdéz 1987). The factors and processes responsible for its emergence remain uncertain, though the local presence of placer gold and easy access by watercraft[5] were undoubtedly critical. The above dating means that La Tolita metallurgy was contemporaneous with, if not somewhat earlier than, the development of Mochica metallurgy in the Lambayeque-Vicús region to the south.[6] Long-term investigations by J. F. Bouchard and colleagues (Bouchard 1984; Patiño 1988; Scott and Bouchard 1988) and the Pro Calima Foundation in the neighboring Colombian coastal regions of Tumaco and Calima, respectively, have also shown coterminous developments in gold working. In the Calima region, sheet gold working that began during the Ilama phase (the first millennium B.C.; Cardale de Schrimpff 1992) reached its height during the succeeding Yotoco phase (ca. A.D. 1 to 800) with mastery of sheetmetal and lost-wax casting of gold and *tumbaga* (Bray 1992).

As a whole, the above clustering of dates, the abundance of placer gold in the Pacific coastal rivers of northern Ecuador and southern Colombia, and the scarcity of gold on the north coast of Peru suggest that the Mochica metalworkers obtained at least some of their gold from the Ecuadorian and Colombian coast and engaged in a creative exchange of technical and technological information. La Tolita and the surrounding coastal region should not be seen as merely a passive receptor of innovations emanating out of the north coast of Peru. It is further suggested here that this interaction was part of the trade (perhaps maritime) for exotic elite and ritual goods that included tropical shells (*Spondylus, Strombus,* and *Conus*) pearls, coral, and emeralds (Shimada 1994). In this regard the discovery of La Tolita style ceramics and gold idols in Frias near Vicús (Jones 1979; Kaulicke 1991b) is not surprising. How the recently reported discovery of gold, gold-silver alloy, and gilded copper objects at Pirincay near Cuenca in southern Ecuadorian highlands (Bruhns 1990) fits in with these coastal developments remains to be seen.

Coastal Ecuadorian metallurgy of later periods has also received much deserved attention. The long-standing effort by O. Holm (1966/67, 1975) to better understand "axe-monies" was taken to another level by Hosler et al. (1990) in their comprehensive documentation and analysis. In spite of some colonial documents pertaining to their uses, the analyzed specimens come from disturbed contexts and their dating is neither secure nor precise. Hosler et al. (1990) suggest similar objects (called *naipes;* fig. 3) that occur primarily in Sicán territory on the north coast of Peru (Lambayeque to Chira; see Estrada 1961) were articles of clothing, basing their interpretation on a single, unique looted specimen. Yet, thousands of excavated *naipes* from intact Sicán graves as well as their limited occurrence on coastal Ecuador suggest they functioned as status items and/or primitive currency (Shimada 1985; Shimada and Montenegro in press). The preceding is an extension of Hosler's (1988) important study of cultural and technological linkages between west Mexican and coastal Ecuadorian metallurgy. She suggests the maritime transmission of copper and copper alloy technology to west Mexico during the last 700 years of prehistory. However, this study too fails to take into account recent field data that bring into question her proposed sequence of metallurgical developments on coastal Ecuador and their transmission to west Mexico (Sutliff 1992; Pollard 1993). Better dating of west Mexican specimens should also help resolve this problem.

Pre-Hispanic Mining

Our understanding of pre-Hispanic mines and mining as well as local mineral resources for much of the Andes is still in its infancy and, alas, advancing at snail's pace. Underlying this situation are conceptual and practical problems. While the cultural significance of metal objects and their production has come to be widely recognized among Andean archaeologists, there is still a strong tendency to treat mining and metallurgy as an-

Fig. 3. Sample of *naipes* placed as part of offerings at the bottom of the Huaca Las Ventanas tomb excavated in 1991. Each *naipe* measures about 5.3 cm by 7.0 cm. Proton-induced x-ray emission analysis of 30 *naipes* from Batán Grande indicates that they are all made of arsenical copper (ca. 1.7 to 5.6 percent As typically around 3.0 to 3.5 percent). Photo by Izumi Shimada.

alytically distinct. Lechtman's general survey of colonial and pre-Hispanic mines and metallurgical sites in the Andes (1976) was significant in treating them in a systemic manner and trying to define their interrelationships. In this way, the mineral bases of the north coast and *altiplano* metallurgical traditions and their respective bronzes became quite clear. In addition, her work effectively complemented the earlier survey and systematic ore and metalworking tool collections made by the nineteenth-century Italian naturalist, Antonio Raimondi. They are today found at the American Museum of Natural History in New York and the National Engineering University in Lima, respectively.

Field surveys of ancient mines are not easy to conduct due to the difficulties in identifying them and gaining access. In her survey, Lechtman (1976) was not able to conclusively identify any pre-Hispanic mines. In a later review of relevant literature, V. P. Oehm (1984, 27) concluded that the Cerro Blanco copper mine in the Batán Grande region on the north coast (Epstein and Shimada 1984; Shimada et al. 1982) was the only credible pre-Hispanic mine in Peru identified up to that time. Lechtman's survey focused on relatively accessible, large and well-known sites that were likely to have suffered post-Conquest mining or other disturbances. In addition, she (1976) suspects that some pre-Hispanic mines were intentionally buried at the time of the Spanish Conquest to prevent the Spaniards from finding and exploiting them, and that many were too small to locate in a macro-scale survey.

For our regional survey of ancient mines in the Lambayeque and adjacent La Leche and Zaña valleys (spanning 1984-1991), we assembled an interdisciplinary team consisting of a geologist (A. K. Craig), archaeometallurgist (J. F. Merkel), local mining engineer (J. L. Suárez), and myself as archaeologist. Our approach had a multi-stage character (Shimada 1984; Shimada and Merkel 1991). Prior to initiating fieldwork, the project historian S. Ramírez (n.d.; also in this volume) was consulted for any archival information on pre-Hispanic and colonial mines, but specific data on copper mines and mining that could guide our fieldwork proved quite limited. Concurrently, geological and natural resource maps available through the Instituto Geográfico Militar and the Oficina Nacional de Evaluación de Recursos Naturales (ONERN), both in Lima, were examined for identification of igneous intrusions and associated mineralizations. Available low-altitude air photographs were also studied to identify ancient roads and workings and associated talus formations. These phases were followed by interviewing of local informants such as animal herders, hunting guides, and prospectors familiar with the local terrain. The last approach proved to be most valuable in locating mines. Encouragingly, the locations of mines identified by informants were often within small areas earmarked for intensive survey through the other avenues described above. Field documentation and ore sampling followed. Active participation of geologists and archaeometallurgists in these mine surveys and examination of excavated ore specimens is crucial to assure that ore samples are appropriate for subsequent analysis (Merkel et al. in press).

We suspect that pre-Hispanic miners were originally guided to exploitable mineralizations by "floats" or weathered ore fragments washed down from sources farther inland (A. K. Craig, pers. comm. 1984). Distinctive colors and even smell could have been used in identifying the ores (Merkel and Shimada 1988).[7]

Overall, the mines we have identified as pre-Hispanic in age (fig. 4) were quite small. With their emphasis on copper oxide ores found close to the surface, their workings were in general quite shallow, with trenches and tunnels generally less than 30 m in length and depth. The size and shape of workings were largely dictated by the nature of the target veins and limitations of primitive stone hammers. Using characteristic, rough hammers (ca. 20-23 cm long and 12-15 cm wide) with blunt points, much like those found with the celebrated Copper Man of the Atacama Desert in

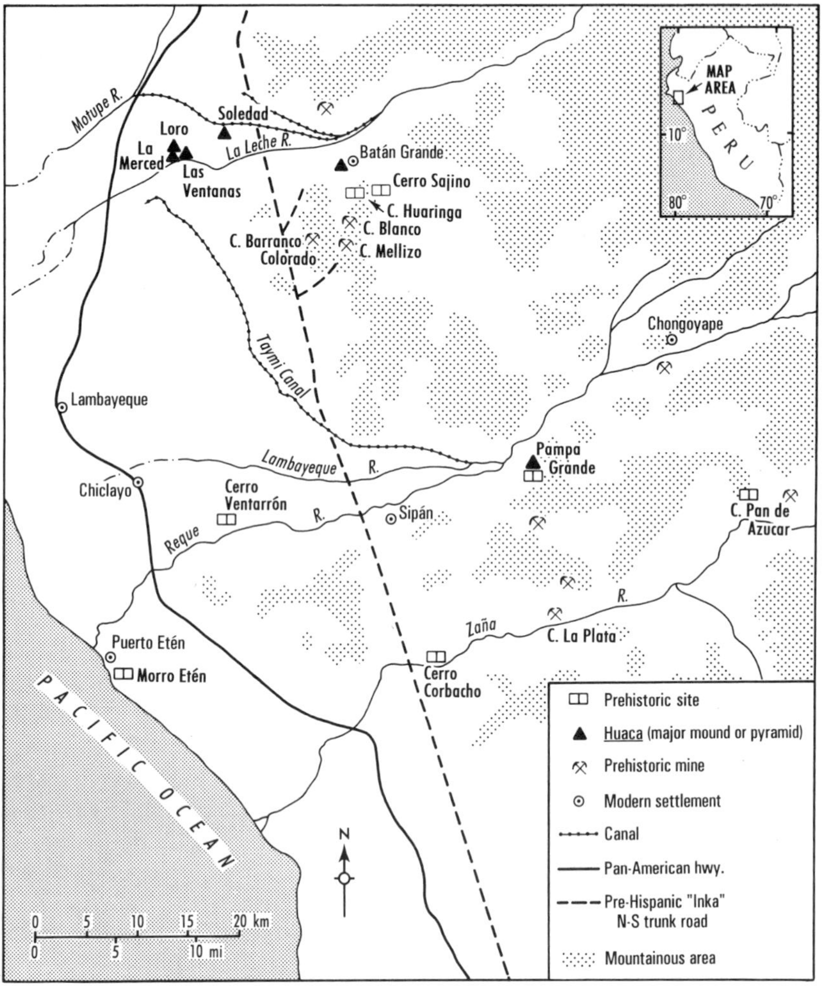

Fig. 4. The Batán Grande region with the locations of pre-Hispanic metallurgical centers and associated copper mines.

north Chile (fig. 5; Bird 1979), pre-Hispanic miners selectively removed only desired high grade ores. We found that shafts or tunnels were usually narrow, sinuous, and showed a fish scale-like pattern of small depressions from battering with stone hammers. In places, these tunnels could be negotiated only by crawling single file (we suspect children assisted in removing the ore). In the case of the Cerro Barranco Colorado, pre-Hispanic miners began removing the richest ore from the surface at the top of the mountain and gradually descended following the vein toward the center and base of the mountain (fig. 6). The narrow trench (ca. 1.5 m wide) ex-

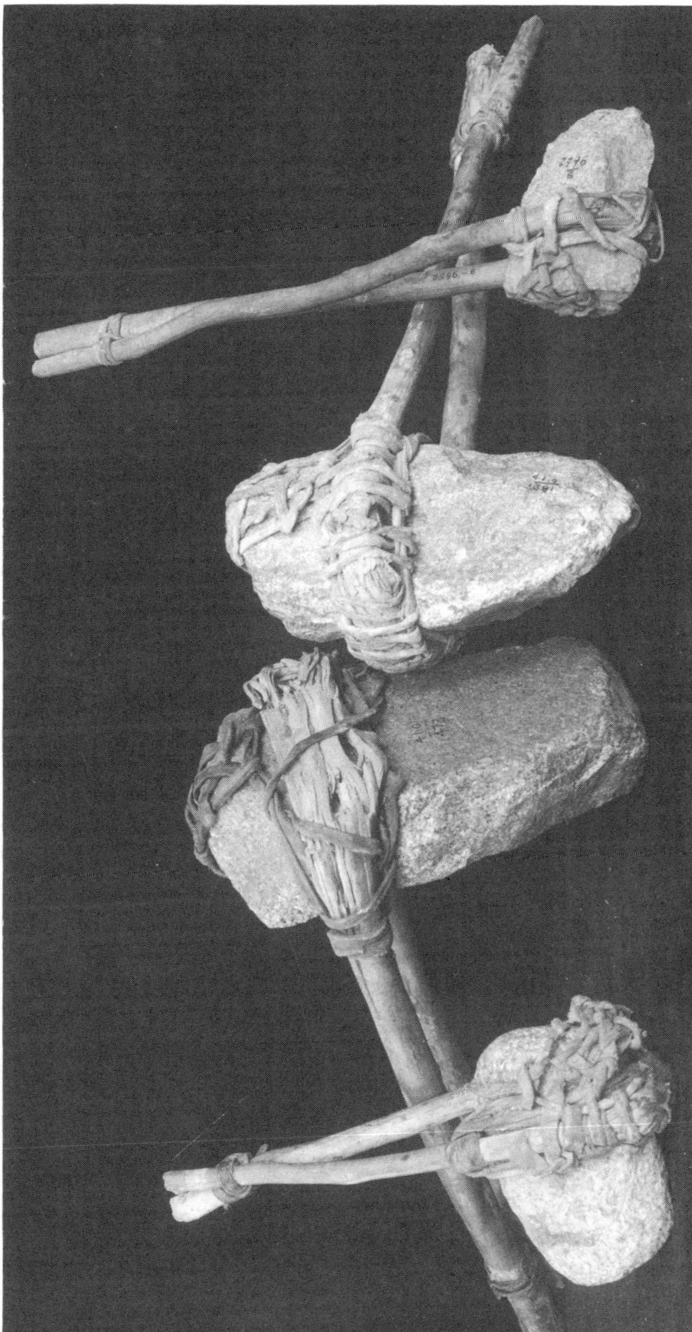

Fig. 5. Well-preserved hafted stone hammers found associated with the "Copper Man" found at Chuquicamata, northern Chile. The hammerstones we found in pre-Hispanic mines surveyed in the Lambayeque region resemble these. Courtesy of the Department of Library Services, American Museum of Natural History, New York. Negative No. 39424.

Fig. 6. Looking toward pre-Hispanic workings on the north face of Cerro Barranco Colorado south of the modern village of Batán Grande. Photo by Izumi Shimada.

tended for about 100 m down the slope. In spite of the narrow opening, some 3 to 5 m below surface near the top of the Cerro, the miners had dug a spacious subterranean stope (2 to 3 m high and 8 to 9 m wide and long) from which a shaft (which we had not yet explored) was excavated. Sometimes, mineralized layers with a shallow angle of dip were exploited as open pits as in the case of Cerro La Plata in the Zaña Valley.

Charcoal smudges found on the ceilings in some workings (Cerro Blanco) may indicate the use of torches for light or even for heat-cracking of ores. Copper or bronze tools may not have been widely used for mining as we found none of the sharp grooves expected from such tools.

The above technique produced little gangue and thus the associated talus near the mouths of mines is *not only quite small in extent but the constituent rocks were also uniformly small in size.* Sometimes, the floors of tunnels are quite clean of ore fragments. Spacious, straight and multi-level (i.e., deep) shafts and tunnels; as well as large mouths and talus are more characteristic of colonial and modern mines dug in much larger scale using different technologies.

Artifact remains within workings are rare and essentially limited to a few discarded stone hammers. However, along ancient roads leading to mines and near the workings we invariably find pre-Hispanic sherds and/

or stone architecture. The cases of Cerro Mellizo and Cerro Barranco Colorado are instructive in this regard. Segments of ancient roads leading to these mines were identified in airphotos and later in the field branching off to the east from the "Inka North-South Trunk Road" that connected the Lambayeque and La Leche valleys. The branch road to Cerro Mellizo had been modified by modern miners who had tested the ancient mine for commercial venture. The Trunk Road has another branch farther north that leads to the Sicán metallurgical centers of Cerro Huaringa (also called Cerro de los Cementerios) and Cerro Sajino near the modern village of Batán Grande (fig. 4; see discussion below). Mid-way between the Cerro Mellizo mine and the Trunk Road (distance of ca. 5 km) on the bank of a seasonal stream, there is a small cluster of crude stone structures with scatterings of copper oxide ores and fragments of utilitarian ceramic vessels (difficult to date being non-diagnostic). Near the base of the main workings at Cerro Mellizo are the remains of two sets of standardized masonry rooms built atop stone terraces (fig. 7). Nearby sherds of Sicán and Chimú (or Chimú-Inka) style mold-made blackware and paddled vessels were recovered. The structures were free of habitational debris and may have been built to store supplies and ores during Chimú or Inka domination of this area in the fifteenth or sixteenth century. The discovery of a cache of-

Fig. 7. Wall foundations of standardized structures built atop a stone-lined terrace near the mouth of the Cerro Mellizo mine south of the modern village of Bátan Grande. Photo by Izumi Shimada.

fering of *Spondylus princeps* shells (presumably imported from coastal Ecuador) and a bowl covered with a plate near the base of the workings suggests the symbolic importance of the mine. Colonial documents also speak of the indigenous perception of mines as *huacas* ("sacred places" in Quechua; Ramírez n.d.).[8]

In the flat area at the north base of Cerro Barranco Colorado we found a small multi-room masonry structure with numerous plainware sherds and a small *batán*.[9] The house is believed to have served as a miners' residence. Nearby circular enclosures with low, coarsely built fieldstone walls (ca. 4 to 6 m in diameter) may have been corrals for llamas used to transport provisions and ores. Similar cultural remains were found at other mines we studied in the Lambayeque and Zaña valleys. Although Sicán, Chimú and Chimú-Inka style ceramics were identified at some of these mines, we have not found any diagnostic Mochica ceramics. The absence of the latter is curious given the abundance of copper and copper alloy objects in the elite tombs excavated at the nearby site of Sipán and the presence of copper workshops documented at another nearby Mochica site, Pampa Grande (Shimada 1978; 1994). At least one Mochica site (phases III-V) in the upper Zaña Valley had surface scatters of copper oxide ores and semi-refined pure copper ingots as a grave offering (fig. 8).

There is evidence of even earlier mining in the Lambayeque Valley. At Morro de Etén, gold may have been mined during the first millennium B.C. Thin-section microscopy of a mineral sample collected by A. K. Craig (pers. comm. 1984) from an ancient mine on the east slope of the Morro (fig. 9), associated with stone walls and sherds, revealed a concentration of adularia, low temperature feldspar typically found with Tertiary gold and silver ores of the bonanza type. In other words, a local vein could have yielded a pocket or mass of gold sufficient to account for the sheet gold objects found at the associated Late Cupisnique site (Elera 1986, 1992). However, no evidence of in situ gold working was found.

As a whole, available data indicate that the prominent pre-Hispanic copper and arsenical copper metallurgical tradition of the Lambayeque region was based on oxide ores from dispersed local mineralized zones associated with Tertiary intrusions. Only high grade ores (ca. 25-30 percent copper) were extracted by carefully following the veins. Ancient miners thoroughly combed the local terrain exploiting numerous small and large superficial ore deposits. We have not found any one disproportionately large mine that could have satisfied the long-term demand for local copper and arsenical copper production. Even for the cluster of three excavated Sicán arsenical copper smelting sites around the town of Batán Grande (see below), ore from the relatively large and close Cerro Blanco mine appear to have been supplemented with ore from the Cerro Mellizo and Cerro Barranco Colorado mines farther south. As noted above, this situation

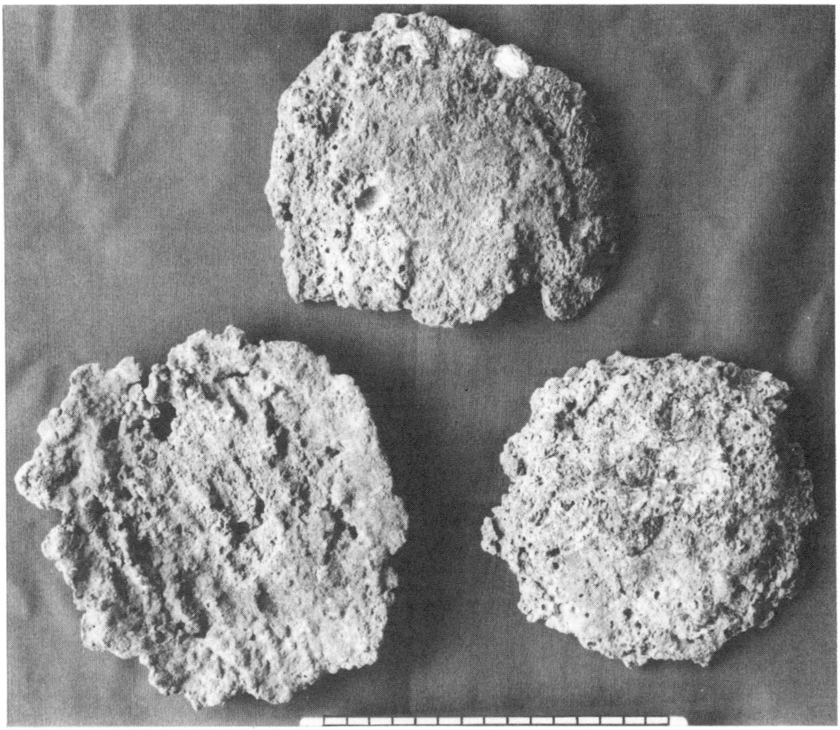

Fig. 8. Three circular ingots recovered from a looted cemetery near Cerro Pan de Azucar on the north bank of the mid-Zaña Valley. The two smaller ingots are shown bottom up. Although the ingots were not highly refined, analysis reveals that the metallic component is nearly pure copper. The scale at the bottom of the photo is 15cm long. Photo by Izumi Shimada.

is likely to reflect the dependence on oxide ores, limited means of ore extraction, and nature of mineralization. However, the documented continuity of Sicán smelting technology over a span of 600 years suggests that this ore supply was adequate.

Holistic Study of Sicán Metallurgy: Approach and Major Findings

The broad anthropological understanding of Andean metallurgy anticipated by Bergsøe (1937; also see Bray 1985) and urged more recently by Lechtman (1979a, 1984a) and others (Shimada 1985; Shimada and Merkel 1991) requires elucidation not only of the technology and products but also their broader cultural and environmental contexts and impacts, akin to the aims of "ceramic ecology" espoused by F. Matson (1965). In this conception, research into the nature of the smelting charge and internal workings of the associated smelters is just as important as the mechanisms

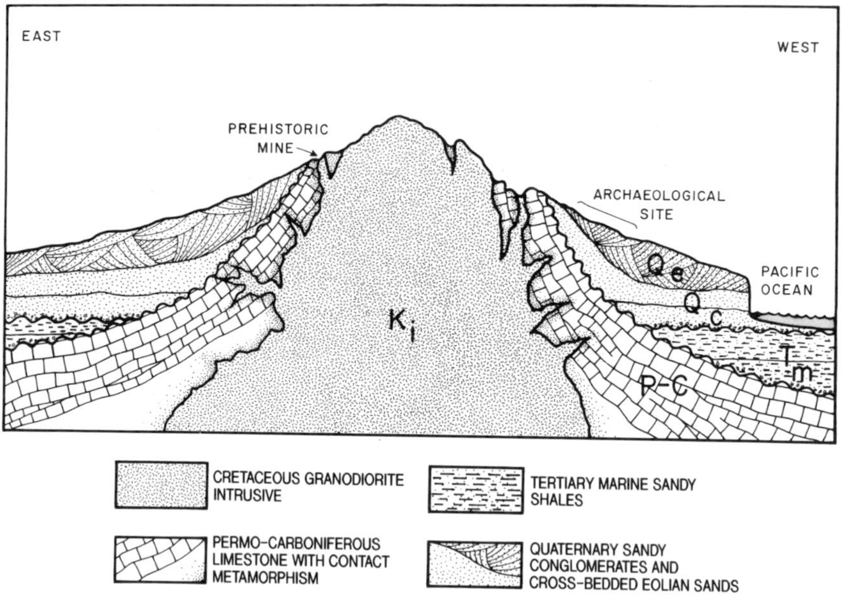

Fig. 9. Schematic drawing of Cerro Morro de Etén with location of prehistoric mine and associated archaeological site. Drawing by Izumi Shimada based on original by Alan K.. Craig.

involved in acquisition of fuel and ore, productivity and associated labor and administrative organization, and social and religious value of the products. Archaeometallurgical studies in both the Old and New Worlds have focused too frequently on a limited phase or facet of the metallurgical process and technology and been conducted too often without adequate anthropological input and concern. Though for heuristic purposes we may distinguish "stages" in metallurgical production, they are in reality closely interdependent and integrated. Moreover, attendant to this fragmented approach, these studies have commonly pursued one or the other of two polarized levels of analysis, either characterizing the elementary composition using high technology and powerful analytical tools or modeling the total production system based on a series of assumptions regarding economic behavior and productivity. Much of the middle ground is left unexplored or unintegrated. What we need are comprehensive studies of all phases and levels (from individual craftsmen to workshops and beyond) of metallurgical production. The behavioral significance of detailed metallographic and compositional data obtained from analyses of end products should be evaluated with the aid of insights from replicative experiments, expert metalworkers, and analysis of tools and their marks.

Further complicating matters, many stylistic, iconographic, and/or technical studies have been carried out on objects in collections without proper understanding of various "conservation" measures taken before they reached collectors or museums. Many of these measures are of questionable value and validity, and the present condition and appearance cannot be taken at face value. In other words, study of the original appearance and use, as well as of the techniques and materials used in manufacture should be based on those objects scientifically recovered from intact contexts.

The sort of "holistic understanding" being urged here requires sustained long-term research specifically focused on metallurgy and related mining and close interdisciplinary collaboration (Shimada and Merkel 1991; Shimada and Griffin 1994; in press). Specialists should be involved both in and out of the field in all phases of research. Further, it calls for in-depth regional study and sampling of multiple sites of different era and character (from industrial to residential and ceremonial centers) to assess the short- and long-term impacts of metallurgical production, and to clarify local resources and pertinent sociopolitical developments.

The Sicán Archaeological Project centered in the Batán Grande region of the La Leche Valley serves to illustrate the productiveness of this approach. Since its inception in 1978, we have been interested in reconstructing the technology and organization of metal production and defining the meaning and roles of metal products within the Middle Sicán culture that dominated much of the north coast during A.D. 900-1100. Our interest encompasses the perception and procurement of raw materials and actual production of metal objects as well as the mundane and ritual uses of the products.

To achieve this goal, we need a multitude of perspectives and expertise and thus our research team has integrated (in and out of the field) archaeologists, archaeometallurgists, conservation specialists, a chemist, an art historian, an historian, and a geologist. In addition to the usual battery of archaeometric analyses (e.g., atomic absorption, electron microprobe, proton-induced x-ray, and neutron activation), we have incorporated a series of replicative field experiments to test and refine our understanding acquired through other means.

From the beginning of our project, the importance of metallurgy in the regional prehistory was apparent. The infamy of the intense grave looting time span from the 1930s to the 1960s, and the folklore regarding the quantity and variety of looted metal objects preceded initiation of the project (Shimada 1981; Carcedo and Shimada 1985). Metallurgical remains are ubiquitous in our study area, while such remains are quite rare in other regions of the Andes. In fact, the name of the principal modern settlement in the valley, Batán Grande, derives from the numerous large *batanes* at the bases of nearby Cerros Huaringa and Sajino.

The pre-Hispanic metallurgical center of Cerro Huaringa was linked by a road (with scatterings of pre-Hispanic sherds) to the copper mine of Cerro Blanco that still evidenced ancient prospecting pits in spite of colonial and modern exploitation. Subsequent excavations conducted at Huaca del Pueblo Batán Grande (fig. 10) and the larger sites of Cerro Huaringa and Cerro Sajino from 1979 to 1983 yielded ample evidence of large-scale smelting of arsenical copper during the Sicán era.

A workshop was typically equipped with a row of 3 to 4 small, bowl-type smelters (each with an operational capacity of ca. 1.5 to 3.0 liters) and a *batán-chungo* set associated with varying accumulations of ground slag (fig. 11). Fuel was charcoal made out of locally abundant *algarrobo* trees (*Prosopis pallida*) and often kept in a shallow pit. In addition, floor debris included some food remains and ore fragments, as well as numerous broken *tuyères* (ceramic tips of blowtubes). The bellows were apparently not known in the New World and the draft was provided by human lung power through long blowtubes, a critical factor that defined the nature of smelting. With this limited draft supply, apparently they were unable to

Fig. 10. Three superimposed sets of Middle Sicán period smelting furnaces excavated at Huaca del Pueblo Batán Grande. Note the excellent state of preservation and ritual offerings made at the time of abandonment in the form of small bowls covered with inverted plates. The bowls probably held some sort of edibles. Photo by Izumi Shimada.

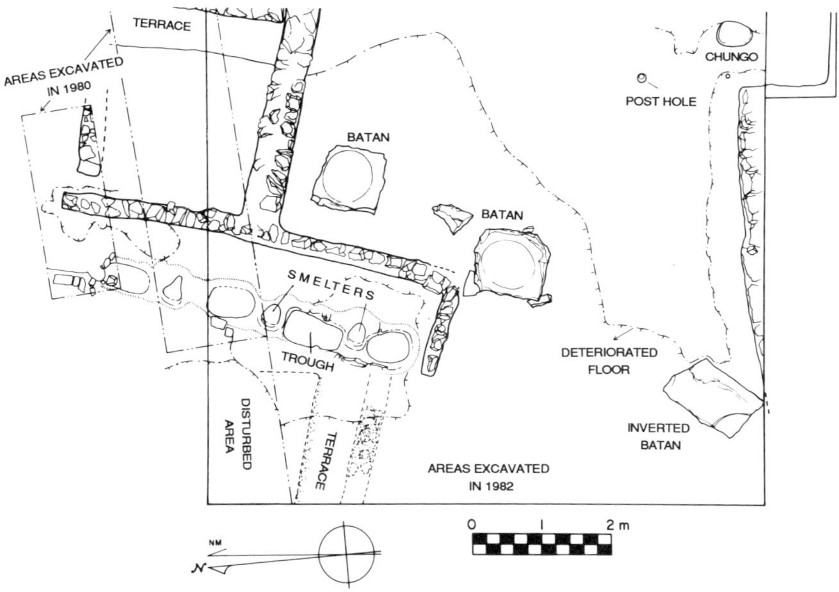

Fig. 11. Floor plan of a Sicán-Chimú (15th century A.D.) smelting workshop excavated in 1982 in the northwestern area of Cerro Huaringa. Note the row of three well-prserved smelting furnaces and two in situ *batanes* surrounded by ground slag. Drawing by Izumi Shimada.

sufficiently melt the charge to produce a usable ingot at the bottom of the furnace; instead, prills or small spherical droplets of metal (typically 1 to 5 mm in diameter) formed and remained trapped in viscous slag. Crushing of the resultant slag freed the prills, which then were collected and remelted to form a plano-convex ingot.

The above reconstruction of the Sicán smelting process was successfully field-tested using both replica furnaces and a 600-year smelter at Cerro Huaringa (fig. 12; Shimada and Merkel 1991; Merkel et al. 1994). The experiments were conducted using local ores, charcoal fuel, and replica *tuyères* inserted into 1.5 meter-long cane shafts. Temperatures at various points inside the furnaces and exiting gases were monitored. The re-enactment provided valuable information on the high labor and material costs of a smelt and how the furnaces operated (Shimada and Merkel 1991).

This "prill extraction" process remained essentially the same over the 600-year span of the Sicán metallurgy. Even after the successive conquests of the Sicán territory by the Chimú and Inka (ca. A.D. 1375 and 1460, respectively), arsenical copper production using the same process continued, albeit on an expanded scale (Shimada et al. 1982; 1983; cf. Lechtman 1979a; 1980)). The additional excavation of a Spanish colonial furnace at

Fig. 12. Replicative smelting experiment conducted in 1986 in the northwestern area of Cerro Huaringa using a 600-year old pre-Hispanic furnace (Sicán-Chimú period). This experiment showed that the size and form of the furnace mouth accommodated up to four *tuyères* simultaneously. Photo by Izumi Shimada.

Huaca Soledad was conducted in an attempt to clarify the fate of this long-standing indigenous bronze tradition upon the introduction of Old World bronzes and associated technology and cultural values (see Shimada and Merkel n.d.).

At the same time, over the 600-year span, a series of improvements were made in the design and construction format of the smelters and tuyeres (Shimada and Merkel 1991). In general, we see a trend toward an increasing efficiency and standardization, as well as greater productivity. In the early stages of the Sicán arsenical copper industry (ca. A.D. 900), smelting apparently had special symbolic significance; e.g. ceramic vessels covered with inverted plates were carefully placed as offerings in the mouths of smelting furnaces at the time of their abandonment (Shimada 1985; Shimada and Merkel 1991, fig. 10). The observed compositional variability in speiss and prills suggests that at this time metalworkers had not yet gained sufficient control over smelting to produce consistent products (Merkel et al. 1994). Within 100 years, however, there are indications that they consciously produced high arsenic prills (up to 30 percent) that are distinguishable for their silvery color. They may have served as stock to produce alloys of different colors, eutectic, and/or malleability. With rapidly improving understanding of the interrelationship between alloy composition and the manufacturing techniques and use of Sicán metal objects, we would indeed expect that Sicán metalworkers had such stock and created a range of alloys according to their needs. During the span of A.D. 900-1100, we witness a permanent shift from earlier reliance on copper to arsenical copper for utilitarian implements throughout Sicán territory that encompassed much of the north coast of Peru. In addition, at Huaca del Pueblo Batán Grande which had a long Sicán occupation, stone and bone tools were increasingly replaced with those of arsenical copper. Thus, the Sicán polity became the harbinger of the "bronze age" (in a generic sense) for northern coastal Peru.

The above work has shed much light on the long standing debate on arsenical bronze (see Lechtman 1979a and 1991 for summary discussions). Much of the debate has focused on the nature of the ores supplying arsenic for bronze and how ancient metalworkers might have coped with the roasting and reduction regimes if they had smelted sulfide ores. Our approach to the problem of defining the nature of the smelting charge involved a coordinated analysis of ore samples excavated at smelting loci and collected from local pre-Hispanic mines (Merkel and Shimada 1988; Merkel et al. 1994). For example, ore specimens recovered from the Huaca del Pueblo Batán Grande workshop included copper oxides, hematite, and scorodite ($FeAsO_4$-H_2O), a common weathering product of arsenopyrite (FeAsS). At the Cerro Mellizo mine, arsenic-rich scorodite was found, although the ore seemed to have been largely exhausted. Overall, the available evidence indicates that the Sicán metalworkers smelted a deliberate mixture of local copper, iron (flux), and arsenic-bearing ores. To date we have no evidence for the roasting of arsenic-bearing sulfide ores at the known Sicán metallurgical sites.

The multi-site sampling and regional perspective of the project have clearly documented that the entire Batán Grande region was a major metallurgical center for at least the last 600 years of prehistory. Whether small or large, residential or ceremonial, sites of these periods invariably have metallurgical remains. It is just as obvious that the primary copper-alloy smelting activities occurred at industrial sites relatively close to copper mines. Other sites were probably engaged in small-scale copper-alloy metalworking.

The Sicán culture was also known for precious metal ceremonial objects such as gold masks and *tumi*-knives. Following the 1991-2 excavations of two major Middle Sicán elite tombs (Shimada 1992; Shimada and Merkel 1993; Shimada and Griffin 1994; in press 1, 2; Shimada and Montenegro in press), the focus of our metallurgical research shifted to issues surrounding the production, use, and social and symbolic values of precious metal objects. The organization, composition, approach, and conceptual underpinnings of the project team that earlier tackled research into Sicán copper alloy metallurgy remained essentially the same, except that the roles of conservation specialists and Peruvian participants were expanded. Inclusion of Jo Ann Griffin, an expert goldsmith and specialist in the conservation of pre-Hispanic metal objects with 30 years' experience, was important in making inferences on the techniques and decision-making processes of ancient goldsmiths. Most archaeologists and archaeometallurgists have little personal expertise or knowledge of the craft under study. Insights of professional metalsmiths working with traditional techniques and tools are essential. Griffin, on her part, conducted a series of replicative experiments, such as producing sheet gold with stone hammers and depletion-gilding *tumbaga* sheets.

In the case of the elite tomb at Huaca Loro, a 3 m-by-3 m burial chamber at the bottom of an 11-m vertical shaft was packed with an estimated 1.2 tons of grave goods. Much of the goods were metal objects, including an estimated 500 kg of scrap metal (overwhelmingly *tumbaga* sheets), 250 kg of cast arsenical copper implements (fig. 13; Vetter 1993), and some 80 major gold alloy objects (e.g., crowns, head ornaments, ear spools, and rattles).

Many objects found in the Huaca Loro tomb can be described as prime examples of the Andean sheetmetal tradition and tours de force of

Fig. 13. One of 14 bundles of cast arsenical copper objects found in the Huaca Loro tomb. They may have been intended as spear points or digging stick tips but many of them were buried following their casting without any forging and shaping. Nearly all of the bundles contained around 30 pieces that typically measured 34 to 36 cm in length and 0.4 to 0.9 kg in weight. Photo by Izumi Shimada.

goldsmithing. However, contrary to a common supposition, the fact that objects are made of gold does not guarantee that they are flawless or carefully manufactured. Many objects display errors and rough finishing (e.g., jagged edges and poorly executed chasing and repoussé). It is the less-skilled or -thorough goldsmith who leaves the clearest evidence of manufacturing techniques and processes (Shimada and Griffin 1994). Museum collections usually select against such pieces and thus limit our technical understanding.

Our interdisciplinary team has also been able to document many clues to the original use and appearance of precious metal objects (Shimada and Griffin 1994; in press). For example, the plainness of cylindrical gold crowns is understandable given that their exterior was covered by interchangeable sheet ornaments with different cut-out designs sewn onto textile backing. One crown was equipped with built-in sockets on its upper rim for *tumi*-shaped ornaments to be inserted. A rather stout liner and additional support mechanisms were probably needed to support larger ornaments placed in front of and above the crown. One such object was an inverted-U shape with 32 regularly placed bangles and 90 gold feathers decorating its perimeter, with overall dimensions of ca. 56 cm and 98 cm in width and height, respectively (fig. 14). These gold objects were characterized by their movement, sound, color, and brilliance.

In many ways the most informative artifacts for understanding Sicán precious metal production are the large piles of scrap metal placed mostly along the edges of the burial chamber and the largest of the seven wall niches. Essentially, the scrap consists of the small pieces that result from any metalworking that involves cutting, filing, and other steps. In fact, these piles contained partially used *tumbaga* ingots, rejects from mishaps in manufacturing such as gold foil squares with poorly executed perforations, and sheetmetal trimmings that still retained the outline of cutout pieces. For any successfully completed object, there would have been a considerable quantity of scrap generated that was carefully saved to be recycled; in other words, it was saved to be re-melted. What is clear is that the scrap represents an enormous investment of manpower and materials. In a recent replicative experiment, Griffin used ancient stone hammers to produce a uniformly thin (approaching 0.1 mm) 10x15-cm gold sheet from a 30-gram nugget in about a day and half (Shimada and Griffin in press 2). Ancient Sicán metalworkers added the additional step of depletion-gilding the *tumbaga* sheets.

The impressive quantity of *tumbaga* sheetmetal scrap in the Huaca Loro tomb and at least 100 m^2 of *tumbaga* sheets that lined the interior of the other excavated elite tomb at Huaca Las Ventanas (fig. 15) forces a major revision of conventional thinking regarding the cultural significance of *tumbaga* and the scale of Middle Sicán precious metal production in gener-

Fig. 14. Middle Sicán ceremonial headdress assembled from a large mask with diagnostic upturned ears and eyes (46.0 cm wide, 29.0 cm high), a large parabolic head ornament with 90 small gold feathers around its perimeter, and a forehead ornament with a sculptural representation of a bat head (42.2 cm wide, 23.6 cm high). The headdress illustrates the originality and impressive sheetmetal working skills of the Sicán goldsmiths, who effectively captured characteristic bat features such as the prominent snout and bony ridge around the eyes. The long, slender tongue was suspended to move from side to side. This headdress was an essential part of ceremonial display.

Fig. 15. A layer of cloth-covered *tumbaga* sheetmetal exposed inside the Huaca Las Ventanas tomb in 1991 at a depth of some 10 m below the surface. Photo by Izumi Shimada.

al. *Tumbaga* sheets were even used to wrap fine pottery and wooden objects.

The new data from Batán Grande suggest a sizeable corps of master sheet makers, probably assisted by an even larger number of apprentices, producing large quantities of sheet goods in different gauges for various applications. Lechtman (1984a, b) has argued that the persistent and widespread use of depletion gilded *tumbaga* relates to the importance of gold color that was so valued in pre-Hispanic Peru. She suggests that this importance reflects cultural attitudes and ideologies; though the depletion gilded *tumbaga* has only a very thin gold-rich layer on the surface, what was important, she argues (1984a, 30), was the knowledge that the "essential quality," gold, even only minimally present, was diffused or contained within the *tumbaga*. The documented scale of production brings into question the applicability of Lechtman's "essence of the object" thesis to the Middle Sicán situation (Shimada and Griffin 1994; in press). In terms of the scale of production and range of use, *tumbaga* appears to top the list of Middle Sicán precious metals. Yet, our comparative study of metal objects found in some two dozen Sicán graves of different size, form, and context indicates that *tumbaga* was clearly secondary to gold in the Sicán elite perception (Shimada and Montenegro in press). Gold objects were reserved for the personal use of the highest echelon elite, while *tumbaga* decorated other items associated with them, as well as personal and non-personal items associated with lower echelon elite. *Tumbaga* allowed the lower echelon elite to emulate the highest elite. Overall, given the limited supplies of gold and silver, gilded *tumbaga* would have been a most practical way of supplying gold-colored sheetmetal on demand (Shimada and Griffin 1994; in press). Economic and pragmatic considerations may well have outweighed ideological factors in the production and use of *tumbaga*.

Manufacturing techniques of Sicán precious metal objects from the Huaca Loro tomb are being examined from a variety of analytical perspectives. Griffin has documented what she has termed "proto-brazing" (often misidentified as simple brazing or fusion welding) as the principal chemical joining technique among them. Proto-brazing is an elegantly simple method of joining gold or silver alloys. It utilizes the copper in the alloy or adds verdigris (copper acetate) to an organic glue, which is then used to hold the pieces in contact during heating done in a reduction atmosphere (Griffin 1986; Shimada and Griffin 1994; in press). Brought to the correct temperature, a new alloy is created at the exact spot where the metals touch. This yields very crisp jointure, with no evidence of filleting. Concurrent with the detailed study of sheetmetal working (e.g., hammer and chisel marks, scribed guidelines, etc.), metalworking stone tools in the National Museum of Anthropology and Archaeology in Lima were studied by Griffin and P. Carcedo. In addition, some 1000 rubbing samples were collected by A. Gordus from nearly all the precious metal objects from this tomb for neutron activation analysis to determine the range of alloys used and how different alloying related to varied manufacturing techniques, styles and inferred uses. Metallographic study of other samples by J. Merkel will provide an additional comparative perspective.

In sum, the approach and perspective of the Sicán Archaeological Project have proven to be quite effective in gaining a holistic understanding of Sicán metallurgy. We have been able to document diverse phases of Sicán metallurgy together with the roles of its products. Concern with metals pervaded this culture to a degree undocumented for any other in the New World.

Conclusion

This paper summarizes major advances in our understanding of pre-Hispanic Andean metallurgy and mining over the past decade. For the sake of brevity, many available details were not presented here. While we have seen notable advances in respect to various facets and phases of north Peruvian metallurgy, in general the pace is disturbingly slow. The number of archaeologists and other specialists actively engaged in original research on pre-Hispanic Andean mining and metallurgy is quite small. Further, limited communication and collaboration among them have led to erroneous conclusions and entrenchment of questionable views. A more open exchange of ideas and data is urged here.

In addition, many museums in the Andes are replete with metal objects that await conservation and analysis. Numerous metalworking implements, whether made of stone or metal, have been stored away without their significance being recognized. Too much public and professional attention has focused on eye-catching precious metal objects that are too often without proper contextual data. If we were to achieve the sort of comprehensive, anthropological understanding of pre-Hispanic Andean metallurgy urged in this paper, we need more sustained, coordinated fieldwork and laboratory analysis specifically focused on this topic. Work in and out of the field should be conducted by the same interdisciplinary team. With the advances made in basic documentation of the *altiplano* metallurgical tradition, the stage is now set for initiation of similar research in northwest Argentina. It is hoped that the recently initiated research at Ramaditas in northern Chile will also be placed in a similarly comprehensive, long-term framework. When these advances are made, we will be in a better position to determine the cultural processes responsible for the development and distribution of certain metals and techniques. For example, was it a matter of ore availability that arsenical copper occurred in northwest Argentina, on the north coast of Peru, and south coast of Ecuador? I suspect that metalworkers in distinct regions maintained much closer contact than we think.

This paper also emphasizes the importance of seeing and treating a given metallurgy and associated mining as an integrated whole. In the case of Sicán metallurgy, there is a clear feedback loop between its smelting technology and mining. Oxide ores found at or close to the surface could be readily mined with primitive tools and smelted without the pretreatment required for sulfide ores. Though it was labor intensive to work with stone hammers, the extracted ores were already carefully selected and largely crushed in the process, necessitating little or no subsequent ore beneficiation. Use of hand-sorted, high grade ores most likely long pre-dated Sicán metallurgy.

Mining is certainly not limited to the extraction of ores for metallurgical ends. Research on ancient mining has focused too narrowly on useful ores for metallurgy; minerals for beads and other lapidary items have received scant attention. Not only are some minerals consistently misidentified (e.g., sodalite as lapis lazuli) in the archaeological literature, but their sources and trade have been little studied (Shimada 1994; Shimada and Griffin 1994; in press). The search for minerals for lapidary or ritual uses (e.g., malachite, chrysocolla, turquoise, azurite, pyrite, quartz crystals, hematite, amber, and even cinnabar) may well have led to the identification of metals in their native form or useful ores for metallurgy or vice versa.

Thus, broadening of our conception of mining should prove more productive.

Though archaeologists commonly speak of the resiliency and persistence of indigenous cultures in spite of the destructuring efforts of colonial institutions, Andean archaeologists have largely excluded the colonial era in their studies. Schaedel (1992, 223) recognizes the tremendous potential archaeology holds in clarifying the transfer process of imported Old World technologies during the early part of the colonial era. But, did indigenous metal production and use really die out as rapidly or thoroughly at the beginning of the colonial era as commonly assumed? We need to consider its survival and perhaps even its fusion with the imported features. It seems there is much to be learned about pre-Hispanic metallurgy and mining even from the small-scale and intermittent mining (including placer mining using traditional methods; Recharte 1988) we see today in various areas of Andean highlands.

Acknowledgements

I am indebted to members of the Sicán Archaeological Project for their collaboration. In particular, I benefitted from the knowledge and expertise of Paloma Carcedo, Alan Craig, Jo Ann Griffin, John Merkel, and Susan Ramírez. Richard L. Burger provided unpublished information on his recent discovery at the site of Mina Perdida. Yutaka Yoshii generously gave permission to use a number of his photographs for this paper. Walter Alva kindly authorized the use of a photo of metal objects from his excavations at Sipán taken by Y. Yoshii. The Department of Library Services, American Museum of Natural History, New York, provided permission to publish the photo of stone hammers from Chuquicamata, Chile.

Endnotes

1. *Tumbaga* refers to gold-copper or gold-silver-copper alloy which approximates 10 to 12 k gold. Depletion gilding is a technique in which acid(s) are applied to the surface of *tumbaga* to selectively remove the baser metal(s) of the alloy to create an enhanced gold appearance. This is followed by burnishing to impart a high gloss and to compact the porous layer of gold left on the surface by the depletion of baser metals.

2. Arsenical copper is an intentional alloy, a type of bronze, that has a slightly yellowish to silvery color, good casting ability, hardness, and durability much like the copper-tin bronze produced in the *altiplano* region to the south.

3. Lost wax casting is a technique by which the matrix or the original object is made in wax and then encased in a refractory material (usually clay). The mold with the matrix inside is then heated to melt the wax, which runs out the open end of the mold. Then, molten metal is poured in to replace the void formerly occupied by wax. When cooled the mold is broken to extract the cast object.

4. The past decade has seen an active debate regarding the identity of the associated iconography. Traditionally, it has been related to the north highland-based Chavín (Janabarriu Phase in R. Burger's [1988] seriation). With advances in the understanding of the north coast-

based Cupisnique and documentation of the strong Cupisnique influence at the type site of Chavín de Huántar (Kaulicke 1991a), various scholars now identify the associated iconography as Late Cupisnique.

5. Dense tropical rainforest vegetation forces the use of watercraft for movement.

6. However, the dating of these Colombian and Ecuadorian developments is still considerably less precise and certain than those of the Mochica.

7. Identification and composition of the ores from our excavations and mine survey are presented in Merkel et al. 1994.

8. J. Nash's (1979) ethnographic study of modern Bolivian miners also reveals that mines are considered *huacas* and that offerings are made to the patron deity of the mountain where the mine is situated to appease him and obtain his permission before ores may be extracted.

9. The *batán* is a large, tabular, stable anvil with central conical depression fashioned out of hard, dense igneous rock such as diorite, andesite or basalt. It was used together with a large rocking stone called *chungo* to grind ores and slag.

References

Alva, W. 1988. Discovering the New World's richest unlooted tomb. *National Geographic Magazine* 174(4): 510-550.

——. 1990. New tomb of royal splendor. *National Geographic Magazine* 177 (6): 2-15.

——. 1992. Orfebrería del Formativo. In *Oro del antiguo Perú*, ed. José Antonio de Lavalle, 17-116. Arte y Tesoros del Perú XIX. Lima: Banco de Crédito.

Alva, W. and Donnan, C. B. 1993. *Royal tombs of Sipán*. Los Angeles: Fowler Museum of Cultural History.

Alva, W., Fecht, M., Schauer, P., and Tellenbach, M. 1989. *La tumba del Señor de Sipán: Descubrimiento y restauración*. Mainz: Verlag des Römisch Germanischen Zentralmuseums.

Bergsøe, P. 1937. The metallurgy and technology of gold and platinum among the pre-Columbian Indians. *Ingeniørvidenskabelige Skrifter*, Nr. A 44. Copenhagen: Danmarks Naturvidenskabelige Samfund i Kommission Hos G.E.C. Gad.

——. 1938. The gilding process and the metallurgy of copper and lead among the pre-Columbian Indians. Translated from Danish by C. F. Reynolds. *Ingeniørvidenskabelige Skrifter*, Nr. A 46. Copenhagen: Danmarks Naturvidenskabelige Samfund i Kommission Hos G.E.C. Gad.

Berthelot, J. 1978. L'exploitation des metaux précieux au temps des Incas. *Annales* 5:948-966. Paris.

——. 1986. The extraction of precious metals at the time of the Inka. In *Anthropological history of Andean polities*, eds. John V. Murra, Nathan Wachtel, and Jacques Revel, 69-88. Cambridge, Mass.: Cambridge University Press.

Bird, J. B. 1979. The "Copper Man": A prehistoric miner and his tools from northern Chile. In *Pre-Columbian metallurgy of South America*, ed. E. P. Benson, 105-131. Washington, D.C.: Dumbarton Oaks.

Bouchard, J-F., ed. 1984. *Recherches archéologiques dans la région de Tumaco (Colombie)*. Editions Recherche sur les Civilisations. A. D. P. F. Mémoire 34. Paris.

Bray, W. 1978. *The gold of El Dorado*. London: Times Books.

——. 1985. Ancient American metallurgy: Five hundred years of study. In *Art of Pre-Columbian gold: Jan Mitchell Collection*, ed. Julie Jones, 76-84. London: Weidenfeld & Nicolson.

——. 1988. Resolving the platinum paradox. *Americas* 40(6): 45-49.

——. 1992. El período Yotoco. In *Calima: Diez mil años de historia en el suroccidente de Colombia* by Marianne Cardale de Schrimpff, Warwick Bray, Theres Gähwiler-Walder, and Leonor Herrera, 75-124. Colombia: Fundación Pro Calima, Santa Fe de Bogotá.

Browman, D. L. 1975. Trade patterns in the central highlands of Peru. *World Archaeology* 6:322-329.

Bruhns, K. O. 1990. Excavations at Pirincay, in the Paute valley of southern Ecuador, 1985-1988. *Antiquity* 64:221-233.

Burger, R. L. 1988. Unity and heterogeneity within the Chavín horizon. In *Peruvian prehistory*, ed. Richard Keatinge, 99-144. Cambridge: Cambridge University Press.

Carcedo, P. and Shimada, I. 1985. Behind the golden mask: Sicán gold artifacts from Batán Grande, Peru. In *Art of Pre-Columbian gold: Jan Mitchell Collection*, ed. Julie Jones, 60-75. London: Weidenfeld & Nicolson.

Cardale de Schrimpff, M. 1992. La gente del período Ilama. In *Calima: Diez mil años de historia en el suroccidente de Colombia*, by Marianne Cardale de Schrimpff, Warwick Bray, Theres Gähwiler-Walder, and Leonor Herrera, 25-71. Colombia: Fundación Pro Calima, Santa Fe de Bogotá.

D'Altroy, T. N. 1993. Research summary of the 1990 season of the Proyecto Arqueológico Calchaquí. *Willay* Nos. 39/40: 6-10. Cambridge, MA.

Donnan, C. B. 1990. Masterworks of art reveal a remarkable pre-Inca world. *National Geographic Magazine* 177(6): 17-33.

——. 1992. Oro en el arte Moche. In *Oro del antiguo Perú*, ed. José Antonio de Lavalle, pp. 119-193. Arte y Tesoros del Perú XIX. Lima: Banco de Crédito.

Donnan, C. B., and Mackey, C. 1978. *Ancient burial patterns of the Moche Valley, Peru*. Austin, Tex.: University of Texas Press.

Earle, T., D'Altroy, T., Hastorf, C., Scott, C, Costin, C., Russell, G., and Sandefur, E. 1987. *Archaeological field research in the Upper Mantaro, Peru, 1982-1983; Investigations of Inka expansion and exchange*. Monograph XXVIII, Institute of Archaeology, University of California, Los Angeles.

Elera, C. G. 1986. *Investigaciones sobre patrones funerarios en el sitio formativo del Morro de Etén, Valle de Lambayeque*. Memoria de Bachiller, Especialidad de Arqueología, Pontificia Universidad de Católica del Perú. Lima.

——. 1992. Arquitectura y otras manifestaciones culturales del sitio formativo del Morro de Etén: Un enfoque preliminar. In *Estudios de arqueología peruana*, ed. Duccio Bonavia, 177-192. Lima: FOMCIENCIAS.

Elera, C. and Pinilla, J. 1990. Research summary of the Proyecto Arqueológico Puémape (1987-1989). *Willay*, No. 34, pp. 2-4. Cambridge, MA.

Epstein, S. M. and Shimada, I. 1984. Sicán metallurgy: A reconstruction of copper alloy production at Cerro de los Cementerios, Peru. *Beiträge zur Allgemeinen und Vergleichenden Archäologie* 5(1983): 379-430. KAVA, Deutschen Archäologischen Instituts, Bonn.

Estrada, E. 1961. Correlaciones entre la arqueología de la costa del Ecuador y Perú. *Humanitas* 2(2): 31-61. Quito.

Fester, G. A. 1962. Copper and copper alloys in ancient Argentina. *Chymia: Annual Studies in the History of Chemistry* 8: 21-31. Philadelphia.

González, A. R. 1979. Pre-Columbian metallurgy of northwest Argentina: Historical development and cultural process. In *Pre-Columbian metallurgy of South America*, edited by E. P. Benson, pp. 133-202. Washington, D.C.: Dumbarton Oaks,

——. 1983. Inca settlement patterns in a marginal province of the empire: Sociocultural implications. In *Prehistoric settlement patterns: Essays in honor of Gordon R. Willey*, ed. Evon Vogt, 337-360. Albuquerque , N.M. and Cambridge, Mass. : University of New Mexico Press and Harvard University, Peabody Museum of Archaeology and Ethnology.

——. 1992a. *Las placas metálicas de los Andes del Sur*. Mainz am Rhein: Verlag Philipp von Zabern.

——. 1992b. La metalurgia precolombina de Sudamérica y la búsqueda de los mecanismos de la evolucíon cultural. In *Prehistoria sudamericana: Nuevas perspectivas*, ed. B. J. Meggers, 35-44. Washington, D.C.: Taraxacum.

Gordon, R. B. and Rutledge, J. W. 1984. Bismuth bronze from Machu Picchu, Peru. *Science* 223:585-586.

——. 1987. The work of metallurgical artificers at Machu Picchu, Peru. *American Antiquity* 52:578-594.

Griffin, J. A. 1986. Investigación sobre la unión de los metales. En *Metalurgia de America precolombina*, ed. Clemencia Plazas, 351-366. Bogotá: Banco de la República.

Grossman, J. W. 1972. An ancient gold worker's kit. *Archaeology* 25:270-275.

Holm, O. 1966/67. Money axes from Ecuador. *Folk* 8-9:135-143. Copenhagen.

——. 1975. Monedas primitivas del Ecuador prehistórico. *La Pieza* 3. Guayaquil: Casa de la Cultura Ecuatoriana.

Hosler, D. 1988. Ancient West Mexican metallurgy: South and Central American origins and West Mexican transformations. *American Anthropologist* 90:832-855.

Hosler, D., Lechtman, H. N. and Holms, O. 1990. *Axe-monies and their relatives*. Studies in Pre-Columbian Art and Archaeology. Washington, D.C.: Dumbarton Oaks.

Jones, J. 1979. Mochica works of art in metal: A review. In *Pre-Columbian metallurgy of South America*, ed. E. P. Benson, 53-104. Washington, D.C.: Dumbarton Oaks.

Kaulicke, P. 1991a. Chavín art and iconography. In *Ancient art of the Andean world*, ed. Shozo Masuda and Izumi Shimada, 47-69. Tokyo: Iwanami Shoten, Publishers.

——. 1991b. El Período intermedio temprano en el Alto Piura: Avances del proyecto arqueológico "Alto Piura (1987-1990)." *Boletín del Instituto Francés de Estudios Andinos* 20:381-422. Lima.

Lechtman, H. N. 1976. A metallurgical site survey in the Peruvian Andes. *Journal of Field Archaeology* 3:1-42.

——. 1979a. Issues in Andean metallurgy. In *Pre-Columbian metallurgy of South America*, ed. E. P. Benson, 1-40. Washington, D.C.: Dumbarton Oaks.

——. 1979b. A pre-Columbian technique for electrochemical replacement plating of gold and silver on copper objects. *Journal of Metals* 31:154-160.

——. 1980. The Central Andes: Metallurgy without iron. In *The coming of the Age of Iron*, ed. T. A. Wertime and J. D. Muhly, 267-334. New Haven, Conn.: Yale University Press.

——. 1981. Copper-arsenic bronzes from the north coast of Peru. *Annals of the New York Academy of Sciences* 376:77-122.

——. 1984a. Pre-Columbian surface metallurgy. *Scientific American* 250(6): 56-63.

——. 1984b. Andean value system and the development of prehistoric metallurgy. *Technology and Culture* 25:1-36.

——. 1988. Traditions and styles in central Andean metalworking. In *The beginning of the use of metals and alloys*, ed. Robert Maddin. Cambridge, Mass.: The MIT Press.

——. 1991. The production of copper-arsenic alloys in the central Andes: Highland ores and coastal smelter? *Journal of Field Archaeology* 18:43-76.

Lechtman, H. N., Erlij, A. and Barry, E. J., Jr. 1982. New perspectives on Moche metallurgy: Techniques of gilding copper at Loma Negra, northern Peru. *American Antiquity* 47:3-30.

Lothrop, S. K. 1941. Gold ornaments of Chavin style from Chongoyape, Peru. *American Antiquity* 6:250-262.

——. 1951. Gold artifacts of Chavin style. *American Antiquity* 16:226-240.

MacNeish, R. S., Patterson, T. C., and Browman, D. L. 1975. *The central Peruvian prehistoric interaction sphere*. Andover, Mass.: Philips Academy.

Maguiña, A. 1992. Informe de los trabajos realizados en Huaca La Merced del complejo arqueológico Batán Grande. In *Informe de la temporada de los años 1991-1992 del Proyecto Arqueológico de Sicán*, ed. Izumi Shimada, n.p. Report submitted to the Instituto Nacional de Cultura, Lima.

Makowski, K. and Cornejo, M. 1993. Research summary of the Pontificia Universidad Católica del Perú fieldwork at Tablada de Lurín, 1991-3. *Willay* Nos. 39/40, pp. 3-8. Cambridge, MA.

Mathewson, C. H. 1915. A metallographic description of some ancient Peruvian bronzes from Machu Picchu. *American Journal of Science* 190:525-616.

Matson, F. R. 1965. Ceramic ecology: An approach to the study of early culture of the Near East. In *Ceramics and man*, ed. F. R. Matson, 202-217. Chicago: Aldine.

Mayer, E. F. 1986. *Armas y herramientas de metal prehispánicas en Argentina y Chile.* Allgemeinen und Vergleichenden Archäologie, Band 18. Bonn: Deutschen Archäologischen Instituts.

——. 1992. *Vorspanische Metallwaffen und -werkzeuge in Ecuador/Armas y herramientas de metal prehispánicas en Ecuador.* Mainz am Rhein: Verlag Philipp von Zabern.

Merkel, J., Shimada, I., Swann, C. P., and Doonan, R. 1994. Investigation of prehistoric copper production at Batán Grande, Peru: Interpretation of the analytical data for ore samples. In *Archaeometry of pre-Columbian sites and artifacts*, eds. David A. Scott and P. Meyers, 199-227. Marina del Rey, Calif.: The Getty Conservation Institute.

Merkel, J. F. and Shimada, I. 1988. Arsenical copper smelting at Batán Grande, Peru. *Newsletter of the Institute for Archaeo-Metallurgical Studies.* No. 12, pp. 4-7. London.

Nash, J. 1979. *We eat the mines and the mines eat us.* New York: Columbia University Press.

Nordenskiöld, E. 1921. The Copper and Bronze Ages in South America. *Comparative Ethnographical Studies*, Vol. 4. Göteborg: Elanders Boktryckeri Aktiebolag.

Oehm, V. P. 1984. *Investigaciones sobre minería y metalurgia en el Perú prehispánico: Una visión crítica actualizada.* Bonner Amerikanistische Studien No. 12. Bonn.

Patiño, D. 1988. Orfebrería prehispánica de la costa Pacífica de Colombia y Ecuador. *Boletín del Museo del Oro* 22:17-31. Bogotá.

Plazas, C. and Falchetti, A. M. 1985. Cultural patterns in the pre-Hispanic goldwork of Colombia. In *Art of Pre-Columbian gold: Jan Mitchell Collection*, ed. Julie Jones, 47-59. Weidenfeld & Nicolson, London.

——. 1986. Patrones culturales en la orfebrería prehispánica de Colombia. In *Metalurgia de América precolombina*, ed. Clemencia Plazas, 210-246. Bogotá: Banco de la República.

——. 1992. *Museo del Oro.* Santa Fe de Bogotá, Colombia: Banco de la República.

Pollard, H. P. 1993. Merchant colonies, semi-Mesoamericans, and the study of cultural contact: A comment on Anawalt. *Latin American Antiquity* 4:383-385.

Ramírez, S. n.d. Ethnohistorical dimensions of mining and metallurgy in northern Peru. In *Sicán metallurgy: Holistic perspective*, ed. Izumi Shimada. ms.

Recharte, J. 1988. Qori Llank'ay: Gold mining technology in a community mine, northern region of Puno, Peru. Working Paper No. 2, Research Project "Production storage and exchange in a terraced environment on the eastern Andean environment", PNSAA, Lima and Department of Anthropology, University of North Carolina, Chapel Hill.

Schaedel, R. P. 1992. The archaeology of the Spanish colonial experience in South America. *Antiquity* 66:217-242.

Scott, D. A. 1980. The conservation and analysis of some ancient copper alloy beads from Colombia. *Studies in Conservation* 25:157-164. London.

Scott, D. A. and Bouchard, J. F. 1988. Orfebrería prehispánica en las llanuras del Pacífico de Ecuador y Colombia. *Boletín del Museo del Oro* 22:2-16. Bogotá.

Scott, D. A. and Bray, W. 1980. Ancient platinum technology in South America: Its use by the Indians in pre-Hispanic times. *Platinum Metals Review* 24:147-157.

Scott, D. A. and Doehne, E. 1990. Soldering with gold alloys in ancient South America: Examination of two small gold studs from Ecuador. *Archaeometry* 32:183-190.

Shimada, I. 1978. Economy of prehistoric urban context: Commodity and labor flow in Moche V Pampa Grande, Peru. *American Antiquity* 43:569-592.

——. 1981. The Batán Grande-La Leche Archaeological Project: The first two seasons. *Journal of Field Archaeology* 8:405-446.

——. 1984. Ancient mining and metallurgy on the northern North Coast of Peru. Paper presented at the Third Annual Meeting of the Northeast Conference on Andean Archaeology and Ethnohistory, Amherst, Massachusetts.

——. 1985. Perception, procurement and management of resources: Archaeological perspective. In *Andean ecology and civilization*, ed. Shozo Masuda, Izumi Shimada and Craig Morris, 357-399. Tokyo: University of Tokyo Press.

——. 1988. A metallurgical survey in the Vicús region of the Far North Coast of Peru. Paper presented at the Seventh Annual Meeting of the Northeast Conference on Andean Archaeology and Ethnohistory, Amherst, Massachusetts.

——. 1992. La sépulture inviolée d'un riche seigneur sicán. In *Les Dossiers d'Archéologie*, Special Issue, " Le Pérou de l'origine aux Incas" (Hors-serie No. 2), ed. Daniel Lévine, 46-57. Paris.

——. 1994. *Pampa Grande and the Mochica Culture*. Austin, Tex.: University of Texas Press.

Shimada, I., Epstein, S. M. and Craig, A. K. 1982. Batán Grande: A prehistoric metallurgical center in Peru. *Science* 216:952-959.

——. 1983. The metallurgical process in ancient north Peru. *Archaeology* 36(5):38-45.

Shimada, I. and Griffin, J. A. 1994. Precious metal objects of the Middle Sicán. *Scientific American* 270 (4): 60-67.

——. In press. Manufactura, uso y significado de los objetos de metal precioso Sicán: Algunas luces sobre la tumba de Huaca Loro. *Pachacamac* 2. Lima: Museo de la Nación.

Shimada, I. and Merkel, J. F. 1991. Copper alloy metallurgy in ancient Peru. *Scientific American* 265:80-86.

——. 1993. A Sicán tomb in Peru. *Minerva* 4(1): 18-25.

——. n.d. Colonial bronze furnace in Batán Grande, Peru. unpublished manuscript.

Shimada, I. and Montenegro, J. In press. El poder y la naturaleza de la élite Sicán: Mirada a la tumba de Huaca Loro, Batán Grande. *Boletín de Lima*. (Peru).

Sutliff, M. J. 1992. *El proceso productivo metalúrgico de la cultura Milagro: El caso de Peñon del Río*. Licenciatura thesis, Escuela Superior Politécnica del Litoral, Guayaquil, Ecuador.

Tolstoy, P. and DeBoer, W. 1989. An archaeological sequence for the Santiago-Cayapas River basin, Esmeraldas, Ecuador. *Journal of Field Archaeology* 16:295-308.

University of Tokyo Ancient Andean Civilization Research Project. 1992. *Catalogue of the exhibit, El Oro y Templo de Kuntur Wasi*. Tokyo: Yomiuri Shinbun.

Váldez, F. 1987. *Proyecto arqueológico La Tolita*. Quito: Banco Central del Ecuador.

Vetter, L. M. 1993. *Análisis de las puntas de aleación de cobre de la tumba de un señor de la élite Sicán, Batán Grande, Lambayeque, Perú*. Memoria de Bachiller, Especialidad de Arqueología, Pontificia Universidad de Católica del Perú. Lima.

Copper Smelting in the Atacama: Ancient Metallurgy at the Ramaditas Site, Northern Chile

Gray Graffam, Trent University, Peterborough, Ontario, Canada
Mario Rivera, Field Museum of Natural History, Chicago, U.S.A.
Alvaro Carevič, Universidad Arturo Prat, Iquique, Chile

Abstract

Today, the copper-rich ores of the Atacama attract the continued attention of international mining efforts, where millions of dollars are spent in the annual extraction of ore from one of the world's richest sources. Pre-Hispanic efforts to tap these ores are evident in the archaeological record of the region covering the past millennia. Sites with furnace ruins, slag, and crushed ore are seen throughout the region, spanning from the Late Formative period through the Inka period. Of these, the later sites have received the most attention to date, although much more waits to be done. The archaeological evidence from Ramaditas, when viewed in this context, is significant in that it is the first firm indication of copper smelting in the Atacama region itself during the Late Formative period of northern Chile (500 B.C. to A.D. 300). The results provided here are the earliest for the region so far, and they provide a firm foundation for understanding the nature of these efforts prior to the incorporation of the region within the Tiwanaku state of later times, as well as the nature of early metallurgy in the South Andes in general.
Key words: copper, metallurgy, northern Chile.

The extraction of minerals and the processing of metals has a long history in the Atacama desert of Chile and its adjacent foothills. Our results provide the first direct evidence that copper smelting in the Atacama goes back at least to the Late Formative period of northern Chile (500 B.C. to A.D. 300), and was one of the major factors affecting permanent settlement in this arid zone. It places the smelting of metals in the desert region firmly within a pre-state context, in association with early village settlements, of which Ramaditas is one. Such a view conforms with the information on

In Quest of Mineral Wealth: Aboriginal and Colonial Mining and Metallurgy in Spanish America, edited by Alan K. Craig and Robert C. West, 1994. Geoscience and Man, vol. 33, pp. 75-92. Dept. of Geography & Anthropology, Louisiana State University, Baton Rouge, LA 70893-6010, U.S.A.

metal artifacts from archaeological contexts in northern Chile (Iribarren 1974; Muñoz 1989). Moreover, it accords well with current models of trade and llama caravan movements (Dransart 1991; Dillehay and Núñez 1988; Núñez and Dillehay 1978), and it complements the published literature on economic development and settlement reorganization during the transition from the late Archaic through Formative periods in northern Chile (Núñez 1989, 1982; Núñez and Moragas 1977; Muñoz 1989; Castillo 1986; Dauelsberg 1985; Rivera 1976; 1991; Rivera et al. 1974; Niemeyer et al. 1989; Santoro 1982; and others).

With copper smelting appearing between 1200 and 250 B.C. at the site of Wankarani (stratum 11a) in the nearby Bolivian highlands (Ponce 1970, 42), it is not surprising to see that efforts to win metals in the Atacama desert date back to Late Formative times. The exploitation of metal-bearing ores of the desert dates to the beginning of the first millennium B.C. or earlier, when brochantite and antlerite were traded into the altiplano highlands. Such arid-land ores, which are superior in quality to those of the altiplano, have been found at the site of Chiripa in the Lake Titicaca basin ca. 1200 to 1000 B.C., and may be among those smelted at the site of Wankarani as well (Browman 1991, Lechtman 1980). Permanent settlement of the Atacama valleys begins at about this time (Tarapacá 6 and 7), associated with the development of irrigation agriculture. As these latter sites show evidence of highland (Wankarani) ceramic influence, the claim can be made that such settlement was articulated within a broader trade network, connecting highlands, desert, and lowlands (Núñez and Dillehay 1978). The exploitation of metal-bearing ores of the Atacama, and the eventual development of smelting in that region by the Late Formative period, grew out of a larger economic context, involving interregional trade and the redistribution of natural resources.

Objects of copper, as well as other metals, have been found in tombs that date to the Early and Late Formative periods in northern Chile. Of course, some of these objects may be native copper rather than smelted copper—a matter that deserves more attention. They include plaques, pendants, copper sheet, diadems, breastplates, bracelets, rings, and zoomorphic art works, as well as the more utilitarian items, such as spoons, pins, tweezers, axes, chisels, fishhooks, and needles (Alvarez 1981, 33; Castillo 1986, 178; Dauelsberg 1985, 21; Iribarren 1974, 11; Latcham 1936; Mayer 1986; Meighan 1980, 121; Muñoz 1986, 307; 1989, 116; Rivera 1976, 74; Willey 1971, 215, 218). The total quantity of metal represented is not great, and it would be inaccurate to discuss the evidence here as the welcoming in of a "copper age," e.g. Nordenskiöld (1921), but it is sufficient to base a claim that items of copper manufacture were certainly valued by the north Chilean natives of the time as both prestige goods and utilitarian objects.

The Ramaditas Site

Ramaditas is a small village that lies within close proximity to the ancient village of Guatacondo (fig. 1), and mention of the site is found in published surveys of the Guatacondo Valley (Meighan 1980; Mostny 1970; 1980). Our efforts to reexamine the site have included mapping its extent and conducting an extensive test program to evaluate metallurgical issues. As is typical of Atacama sites, the ruins are in an excellent state of preservation, with many walls standing close to their full original height.

The site does not offer much by way of monumental ruins. All in all, its size is small (core area of 3 hectares) and unimposing (fig. 2), and it represents a small village with a probable population of about 50 to 100 people (11 to 20 families). This is undoubtedly the reason why previous researchers have not called attention to the site, and why virtually no previous work has been done there. Ramaditas appears, at first glance, to be rather inconsequential. In terms of its degree of "urbanism" or "monumentality," this is certainly the case; its significance lies in other areas. It was clearly not a civilization center or urban settlement.

The site consists of three architectural compounds, a large mound, a scatter of individual structures, and a large expanse of irrigated fields (figs. 2 and 3). Scattered over the site are numerous lithic fragments and sherds of Alto Ramírez plainware.

Ramaditas is the earliest of the permanent village sites in the Guatacondo Valley, predating the main *aldea* of Guatacondo by about 150 years. It is predominantly a single component site dating to the Late Formative period, but with some structures showing subsequent marginal use as temporary campsites in the upper fill stratigraphy. There are several pieces of evidence that point to a date of the Late Formative period for site construction and main occupation:

1. artifacts in association with the structural floors and the one to two centimeters of accumulation on top of them are all, without exception, characteristic of the Late Formative period; these include Alto Ramírez plainwares, unfired ceramic figurines, and projectile points typical of the period;

2. architectural style and treatment at Ramaditas is strikingly similar to that of the main *aldea* of Guatacondo and other Late Formative period centers in the region; this not only includes the shape of the semi-subterranean structures, the pattern of agglutination, roof type, and pattern of entrance and ventilation openings, but also the treatment of adding modeled clay faces to the interior walls of some structures;

3. an accelerated (AMS) date taken on charred algarrobo seed from a sealed pit in the floor of a structure (Structure #57) at Compound 3

Fig. 1 Northern Chile, showing location of Ramaditas and Alto Ramírez archaeological sites.

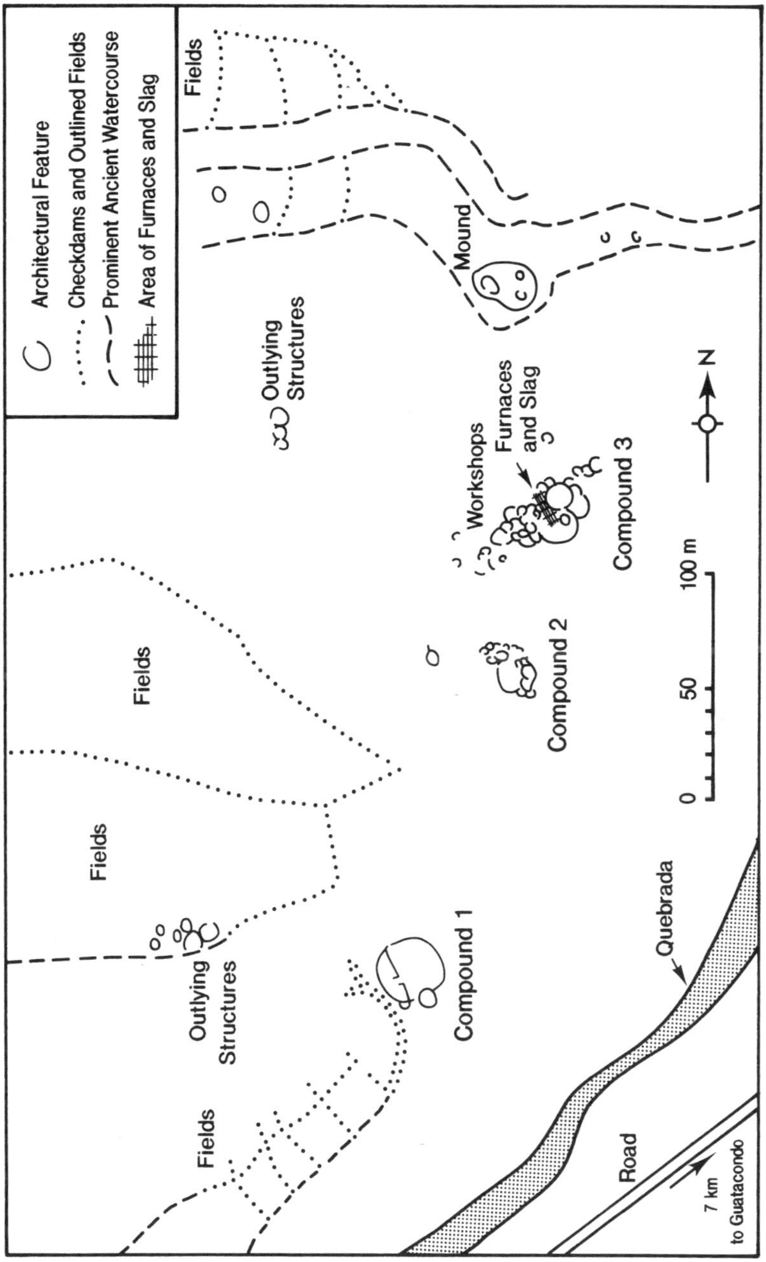

Fig. 2. The Ramaditas archaeological site, northern Chile.

Fig. 3. Portion of Compound 1, Ramaditas site. Diameter of compound is roughly 30 meters, with sections of walls standing about 1.3 meters above desert floor. Andean foot hills, right background.

with a result of 2040 ± 50 B.P. (corrected and calibrated to 50 B.C.; TO-3573); and

4. a standard ^{14}C date taken on a support post from Compound 1 (Room #1) with a result of 2040 ± 60 B.P. (corrected; Beta-48822).

The archaeological evidence is consistent and convincing that site construction and first occupation date to the Late Formative period of Chile (Early Intermediate period of Peru).

Ancient Metallurgy at Ramaditas

The topic of ancient metallurgy at Ramaditas is a theme of some importance. Few sites with evidence of slag, ore, metal, and furnace ruins nearby have been extensively studied so far in the South Andes, particularly for the pre-Inka period. This has been the case, despite the generally held view that ancient metallurgy and the extraction of mineral resources were part of the rationale for permanent human settlement in the Atacama desert early on. The questions that surround the precise nature of the use and extraction of metallic resources, and the way in which these resources affected settlement, have not been rigorously pursued. Part of the problem has been the relatively small amount of research done at Atacama sites, as compared to areas with a much greater concentration of archaeological ruins and dense urban centers. A second, and perhaps more important problem, relates to the relative scale of the early copper industry itself. As shown below (and as thought to be represented by Ramaditas), copper production during the Late Formative period in the Atacama was a different sort of affair from that at the large copper centers of later times elsewhere in the Andes, e.g. Batán Grande in Peru (Shimada et al. 1982). Part of the problem then has been to recognize the relative significance of copper manufacture in a context of marginal permanent settlement, where it was the result of more repeated, dispersed efforts, and not the long-term, centralized production of a single industrial center. As shown by the Atacama sites, settlement in the desert was fragile. The marginal sites of the Late Formative period were organized within a network of far-flung trade relations, including vertical economic ties (Dillehay and Núñez 1988; Núñez 1989). The scale of copper industry at Ramaditas, and the thin veneer of evidence from other pre-Inka sites in the region, seems to reflect this.

It has been all too easy to overlook evidence of copper smelting in the Atacama, despite the rich abundance of copper deposits there. The "thinness" of the evidence at Late Formative period sites, which is more a result of the dispersed and mobile "off-site" nature of the smelting itself than any lack of production, means that when noted, its significance has gone unstudied. It seems to be that researchers have tended to overlook the potential of their evidence, as they have concentrated on other matters, primarily because their findings have not included vast amounts of slag and copper. More to the point, if Ramaditas is representative, there is so little slag to be found at these habitational sites, that it is very easy to miss the evidence altogether if excavation is limited. For example, the only piece of subsurface slag to be found so far at Ramaditas was less than one cubic centimeter. When thinking about metallurgy in the Atacama, it is important to dismiss the idea of a single major production center. As the settlements of the Atacama interior and those along the adjacent coast never developed into large urban centers like those of central and northern Peru, evidence of early smelting should not be expected to fit within a model of centralized, urban development. Populations in the Atacama remained scattered due to the desert's inability to support dense populations, and patterns of mineral extraction and metal production were decidedly different from those where urban settlement was possible over longer periods of time.

It is our claim that the evidence of metallurgy from sites such as Ramaditas is a topic worthy of more extensive and exhaustive research. The pattern that emerges from our study is one where, during the Late Formative period:

1. metal smelting was a process carried out close to small permanent villages;

2. the location of furnaces was a flexible matter dependent upon distance from fuel source and ore;
3. smelting tended to be episodic not continual, meaning that furnaces were constructed when and where needed; and
4. ingots were transported to the villages for subsequent working and refinement.

This pattern, which is also similar to that noted in northwest Argentina during the Inka period (Rodríguez 1986), indicates that metal smelting was a dispersed activity, but not necessarily an unimportant one. It also infers that it could be done independent of direct labor supervision, at least up to the point of ingot production.

The examination of ancient metallurgy at Ramaditas, from a materials science perspective, has been the focus of research by one of us (Graffam n.d.). This study demonstrates conclusively that copper was smelted near to the site, and that copper metal itself was worked there. It also shows that the metal produced was of unusual purity, due in great part to the copper-rich ores selected for reduction. Furthermore, it provides an insight into the nature of the smelting process itself, addressing issues of importance relative to the degree of technical skill involved and the type of copper chemistry associated with copper production from the site. What follows below is a series of observations that relate to the metal industry, as they relate to issues of archaeological and metallurgical importance.

Study Samples

Nineteen samples of crushed ore from both surface and subsurface contexts were submitted to the Mineralogy Section, Royal Ontario Museum, for identification by x-ray powder diffraction analysis. In the case of surface samples, these were collected from known furnace locations at which slag was also collected. These samples form a control group for comparison, primarily for subsurface materials that were found *ex situ* of their primary industrial context.

Numerous pieces of slag were collected and brought back for study. Three samples (one surface slag, one piece of coated furnace block from a surface ruin, and one subsurface slag) were submitted to the Department of Metallurgy and Materials Science, University of Toronto, for identification via x-ray florescence analysis. Again, in the case of surface samples, these were collected from known furnace locations near the site. They form a control group for comparison, which is particularly useful where isolated subsurface slags are concerned. The subsurface sample comes from a sealed floor context dated to 50 B.C. at Compound 2 at the site.

One piece of metal (12 mm. x 10 mm. x 1 mm.), which was discovered *in situ* in an undisturbed floor context of Compound 1 (50 B.C.), was brought back for study. Two samples of it were submitted to the Slowpoke Reactor Facility at the University of Toronto for study by instrumental neutron activation analysis (INAA). The remainder was submitted to the Department of Metallurgy and Materials Science, University of Toronto, where it was sectioned for microscopic study: microstructural analysis and microcompositional analysis via scanning electron microscope (SEM).

Dating the Evidence

One of the concerns when working on sites from the Atacama is a matter of dating. Elsewhere, this is not as worrisome a problem, as carbon is found in some abundance at industrial sites for radiocarbon dating. The issue here, however, is complicated by the fact that wood in the Atacama is exceptionally well preserved, and it may be "mined" from the ground to be used centuries later. This makes it very difficult to date a surface feature from its associated charcoal should it be found "off site," and not within the specific limits of an architectural compound or constructed mound where other means can be employed. Smelting furnaces in the Atacama suffer from such a dating problem. Within a one kilometer radius of Ramaditas, there are more than one dozen surface furnace ruins. Although one may assume that, as a class, these ruins date primarily to the Ramaditas phase of occupation, it remains difficult to ascertain this directly. For example, a furnace ruin (Furnace A) that was found situated on top of a sand-filled structure at Compound 3, and hence stratigraphically later than the site occupation and subsequent abandonment of that structure, dated earlier than the date of site construction! Charcoal from the feature was radiocarbon dated to 2320 ± 60 B.P. (Beta-50834), some 280 years prior to the initial settlement at Ramaditas (2040 ± 50 B.P., TO-3573)—a clear impossibility. Accepting that the date is accurate from the laboratory, it demonstrates the point that "old" wood (wood that died centuries before site construction) can be, and indeed was, collected from the desert for use in smelting operations well after site abandonment—in this case, long after settlement had moved up valley but while smelting operations continued in the immediate vicinity of Ramaditas.

Still, it can be determined that smelting was also carried out during the original phase of occupation at Ramaditas. The best evidence to date is a small piece of copper slag (Slag A) that was found in a sealed context at Compound 2 (table 1). This item was found on the floor of one of the standing structures (Structure #22), below more than a meter of sand overburden and a level of roof collapse, and in association with Late Formative period pottery sherds *in situ* (Martindale n.d.). The evidence indicates that the slag was carried into the structure at the time of original occupation. The slag cannot be intrusive from the top surface, nor can it have fallen

Table 1. X-Ray florescence analysis of Ramaditas slags.

	Slag A	Slag B
$A1_2 0_3$	18.7	14.4
CaO	3.0	9.3
Cu_2O	1.0	2.3
FeO	4.1	3.8
K_2O	1.3	3.3
MgO	2.0	2.0
MnO	0.2	0.0
Na_2O	3.1	3.3
P_2O_5	1.2	0.4
PbO	0.6	0.0
SiO_2	61.2	60.0
TiO_2	0.2	0.4
Total	96.6%	99.0%
Approx. Fluidity	1250°C	1250°C
Metallic Copper	0.9%	2.0%

Note: The analyses identify the percentages of various metals present in the samples, reporting them as oxides. Content of metallic copper is also reported, separating it from the reading of copper oxide.

down onto the floor during archaeological excavation, for there is no slag found anywhere on the top surface of Compound 2. The present interpretation is that the item was probably "carried" to the site adhering to a copper ingot about to be worked. In the course of hammering the ingot, the slag, being brittle, broke off and either fell to the floor where it was found, or temporarily "clung" to the worker's clothing until it fell to the floor later. Ceramic associations, corroborated by radiocarbon dates for site construction and original occupation in general, place this floor context within the Alto Ramírez II time frame of the Late Formative period.

Local Production

A subsequent line of evidence removes the possibility that the ingot and slag, referred to above, were transported to the site from great distance, i.e. that copper production was not carried out near Ramaditas during its original phase of occupation and that the ingot in question was a long-distance trade item. As shown in table 1, Slag A is chemically very consistent with the surface slags (Slag B) found in the immediate vicinity (Ward n.d.). Such consistency argues strongly for the use of the same ore mixture, i.e. ores with the same gangues, and the same additive flux. As the surface slags are known local products from local ores in the vicinity, the comparative evidence argues convincingly that the subsurface slag is also the byproduct of local production.

Smelted Metal

Of course, the presence of slags does not automatically mean that a smelting operation was being run. But, as shown in table 1, slags from the site were sufficiently low in copper content that they were obviously smelting slags, not refining crucible slags, such as those associated with the melting and purifying of native coppers (Rehder n.d.; Franklin, pers. comm.). Analyses of both surface and subsurface slags show them to be the byproducts of copper smelting. As the surface sample (Slag B) had been found in direct association with crushed brochantite and antlerite ores, this was clearly an expected result. As shown by their chemistry, both samples are high silica slags that contain a small percentage of metallic copper, and such slags are consistent with the composition of known copper smelting slags (Rehder n.d.).

Yet, as shown in table 2, INAA revealed that the copper from the site is 99.5 percent pure, and low enough in trace elements to fit the profile of a native copper (Hancock n.d.). This raised the possibility that either the interpretation of the slag results were in error, or that the metal in question was a smelted product of unusual purity. It needed to be demonstrated directly that the metal is a smelted product to remove all doubt.

Turning to the metal itself, the presence of an estimated two to four percent by volume of copper oxide inclusions within the metal demonstrates conclusively that the sample in question is smelted copper. These inclusions are seen under normal microscopy, of an etched cross-section, as small rounded "nodules" throughout the metal. Subsequently SEM analysis shows them to range in size from three to five microns, with an oval to somewhat flattened shape (Nueb n.d.). All of them, without apparent exception, are inclusions of copper oxide, the result of oxygen in the molten metal being forced out as the metal solidified (Rehder n.d.). Such inclusions are not characteristic of native copper in its natural state (oxygen is not present when native copper is formed naturally), but are typical inclusions in smelted products (Franklin, pers. comm.). Although it cannot be ruled out that native copper may have been added to the smelting charges along with copper ore, it seems abundantly clear that the metal from Ramaditas is a smelted product.

Table 2. Instrumental neutron activation analysis of Ramditas copper sheet.

Element	Sample #1	Sample #2
Ag	560	600
As	≤2.9	240
Au	0.099	0.090
Br	7.9	7.8
Cd	≤29	≤9
Co	4.3	8.4
Cr	≤14	≤13
Fe	≤540	≤640
Hg	14	11
In	≤1.6	≤1.2
La	≤1.6	≤1.6
Mn	23	20
Na	1000	970
Ni	≤100	≤100
Sb	≤0.43	≤0.19
Sc	0.18	0.06
Se	≤14	≤13
Sm	≤0.22	≤0.21
Sn	≤270	≤240
Th	≤0.91	≤0.81
Zn	≤41	≤40
Total	0.26%	0.29%

Technological Efficiency

The issue of smelting temperature and slag fluidity is important to consider from a technological standpoint. As shown in table 1, both subsurface and surface slags contain high silica and a small percentage of metallic copper. Slags with similar composition are known to have an approximate temperature of fluidity of 1250°C, and would be somewhat more viscous at lower temperatures as one approaches the lower limit for copper smelting (Rehder n.d.). The low percentage of metallic copper in the slags, however, indicates that the slag was sufficiently fluid when molten (in operation) to achieve good separation of metal, and that furnace operation was close to a 1250°C temperature (Rehder n.d.). Such an interpretation was further supported by the glossy and "drippy" appearance of the slags associated with the surface furnaces. Slag fluidity is an important observation, as it infers technological efficiency.

The Ramaditas evidence is in striking contrast to the kind of process reported from Batán Grande in Peru. In the latter case, adequate separation of metal was not achieved during the smelting. Copper prills embedded within the slag had to be removed laboriously by hand after the slag had cooled and subsequently been crushed, and the prills then remelted to form ingots (Shimada et al. 1982; Shimada et al. 1983). Blowpipes were used to achieve combustion, as was the general practice in pre-Hispanic New World, bellows apparently not being known.

Such a process was not employed at Ramaditas. Here, higher slag fluidity had been achieved so that most of the reduced copper had separated out to form a pool in the hearth of the furnace. Higher slag fluidity implies higher furnace temperatures, or better fluxing, or both. Since there is no evidence of blowpipe nozzles at Ramaditas, it may be suggested that natural draft furnaces were used. These can achieve demonstrably higher temperatures than blowpipe furnaces, the latter being affected by the moisture and reduced oxygen content of human breath (Rehder n.d.). Unfortunately, all of the surface furnaces near the site are smashed apart, presumably to remove every gram of metallic copper from the smelting, and none show intact architecture that could help determine whether natural draft may have been employed. Still, as not one fragment of a single blowpipe is seen in association with any of the surface furnaces, it points to a probability of natural draft over other techniques of smelting, and it suggests that the well-known *huayra* (wind-blown) furnace described in historic accounts may date back as early as the Late Formative period in the South Andes.

Sulphate Ores

The issue of type of ore used in smelting is also important from a straightforward technological standpoint. Copper ores containing sulphur are more difficult to smelt than other ores, as the sulphur must first be removed (and replaced by oxygen by roasting in air) in order to obtain a satisfactory copper product. Smelting of such ores is more difficult, as they have proven to be problematic to work.

Crushed brochantite [$CuSO_4+3Cu(OH)_2$] and antlerite [$CuSO_4+2Cu(OH)_2$] ores are found in association with the surface slags and furnace ruins (Back n.d.), which raises several interesting points. Firstly, it provides abundant proof that sulphate ores were the source of metal for the surface operations. Secondly, given the striking similarity of the surface slag (Slag

B) to the subsurface slag (Slag A), it can be suggested that sulphate ores were processed here during the original phase of occupation as well. Slag is the product of both the gangue in the ore, and the fluxing agents and additives put into the charge. While it cannot be stated with certainty that sulphate ores were used during the original occupation of the site—the use of relatively pure ores would mean that the ratio of gangue to silica flux and added base in the slag would be low, which would mask otherwise obvious differences—the striking similarity in slags makes it possible to suggest continuity in both cases. It infers that brochantite and antlerite ores were used at about 50 B.C. as well. This evidence would tend to agree with the interpretation that sulphate ores, particularly brochantite, have a long history of use in smelting in the South Andes; such ores, which originate only in hyperarid regions such as northern Chile, have been reported from sites in the Lake Titicaca basin as early as 1250 to 850 B.C., as well as elsewhere in the Bolivian highlands and eastern slopes in Formative period contexts (Browman 1991; Lechtman 1980).

Direct evidence of ore smelting is seen at Ramaditas in association with Furnace A (Compound 3 surface furnace with erroneous date). Adjacent to this surface ruin is a scatter of crushed brochantite and antlerite ore on top of a sand-filled structure (Structure #55). The size of the ore pieces ranges from 0.2 to 1.0 $cm^2$and a thickness of 3.0 to 4.0 mm. Such a size will permit roasting to oxidize the pieces fully, and is still not too fine to prevent air circulation among the pieces. Excavation of the top five to ten centimeters of this feature revealed indications of irregular burning, substantiating that roasting of the ore had been carried out—a matter that would have had to have been done to achieve adequate copper reduction.

Turning again to the metal, it is not possible to evaluate whether or not a sulphate ore was used in its production. Substantial inclusions of copper oxide throughout its matrix signify that there was more than sufficient oxygen within the metal when molten to drive off any traces of sulphur that otherwise may have been left.

Metal Processing

At present, there is only very limited evidence of metal processing at the site. In the area opened, copper metal has been discovered with very low frequency, as would be expected where prized metal was being fashioned into valued items and where metal scrap could be collected for remelting. In fact, metal has only been found on the sampled floor of a room in Compound 1—a sampled area of about 3 m^2. This area, which represents less than five percent of the excavated area of the site to date, is the only context that provides firm evidence of metal processing. It may be that it was carried out in other areas of the site as well, but there is little by way of direct evidence of this, as yet. Given the current circumstances, metal processing remains a difficult topic to address.

Still, the presence of small rectangular pieces of copper sheet from the floor of Compound 1 indicates that some degree of metal processing was carried out here. The presence of the copper sheet argues for hammering to some degree, as does the thinness of the pieces themselves. As would be expected, a microstructural analysis of the metal bears this out, as it shows the metal to have a distorted (downward) curvature at its edge, indicative of a shearing or chisel action. SEM analysis also shows some degree of distortion of the copper oxide inclusions, which again indicate hammering, although not extensive in this case. It is tempting to think that the regular shape of these rectangular pieces means that they were not metal scrap, but instead metal blanks waiting to be formed into their final shape, e.g. small tubular beads. Exactly what items were produced here remains unknown, but status items as well as utilitarian objects were part of the Alto Ramírez copper assemblage, and it seems probable that such items were made at Ramaditas as well.

Conclusion

Evidence from Ramaditas shows conclusively that the smelting of copper ores was an activity that dates at least to 50 B.C. in the Atacama desert. It provides the earliest direct evidence that copper metallurgy was a major factor affecting permanent settlement in the zone during the Late Formative if not before, and it provides direct support for the current models of economic and cultural development in the South Andes. For these reasons alone, the results reported here are highly significant.

The results demonstrate clearly that the metallurgical process involved the smelting of ores, and not simply the melting and purification of native copper. The examination of the slags show them to fall within the range of smelted byproducts, which indicates the process of ore reduction, not metal refinement. The results also indicate that copper-rich ores of brochantite and antlerite were most probably those selected for use. As shown by slag analysis, these ores were smelted at a temperature of about 1250°C. The temperature achieved was sufficient to effect good separation of metal from slag, which indicates a good degree of slag fluidity. This point is an extremely important one, for it means that the technology did not rely on the collection of prills by hand, as has been described for Batán Grande in Peru. Lastly, the results show that some degree of metal processing was carried out at the site, although this is an area that is clearly in need of more attention.

A picture emerges of inhabitants at small villages in the Atacama desert valleys with a viable and efficient technology during the Late Formative period. This view accords well with the archaeological evidence of copper artifacts in northern Chile in general (Iribarren 1974; Muñoz 1989), and it helps complete our understanding of the ancient economy of mineral exploitation in this region. By this period, small villages that were scattered among the desert valleys were engaged in metal smelting, and in the manufacture and supply of desired goods. Copper smelting in the Atacama grew out of a larger context of resource exploitation and interregional trade that connected highlands, desert, and coast.

Acknowledgments

This work was made possible by research grants from the Social Sciences and Humanities Research Council of Canada, Trent University, Cominco Ltd., Cominco Resources International Ltd., and Teck Corporation. The research has benefited greatly from the assistance of Prof. J. E. Redher and Prof. Ursula Franklin, both of the University of Toronto, who gave a great deal of their time and expertise to the study; from conversations with Profs. John and Theresa Topic, both of Trent University, and Prof. Ron Hancock of the University of Toronto; and from the assistance of Dr. J. Mandarino and Dr. M. Back, both of the Royal Ontario Museum, for help with ore identifications. We thank Andrew Martindale, Craig Bissell, and Emily Stovel for their work during the 1992 season at Ramaditas which helped bring this study to fruition, and for the assistance of the 1991 field school students in getting the site mapped and recorded. We also thank the museum staff at Universidad Arturo Prat for their help and assistance with the project, and the staff at the Department of Metallurgy and Materials Science, University of Toronto, for help in conducting the research.

References

Alvarez Miranda, L. 1981. Metalurgia prehispánica en el extremo norte de Chile. *Revista Parina* 1:33-38.

Back, M. n.d. Results of x-ray powder diffraction analyses of copper ores from the Ramaditas site, northern Chile. Technical report on file, Trent University.

Browman, D. L. 1991. The dynamics of the Chiripa polity. Paper presented at the 47th International Congress of Americanists, New Orleans. Manuscript on file, Trent University.

Castillo G., G. 1986. Sociedades agropecuarias tempranas y control de recursos en el ambiente semiárido de Chile. *Chungará* 16-17:173-180.

Dauelsberg H., P. 1985. Faldas del Morro: Fase cultural agro-alfarera temprana. *Chungará* 14:7-44.

Dillehay, T. D. and Núñez A., L. 1988. Camelids, caravans, and complex societies in the south-central Andes. In *Recent studies in pre-Columbian archaeology*, eds. N. J. Saunders and O. de Montmollin, 603-634. Oxford: BAR International Series 421 ii.

Dransart, P. 1991. Llamas, herders and the exploitation of raw materials in the Atacama. *World Archaeology* 22(3):304-319.

Graffam, G. n.d. Metallurgical observations from the Ramaditas site, Atacama desert, northern Chile. Technical report on file, Trent University.

Hancock, R. n.d. Instrumental neutron activation analysis of metal from the Ramaditas site, Atacama desert, northern Chile. Technical report on file, Trent University.

Iribarren Charlín, J. 1974. *La metalurgia en Chile en época precolombina.* Trujillo: Trabajo expuesto en el segundo congreso Peruano del hombre y la cultura andinos.

Latcham, R. 1936. Metalurgia Atacameña. *Boletín del Museo Nacional* 15:107-151.

Lechtman, H. 1980. The Central Andes: Metallurgy without iron. In *The coming of the age of iron*, eds. T. A. Wertime and J. D. Muhly, 269-334. New Haven: Yale University Press.

Martindale, A. n.d. Excavations at Compound 2, Ramaditas site, northern Chile. Technical report on file, Trent University.

Mayer, E. F. 1986. *Armas y herramientas de metal prehispánicas en Argentina y Chile.* Bonn: AVA Materialien Band 18.

Meighan, C. W. 1980. Archaeology of Guatacondo, Chile. In *Prehistoric trails of Atacama*, eds. C. W. Meighan and D. L. True, 99-126. Los Angeles: Monumenta Archaeológica 7, University of California.

Mostny, G. 1980. The archaeological zone of Guatacondo. In *Prehistoric trails of Atacama*, eds. C. W. Meighan and D. L. True, 91-97. Los Angeles: Monumenta Archaeológica 7, University of California.

——. 1970. La subarea arqueológica de Guatacondo. *Boletín del Museo Nacional de Historia Natural* 29:271-287.

Muñoz O., I. 1989. El período formativo en el Norte Grande (1000 a.C. a 500 d.C.). In *Culturas de Chile: Prehistoria desde sus orígenes hasta los albores de la conquista*, eds. J. Hidalgo L., V. Schiappacasse F., H. Niemeyer F., C. Aldunate del S., I. Solimano R., 107-128. Santiago: Sociedad Chilena de Arqueología.

——. 1986. Aportes a la reconstitución histórica del poblamiento aldeano en el valle de Azapa (Arica-Chile). *Chungará* 16/17:307-322.

Niemeyer F., H., Castillo G., G., Cervellino G., M. 1989. Los primeros ceramistas del Norte Chico: Complejo Molle (0 a 800 d.C.). In *Culturas de Chile: Prehistoria desde sus orígenes hasta los albores de la conquista.* eds. J. Hidalgo L., V. Schiappacasse F., H. Niemeyer F., C. Aldunate del S., I. Solimano R., 227-263. Santiago: Sociedad Chilena de Arqueología.

Nordenskiöld, E. 1921. *The copper and bronze ages in South America.* Comparative ethnographical studies, vol. 4. Göteberg: Elanders Boktrykeri Aktiebolag.

Nueb, F. n.d. SEM analyses of metal from Ramaditas, northern Chile. Technical report on file, Trent University.

Núñez A., L. 1989. Hacia la producción de alimentos y la vida sedentaria (5.000 a.C. a 900 a.C.). In *Culturas de Chile: prehistoria desde sus orígenes hasta los albores de la conquista.* eds. J. Hidalgo L., V. Schiappacasse F., H. Niemeyer F., C. Aldunate del S., I. Solimano R., 81-105. Santiago: Sociedad Chilena de Arqueología.

——. 1982. Temprana emergencia de sedentarismo en el desierto Chileno: proyecto caserones. *Chungará* 9:80-123.

Núñez A., L. and Dillehay, T. 1978. *Movilidad giratoria, armonía social y desarrollo en los Andes meridionales: Patrones de tráfico e interacción económica.* Antofagasta: Universidad del Norte.

Núñez A., L. and Moragas W., C. 1977. Ocupación con cerámica temprana en la secuencia del Distrito de Cañamo (Costa Desértica del Norte de Chile). *Estudios Atacameños* 5:21-49.

Ponce S., C. 1970. *Las culturas Wankarani y Chiripa y su relación con Tiwanaku.* La Paz: Editorial "Los Amigos del Libro".

Rehder, J. E. n.d. Metallurgical materials from Ramaditas, Chile. Technical report on file, Trent University.

Rivera, M. 1991. Prehistory of northern Chile: A synthesis. *Journal of World Prehistory* 5(1): 1-47.

——. 1976. Nuevos aportes sobre el desarrollo cultural altiplánico en los valles bajos del extremo Norte de Chile, durante el período intermedio temprano. In *Homenaje al Dr. Gustavo le Paige, S.J.* Arica: Universidad del Norte.

Rivera, M., Soto, P., Ulloa, L., and Kushner, D. 1974. Aspectos sobre el desarrollo tecnológico en el proceso de agriculturación en el norte prehispánico, especialmente Arica (Chile). *Chungará* 3:79-107.

Rodríguez O., L. 1986. Pre-Columbian metallurgy of the southern Andes: A regional synthesis. In *Pre-Columbian metallurgy*, 402-417. 45th International Congress of Americanists, Banco de la Republica. Bogotá.

Santoro V., C. 1982. Formativo temprano en el extremo norte de Chile. *Chungará* 8:33-62.

Shimada, I., Epstein, S. M., and Craig, A. K. 1982. Batán Grande: A prehistoric metallurgical center in Peru. *Science* 216:952-959.

——. 1983. The metallurgical process in ancient north Peru. *Archaeology* 36(5):38-45.

Ward, M. n.d. Results of x-ray florescence analyses of slags from the Ramaditas site, northern Chile. Technical report on file, Trent University.

Willey, G. R. 1971. South America. Vol. 2 in *An introduction to American archaeology*. Englewood Cliffs: Prentice-Hall, Inc.

Ethnohistorical Dimensions of Mining and Metallurgy in Sixteenth-Century Northern Peru

Susan E. Ramírez, DePaul University, Chicago, Illinois, U.S.A.

Abstract

Original manuscript sources describe technological aspects of mining which have long been neglected in studies of sixteenth century northern Peru. The socio-economic organization of indigenous production featured a redistributive system that was destroyed by the arrival of the Spanish who brought in the amalagamation process and appropriated Indian labor into an expanding market economy.
Key words: mining, metallurgy, aboriginal, early colonial, northern Peru.

Introduction

Despite headline attention to rich archaeological finds of hundreds of gold, silver, copper, and other metal objects over the last few decades, ethnohistorians have long neglected the study of mining activity in the North of Peru. One reason for this inattention was the "bullionist" bias of the first waves of Spanish conquerors and settlers. They equated quantities of gold and silver — whether it be of a kingdom, nation or individual — with strength and wealth, and, consequently, given their own aspirations, were mostly uninterested in the more utilitarian metals, such as copper or bronze. The discovery of the "mountain of silver," the Cerro Rico of Potosí, in 1545 with the initial yields of up to one mark (8-8.5 ounces each) per pound of silver ore, quickly shifted Spanish attention and colonization efforts to the South. At a comparable time, the North — whether speaking of the Lambayequen coast alone or the coast and its highland hinterland of Cajamarca (and even on north into Cañar country) (fig. 1) — offered few such opportunities for quick enrichment. (Lohman Villena 1970, 644; Shimada, Epstein and Craig 1983, 38; Bargalló 1955, 74).

In Quest of Mineral Wealth: Aboriginal and Colonial Mining and Metallurgy in Spanish America, edited by Alan K. Craig and Robert C. West, 1994. Geoscience and Man, vol. 33, pp. 93-108. Dept. of Geography & Anthropology, Louisiana State University, Baton Rouge, LA 70893-6010, U.S.A.

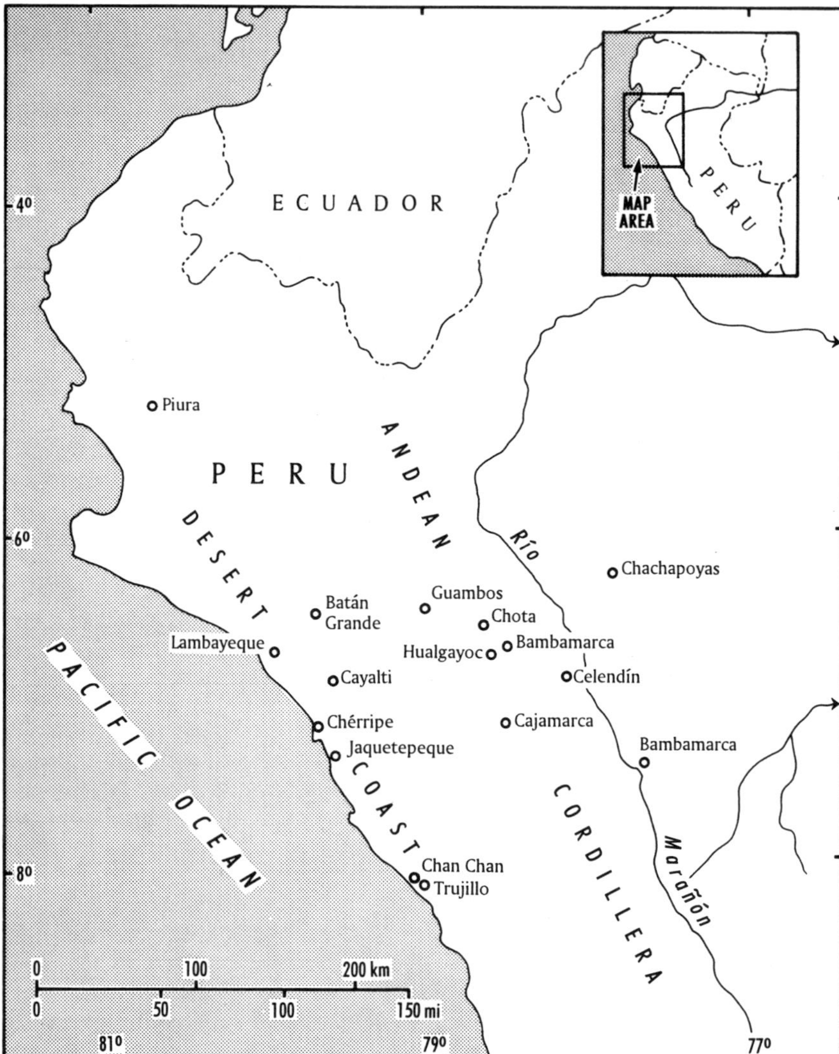

Fig. 1. Locale of late Inka and early Spanish mining and metallurgy in northern Peru. Present-day international boundaries are indicated.

Furthermore, during the first few decades after the Conquest, the Spanish assumed that prospecting, mining, and even some metal working (as will be shown below in the discussion of *plateros*, silver smiths) were the jobs of Indians. As a result of these attitudes and the inordinate attention to Potosí and later to Huancavelica and other rich, southern and usually highland sites, the Spanish left comparatively few early written records on the topics of mining and metallurgy for the North. Chroniclers, some of whom will be cited below, mention the North, but do not devote the same space to describing the mines and technological processes there as they do for the South — on which long, detailed, and often elaborate accounts are included in their writings. Other colonial manuscript sources, such as tribute and census records, court cases, *visitas* (diaries or reports of inspection tours) and other administrative accounts, refer to mining, if at all, in a cursory and incidental manner, until the latter part of the seventeenth century.

Indigenous Metallurgy in the Sixteenth Century

To the Indian inhabitants of northern Peru, like those elsewhere, mines were objects or locations of veneration. Cristóbal Sánchez de Albornoz, writing about *huacas* (shrines) in Lucanas, Angaraes, Jauxa and other provinces of Chinchaysuyo, states:

> They [the Indians] choose the most beautiful piece and they save it ... and this they do from all the gold and silver ores and quicksilver that they have discovered over many years. They have chosen the most beautiful metal ores and they have kept them and they keep them and they worship them calling them mothers of the said mines. And, before they go to work them [the mines], the day that they are to work, they worship and they toast the said rock, calling it mother of what they work.

> Escoxen el más hermoso fruto y le guardan...y desta forma de todos los minerales de oro o plata o azogue que antiquísimamente se han descubierto. Han escogido las más hermosas piedras de los metales y los han guardado y guardan y los mochan llamándolas madres de las tales minas. Y, primero que los vayan a labrar, el día que han de travajar, mochan y beven a la tal piedra llamándola mama de lo que travajan. (Duviols 1967, 18)[1]

That the Indians of the North Coast followed this practice is clear from an administrative report on the district of Piura, directly north and adjacent to the later colonial district of Lambayeque, which until a few years after the middle of the sixteenth century included the Jayanca (North Coast) district as well. The anonymous[2] report, titled "Relación de la Ciudad de San Miguel de Piura," states that the inhabitants offered gold and silver to their shrines and *adoratorios* (places of worship), like their southern counterparts (Anonymous 1925, 96).

Although the "Relación" does not directly relate the known mines in the district to huacas and offerings, archaeological evidence shows the existence of such practices in the Lambayeque region. During a five-week site survey in 1984, a team consisting of Izumi Shimada, Alan Craig and Ing. José Luis Suárez found *Spondylus* shells, revered and precious commodities among the Indians, just a few meters from an ancient mining pit at Cerro Melliso. They interpreted these as ritual offerings similar to the ones described by Albornoz (see below) (Shimada 1984, 4,7). Shimada also reports excavating fourteen sacrificial camelids at the Huaca del Pueblo Batán Grande, a relatively small residential and metallurgical site. They were associated with metallurgical remains, dating to about A.D. 900 (Shimada and Shimada 1985, 14-15).

Spondylus shells and camelid offerings are mentioned by Albornoz in a discussion of a method used to locate *huacas*. He says that

> in and around all the shrines that are in the mountains and on the coast are some markers that they call cachauis that are signs of the offerings that they made to these shrines and each marker takes the name of the person who there offered a child or llama, gold or silver or conch shell. You will find the offerings at the said markers.

> en todas las más guacas que están en los cerros y en llanos, tienen al rededor de si unas señales que llaman cachauis, que son señales de los ofrescimientos que a las tales guacas hazían y tienen sus nombres en nombre cada señal del que allí, ofresció hijo o carnero [,] de oro o plata o de mollo. Hallarán los ofrescimientos en los tales ceques o cachauis (Duviols 1967, 38).

These *cachauis* or *ceques*, he argues, could help locate the huacas and, by implication, other mines.

Apparently in the North, the Spanish did not heed this advice. Although the first Spaniards soon equated the pyramids and temple structures with buried treasure, they did not discover many natural mines (CLDHP 1925a, 129; AGI J 458, 1468; J 1065, 5-VIII-1535, the latter courtesy of J. V. Murra). Early colonial reports of mines on the North Coast are few. Felipe Guamán Poma de Ayala wrote that Saña was "very poor in silver and little gold..." (1980, 1020[1028]). His information is confirmed by a copy of Presidente Pedro de la Gasca's 1549 tribute list for Saña that reads "they have no type of mines in their lands" (BAH 9-4664, 23^{v}). In 1562, however, there is a reference to a mine in Saña, although it is not known what ore was being exploited there or, indeed, if it was a natural mine or a sacred tomb ("*guaca entierro*"), that was sometimes referred to as a "mine" by the Spanish (ART LC 24-IX-1562). In the same year, a few kilometers to the north, Spaniards reported gold and silver deposits in the Túcume Valley (ART Mata 1-VII-1562). The aforementioned early, but undated, "Relación" describing Piura also reports that gold and silver mines existed within sight of the city, but were not then being exploited as uneconomical (Anonymous 1925, 79).

Both the archaeological and colonial records indicate, in contrast, that the northern highlands were richer in mineral resources than the coast. Parts of the general Cajamarca region were known to contain numerous mineral deposits, especially of silver. Chilete, as mentioned above, was most probably an important silver mining center even before its Spanish discovery and exploitation, as early as 1540 (AGI J 1063, No. 3, 1570, 5^{v} and J 415, 194^{v}).[3] Elsewhere in Cajamarca, silver and gold were extracted in the Conchucos area (Bargalló 1955, 38 and 76). In fact, as early as 1535, Pizarro gave Captain Martín Estete, his lieutenant in the Valley of Trujillo, permission to *rescatar* (exchange) with the *caciques* (lords) Chimo and Conchuco and their subordinates and subjects for the gold and silver in their possession (AGI J 1065, 1535 courtesy of J.V. Murra; Bargalló 1955, 76).

The Guamachuco district was also a region rich in ores (Loredo 1958, 255), as was the area near Chota, where, in the 1560s, Indians from Conduzmarca and Bambamarca worked mines (AGI J 461,1443^{v}-44). Later, on an inspection tour of the northern area, Archbishop Don Toribio Alfonso de Mogrovejo ([1593] 1920-21, 67) encountered four Indian families with sixteen members working mines in the 1590s. Another mine he visited at the same time near Celendín was worked by black slaves and an undetermined number of Indians (Mogrovejo ([1593] 1920-21). Only the native inhabitants of Guambos, the area that included the headwaters region of the Río Lambayeque or Chancay, directly inland from the town of Illimo, categorically and repeatedly denied having access to mines: "...the natives of Guambos are poor. They have no mines (AGI J 457, 744)."[4]

Further afield, silver was known to originate in Cañar country, to the north of Cajamarca (Bargalló 1955, 38). Chachapoyas, further inland from Cajamarca, was also reported to be rich in gold and silver deposits as early as 1538, almost assuredly indicating pre-Columbian mining activities (Romero 1937, 185 and 187; Loredo 1958, 259-64).

The references are curiously silent in regards to the rich gold mining areas of Pataz and the silver deposits of Hualgayoc near Chota[5] that became known in later centuries. Also, copper deposits that archaeological finds suggest were so important to the indigenous economy are rarely mentioned, almost certainly reflecting the Spanish bullionist myopia referred to above.[6]

The pre-Hispanic inhabitants of the region, however, certainly knew where the mines were. Craig (in Shimada, this volume) argues that ore sources could have been located by finding mineralized "float" in dry stream beds and following the trail upstream or by using plant indicators. As a result, an advanced metallurgical industry had flourished on the North Coast under local authorities for centuries. Later, upon their conquest of this and other areas, the Inkas may actually have taken measures to expand the scope of the industry. Guamán Poma de Ayala suggests that Capac Yupanqui and presumably his successors initiated a systematic search for mineral resources upon securing their expanding domains.

Archaeological and local-level manuscript sources indicate that local lords were entrusted with overseeing the actual procurement and extraction of the ores (Shimada 1985; Carranza 1923, 53; Rodríguez-Rivas 1970, 659). In general, the available historical sources suggest that the *curaca* (paramount lord) assigned *mineros* (commoners) (BAH A 92, 17) to work in the shallow pits and mines where most of the ores (except gold) were removed. Commoners worked as miners on a seasonal and temporary basis, perhaps as part of their *mita* (rotating tribute labor service) obligations (AGI/P 188, R. 22, 1561, 10). Numbers of workers assigned to such tasks were small.

The exact means of access to or procurement of the ore-bearing resources is still a matter of some debate. Evidence exists to support several different mechanisms. The most evident is direct exploitation of resources close to the *curaca's* community. Both the archaeological and historical records support this explanation. Recent geological surveys report that both the copper and arsenic-bearing ores necessary to produce the archaeologically-found ceremonial and utilitarian artifacts for which the region is becoming famous were available on the coast (Shimada 1984; and 1985, 372-77; and Shimada, pers. comm., 22 March 1985).

A growing number of primary sources, including the inspection tour of Jayanca in 1540, the *encomienda* (grant of Indians in trust to a conqueror in return for protection and religious instruction) grant of Túcume of 1541, and numerous others dating from before the major *reducción* (resettlement) efforts of the visitor Dr. Gregorio González de Cuenca in 1566 and Viceroy Francisco de Toledo in 1572-73 (Espinoza Soriano 1975; ART CoO 13-VII-1570, 96^{v}-99; Ramírez 1979), indicate that it was not uncommon for coastal lords to have subjects in the highlands (Loredo 1958, 271-73; AGI J 457, 1013; and AGI P 189, R. 11, 1566). One logical site of highland-coastal interaction in the North are the mines of Chilete at an accessible 1,000 meters above sea level, inland from the coast, following the Jequetepeque River, between Pacasmayo and Cajamarca. Chilete is known to be a source of the arsenical ore needed to make the bronze or copper alloy so common in archaeological artifacts found on the coast (Bargalló 1955, 41; Malaga 1905, 24-25).

One document states that at least two groups of Indians worked there under separate lords in the 1540s (AGI J 415). José Antonio del Busto (1959), citing early colonial records, says that the *curacas* of Cajamarca *made* Indians from the coast exploit certain highland mines under their jurisdiction or control (emphasis mine, AGI AL 568 as cited by Busto 1959, 369 and 386). This suggests that their own coastal lords did not manage the mineral sources themselves. Hence, evidence for direct exploitation of mineral resources in the highlands by coastal colonists remains circumstantial, but argues forcefully for some type of resource sharing between the peoples of the highlands and the coast.

Exchange was another mechanism for procurement which dates to before the Inkas conquered the coast. Foodstuffs and *bastimientos* (cloth) were traded for minerals by coastal peoples and highlanders, respectively. After the Inka, these exchanges diminished in number and importance, perhaps coinciding with the notorious resettlement of peoples and the redistribution of resources that followed Inkan expansion. The scale of exchanges both before and after the Inka conquest argues for state administration.

The evidence on hand, in sum, leaves unresolved the exact location of the sources of ores needed for coastal metallurgy (Shimada 1985, 357-58), but does help resolve the debate over the mechanism of procurement. If ores were available locally, direct exploitation would probably have been the rule. If not, commodity exchange between polities was another mechanism, which, according to our current knowledge, predated the Inka domination of the coast.[7] Subsequently, this state-administered exchange was apparently largely replaced by resource sharing by colonists or *mitayos* (draft laborers) doing temporary duty away from home.

Once at the processing center, the ores were smelted and the resulting metals later worked into artifacts by small groups of specialized workmen. Archaeological finds and the works of the early chroniclers suggest that the ores were ground with stone hammers and/or *batanes* (Andean grinding stones) and then smelted in small clay furnaces, which used natural wind power or drafts produced through blow tubes to obtain the hot flames and high temperatures needed to smelt the metal ore.[8] Although Morúa's description is of the process practiced in the southern highlands, it still gives us some insights into the probable northern tradition.

> Only the very rich metals were smelted all alone, if they contained lead, and if dry, [they were smelted] with a collector that made them soon melt and run .... To melt them they put them on the hills and slopes, where winds blow with more force in the large clay brasiers that they call guayras, with burning charcoal and the ore inside; and as it melts, the fire consumes the slag and purifies the silver. All that the Indians of Peru extracted in old times was by this method of smelting, because they knew no other method; and because of this they only used rich metals.

> Guayréanse solamente los metales muy ricos, por si solos, si son plomizos; y si secos, con alguna liga que los haga luego derretirse y correr .... Para derretirlos los ponen en los collados y laderas, donde con mas fuerza soplan los vientos en unos braseros grandes de barro que llaman guayras, con carbón encendido y el metal dentro; y como se va derritiendo, va consumiendo el fuego la escoria y purificando la plata. Toda la que sacaban los indios del Perú antiguamente era por este modo de fundición, por que no supieron otro beneficio; y a esta causa no aprovechaban sino los metales muy ricos. (Morúa [1590] 1946, 278-79)

On the coast, rather than place the clay furnaces on the hillsides to take advantage of the wind, it was apparently more common to elevate the internal temperature with the use of blow tubes. Garcilaso de la Vega (1943) describes the practice.

They smelted with the power of their breath [delivered] through copper tubes, long ones of a half fathom, more or less, depending on the size — large or small — of the task or fusion. Eight, ten, or twelve got together, as were necessary for the smelting; they circled the fire, blowing with the tubes, and today [circa 1609] they continue to do this because they have not wanted to change their custom.

Fundían a poder de Soplos con unos cañutos de cobre, largos de media braza, más o menos, como era la fundición, grande o chica. Juntávanse ocho, diez, y doce, como era menester para la fundición: andaban al derredor del fuego, soplando con los cañutos, y oy [hacia 1609] se están en lo mismo que no han querido mudar costumbre. (as cited by Bargalló 1955, 40)[9]

Subsistence for the metal smiths, called by the Spaniards "*plateros*," as well as the raw materials needed for the processing itself were supplied by the *curacas*. Small[10] groups of these smiths served at the courts of the lords of Lambayeque, Collique and Jayanca (AGI J 458, 1800v; J 461, 1456-56[v]; and J 418, 1573, 217); and received from them food, cloth, and other items necessary for subsistence (Morúa [1590] 1946, 329). According to Morúa, they were also exempted from other tribute service (Morúa [1590] 1946, 329; AGI J 461, 1514v-15; and J 457, 848[v]).[11]

Indians also fabricated *naipes* in prehistoric times, as well as slabs, tokens and larger pieces of metal, described in various Spanish texts as *barretillas* (small bars), *tejuelos* (small ingots or pieces), *planchas* (slabs), *ojas* (sheets) and *platos* (plates) (Alcina Franch 1970, 319), whose form and distribution have stimulated widespread debate as to their use.

Metallurgy on the North Coast During Early Colonial Times

The Spanish invasion and settlement seriously disrupted this procurement and production pattern. At Cajamarca in 1532 and 1533, the Indians delivered unto Pizarro's small conquering band of men 2,475,302 pesos of gold or a total of 600,655,410 *maravedís* of the precious metals, which they had accumulated over decades and even centuries (Bargalló 1955, 68 and 101). Yet, the Spanish insisted on finding more (Jara 1963, 138 and 142).

The first *encomienda* grants reflect this preoccupation, specifically stating that Indian charges could be used to work mines, even in districts where no mines were known to exist (AGI AL 201, 1633 for Túcume in 1536). The carte blanche power that the first *encomienda* grants implied gave individuals such as Melchior de Verdugo, the *encomendero* of Cajamarca, the authority to send hundreds into the pits. Verdugo, who later claimed that he was the first Spaniard to discover and exploit silver mines in Peru (Busto 1959, 330 and 364), pressured the lords of Cajamarca with threats of death and by other means, to send 100 men each to the Chilete mines as early as 1540 (AGI J 415, 199).

The Memoria of 1548, published by Rafael Loredo (1958), estimates that the 3,000 encomienda Indians Verdugo still controlled at that date produced 6,000 pesos per year of income for him in the mines.[12] Apparently, the 1,000 Indians that Verdugo lost to Hernando de Alvarado from his original encomienda of 4,000, during the Crown's adjustment and redistribution of encomiendas, could also produce almost as much in the mines as the 3,000 Indians that still served Verdugo. The Memoria estimates Alvarado's Indians could produce 5-6,000 pesos per year (Loredo 1958, especially 255-56). The Memoria goes on to list the other encomiendas where the Indians had access to mines and how much such production was worth. In the highlands, then, as compared to the coast, most of the encomienda Indians could have had and perhaps did have access to mines (Loredo 1958, 255-56 and 259-64).

In areas without known mines, the *encomenderos* likewise forced the *curacas* and *principales* to procure the precious metals. In Jayanca, for example, the coastal lords had to seek gold and silver outside their communities. According to the 1540 testimony, the *principal* Labamy and his approximately 200 subjects who lived on lands of Túcume in the highland district of Guambos, gave nine silver plates every three lunar months to their *curaca* (Espinoza Soriano 1975, 268).[13] The *curaca* of Jayanca delivered to his *encomendero*, Francisco Lobo, ten *tejuelos* (pieces) of gold and *barretillas* (little bars) of silver. In addition, two lesser lords gave him three *tejuelos* and *barretillas* of silver and one of gold. The *principal*, Facollapa, who was entrusted to the encomendero Diego Gutiérrez, gave him five little bars of silver every three lunar months or four or five *tejuelos* of gold. Two of his *principales* also gave Gutiérrez two *tejuelos* of silver, apparently in the same time period.[14]

Other coastal communities, like Saña, were forced to find silver for their *encomendero*, too. La Gasca's tribute list for Saña, dated 1549, states that the 1300-1500 tribute payers had to pay "two hundred pesos of perfect purity [*ley*] in gold or silver" (BAH 9-4664, 23^{v}) — even though there were no known mines under their control.

It is not known whether or not this demand was designed to force the Indians into the money economy and/or whether they were expected to obtain the pesos through labor or goods exchange. In favor of the first explanation is the fact that already by 1566, the Spaniards experienced a labor shortage. Indians did not want to work for the Spanish. They had to be forced.

Likewise, the increasingly severe labor shortage probably prevented *curacas* from dispatching Indians to mine ore, even that found on the coast, as often as before the Conquest. Initially, however, some copper mining was apparently carried on, because a few of the native inhabitants of the area used copper and lead as currency in 1566. Others used "*tejos*" (pieces) or "*pancitos*" (little cubes or pieces) of silver. Silver currency did not circulate regularly until decades later (AGI J 457, 719, 837^{v} and 879). Implements of their own manufacture were also rapidly replaced with those fashioned from copper, imported early from the Old World and later from Chile.

Nevertheless, Indian workers dominated the process of smelting and working precious metals in Peru until the 1570s (Bargalló 1955, 91). Because of the technology, production remained dependent on high grade ores and was relatively small-scale and labor intensive, a pattern that apparently dates back to the Middle Sicán period (Morúa [1590] 1946, 279; Cobo 1956, 151).

In 1572, Pedro Fernández de Velasco, backed by Viceroy Toledo, experimented successfully with the amalgamation process (that had first been perfected in New Spain) using mercury and salt (Brading and Cross 1972, 552-55). In the South where the process was first tried, ores were mixed with mercury, salt and sometimes other ingredients, in stone vats or tiled containers. Heating hastened the actual amalgamation process. This technological development allowed poorer grades of ores to be used and also increased the scale of production.[15]

In the North, the amalgamation process took place in vats or *cajones* of 50 *quintales* (hundredweights) of ore each. By the end of the sixteenth century, the process was accomplished without heating. This technological breakthrough, meant that Spaniards took control over the refining process, although they always remained dependent on Indians for most of the manual labor.

That smelting was established on the North Coast is evident from a document found in the archive of Cayaltí, a sugar plantation (now an agricultural cooperative) that refers back to the sixteenth century. The document records that Gaspar de Coria (b. 1545?), an early resident and rancher of the north coastal valleys, had a functioning "mill to crush metal [ores] with four little ovens [*hornillos*]" (ACayaltí, 1920) near the site of Sipán. A geologist and this author visited Sipán (in the Lambayeque Valley, 13 kilometers from Saña), near one of Coria's known landholdings, in the early 1970s. The bases of furnaces were still visible. Lechtman mentions Sipán as a site where several large cakes or ingots of cast copper have been found (ART IC 16-111, 1785; ANP R 1. 22, c. 57, 1611, 185^{v}; and RA 1. 23, c. 123, 1614, 129; Mogrovejo [1593] 1920-21, 237; ART CoO 27-VII-1580 and 30-IX-1582; and Lechtman 1976, 14). It should be noted, however, that throughout colonial times most of the copper used on the coast to make cauldrons and mills for the sugar industry was imported (Ramiréz 1986, chap. 6).

A second type of mining activity in the sixteenth century was *huaca* looting. As early as 1559, the "*huaca*" Yomayoguan, a structure in Chan Chan, was mined for buried treasure (Ramiréz, forthcoming). Other "*huacas*" were also sources of big finds in the same middle years of the sixteenth century. City officials, in fact, subsidized such "mining" activity as early as 1562 by assigning *mitayo* laborers to aid the digging and dismantling of the structures. Of a total of 882 Indians allocated to all *mita* activities, 45 percent were specifically assigned to digging at *huacas* (table 1). This assignment of labor indicates that excavating for buried treasure was big business, important to the local citizens and to the Crown (for the tax revenues rich discoveries would yield) (ART 1562) (fig. 2).

Conclusions

This work is a brief first attempt to describe the significance of mining and metallurgy within the context of the socioeconomic organization of the North Coast during the sixteenth century. The picture that emerges from the records and findings on hand, for the pre-Spanish period, is one of an

Table 1. Distribution of the *mita* for *huaca* looting

Community	Total Obligation	For *Huaca* Looting	Percent
Chimú	46	0	0
Guamán	12	0	0
de Don Alonso and de Don Xpoual	8	0	0
Guanape	12	0	0
Chao	15	0	0
Chapiyungas	6	0	0
Chicama	100	40	40
Licapa	35	15	43
Xeequetepeque	60	30	50
Chérrepe	50	30	60
Moro	40	20	50
Chepen	18	10	56
Lambayeque	100	50	50
Saña	55	30	55
Sinto	35	20	57
Collique	35	20	57
Chuspo	35	20	57
Ferriñafe	40	20	50
Raco	40	20	50
Túcume	100	50	50
Reque	40	20	20
Total	882	395	45

Source: ART 1562.

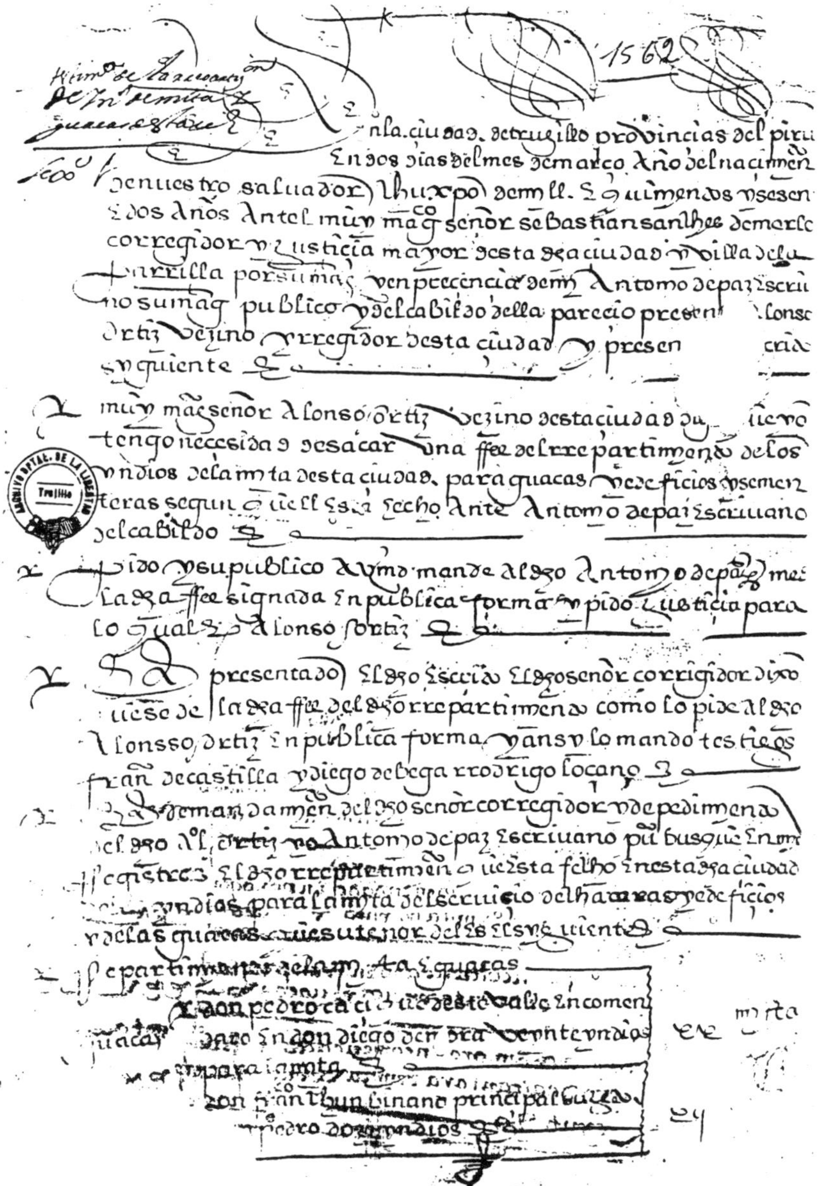

Fig. 2. Sample folio from 1562 Spanish manuscript requesting *mita* labor to exhume artifacts from graves near the site of Chan Chan, northern Peru (ART 1562).

indigenous procurement and production system that served the needs of the community and state. A lord assigned a relatively small proportion of his subjects, on a short-term or rotating basis, to procure ores — both on the coast and probably also in the highland hinterlands. Ores were transported to central workshops where workers smelted the material and smiths fashioned the finished goods in return for subsistence and recognition. Apparently under the Inka, the objects of gold and silver entered into the redistributive system of the imperial state. Utilitarian objects, rarely mentioned in the Spanish manuscript sources, probably entered the local redistributive networks and thus were allocated by the lord to those who needed or merited them, e.g., copper or bronze fishhooks to fishermen; agricultural implements, such as hoe blades, to farmers; and domestic tools, such as needles and *tumi* knives to weavers and cooks, respectively.

This general picture breaks apart rapidly under the onslaught of the Spanish, who apparently did not understand the indigenous system, so foreign was it to their own manner of interaction and thinking. The first Spaniards in the North had risked life and limb for the opportunity to obtain riches defined, first, as gold and silver and, second, as Indian subjects to do their bidding — as done in the society they had left on the Iberian peninsula. Success was defined by bullion and the position and status it could buy. The very presence of the conquerors, given their objectives, resulted in the rapid, and perhaps unwitting, dismantling of the indigenous political economy. Disease decimated Indian communities. The surviving population was concentrated in nuclear villages to facilitate tribute exaction and control. This resettlement meant not only loss of their homes, but also a loss of their traditional lands and other resources that the Spanish subsequently insisted were too far away to be effectively utilized. Tribute demands of labor and exotic items — such as gold for many of the coastal communities — took a disproportionate amount of labor power and time away from other pursuits. Finally, the Crown's proscriptions on tribute, travel and hospitality — as interpreted and implemented by Dr. Cuenca in the 1560s — and the Spanish *encomendero's* meddling with succession and the basis of legitimacy undermined the position and power of the lords to the point where their word and orders were questioned and disobeyed. Simultaneously, Indian labor was gradually appropriated from the control of the lords through the expanding market economy. Some of this labor was directed into the mines, especially silver mines, found in close proximity to the Spanish cities in the North. The scale of operations multiplied with the introduction of the amalgamation process in the 1570s. By then most of the traces of the indigenous system had vanished, replaced by a relatively large-scale system that was controlled by the Spanish and designed to meet their objectives. Ironically, the native population continued to provide the bulk of the labor power for the extraction activities. The metalsmiths, too, were forced to leave the courts of the their traditional indigenous lords. They became dependent for their livelihood on serving the Spanish.

Endnotes

1. See also Duviols 1967, 38. Albornoz' statements cover from Quito southward. Duviols dates the manuscript about the end of the sixteenth century.

2. Waldemar Espinoza attributes the manuscript to a person by the name of Céspedes and dates the document at 1586 (1975, 272).

3. Malaga (1905, 21) erroneously reports that the ore deposits of Chilete were discovered by Portuguese in the middle of the seventeenth century.

4.The Memoria, published by Loredo (1958, 256), states that they could have worked in the mines in 1549.

5. Unless these were the deposits that the Indians of Conduzmarca and Bambamarca were working in the 1560s.

6. Guamán Poma (1980, 101) did mention copper.

7. Could this explain the survival of exchange specialists in areas that been incorporated into the Inka empire only briefly (Salomon 1978)?

8. See Shimada, Epstein and Craig (1983) for a more complete description of this process.

9. See also Francisco Jerez for a description of an alternative process called "torrefacción" that he observed in the highlands of Cajamarca and Huanuco, as cited by Bargalló, 39. The numbers in the above quote are too high from an archaeological perspective (Shimada, pers. com., March 1985).

10. Most artisan groups were small. Salt makers numbered 20 to 21 in 1566 (AGI/J461, 1468^{v}). Seven or eight fishermen apparently made up one group that petitioned Dr. Cuenca in 1566 (AGI J 457, 1440^{v}). Documents regarding the encomienda of Túcume in 1541 show 20 "indios tamberos" (inn keepers) and about 100 fishermen (the largest group of specialists referred to in the manuscript) (ART/CoO, 13-VII-1570). Another reference for the central coast suggests that the total of artisans equalled between six and seven percent of a community's population (Ramírez 1981, note 14, 293).

11. Whether or not these artisans were full-time is still debated. Netherly believes that they were full-time (1977a and 1977b, 202). Rostworowski reports that certain artisans had access to land and therefore were part-time (1977, 126), a conclusion shared by Ravines (1978), based on Cobo (1956, 93).

12. Loredo's Memoria (1958) reports that Verdugo lost a third of his Indians. Busto's figure of a fourth is more in accord with the actual numbers given in the visita in AGI/J1063, No. 3, 1570.

13. Espinoza 1975, 257, says that Labamy gave three plates of silver to his lord, when in fact the testimony of the visita (268), that he transcribes and publishes, says nine.

14. Does the fact that only certain *principales* gave tribute silver indicate that they were specialized producers of the substance or that they had some special capacity (location?) to procure it and that the other lords and Indians did not?

15. For a description of the extraction and refining process at Potosí, see Bargalló 1955, 83 and 91-103; Lohmann Villena 1970, especially 644-45, and Craig (this volume).

References

Archival Sources

ACayaltí	— Archivo Cayaltí [Peru]		
AGI	— Archivo General de las Indias [Sevilla]		
	AL	—	Audiencia de Lima
	J	—	Justicia
	P	—	Patronato
ANP	— Archivo Nacional del Perú		
	RA	—	Real Audiencia
	R	—	Residencia
ART	— Archivo Regional de Trujillo [Peru]		
	CoO	—	Corregimiento Ordinario
	CoR	—	Corregimiento Residencia
	IC	—	Intendencia Compulsa
	LC	—	López Cordova
	Mata	—	Mata
BAH	— Biblioteca de la Real Academia de Historia [Madrid]		
CLDHP	— Colección de libros y documentos referentes a la historia del Perú		

Published and Other Materials

Alcina Franch, J. 1970. La producción y el uso de metales en la América precolombina. *La minería Hispana e Ibero-Americana, Ponencias del I Coloquio Internacional sobre historia de la minería*. I:307-31. León: Departamento de Publicaciones, Catedral de San Isidro.

Anonymous. 1925. Relación de la Ciudad de San Miguel de Piura. In *Colección de libros y documentos referentes a la historia del Perú (CLDHP)*, Horacio H. Urteaga, ed., t. 5, 2a serie, 75-98. Lima: Sanmarti y Cía.

Bargalló, M. 1955. *La minería y la metalurgia en la América Española durante la época colonial*. Mexico City: Fondo de Cultura Económica.

Brading, D. A. and Cross, H. E. 1972. Colonial silver mining: Mexico and Peru. *Hispanic American Historical Review* 7(4): 545-79.

Busto, J. A. del 1959. El capitan Melchor Verdugo, Encomendero de Cajamarca. *Revista Histórica* 26:318-87.

Carranza, F. 1923. La metalurgia del Perú Incaico. *Boletín de la Sociedad Geográfica* [Lima] 40(1-2):47-56.

Cobo, B. 1956. *Historia del Nuevo Mundo*. Biblioteca de Autores Españoles vols. XCI-XCII. Madrid: Estades, Artes Gráficas.

Duviols, P. 1967. Un Inedit de Cristobal de Albornoz: La Instrucción para descubrir todas las guacas del Pirú y sus camoyos y haziendas. *Journal de la Société des Americanistes* 56:7-39.

Espinoza Soriano, W. 1975. El Valle de Jayanca y el reino de los Mochica, Siglos XV y XVI. *Bulletin de l'Institut Français d'Études Andines* 4(3-4): 243-74.

Guamán Poma de Ayala, F. 1980. *Nueva crónica y buen gobierno*. 3 vols. Mexico City: Siglo XXI Editores, S.A.

Jara, A. 1963. La producción de metales preciosos en el Perú en el Siglo XVI. In *Anales del III Congreso Nacional de Historia del Perú*, 134-42. Lima: Centro de Estudios Histórico-Militares del Perú.

Lechtman, H. 1976. A metallurgical site survey in the Peruvian Andes. *Journal of Field Archaeology* 3:1-42.

Lohmann Villena, G. 1970. La minería en el marco del virreinato Peruano. In *La minería Hispana e Ibero-Americana, Ponencias del I Coloquio Internacional sobre historia de la minería*, I:639-55. León: Departamento de Publicaciones, Catedral de San Isidro.

Loredo, R. 1958. *Los Repartos: Bocetos para la nueva historia del Perú*. Lima.

Malaga Santolalla, F. 1905. Importancia minera de la Provincia de Cajamarca. In B*oletín del Cuerpo de Ingenieros Mineros del Perú* (BCIMP) 31:9-83.

Mogrovejo, Don T. A. de. [1593] 1920-21. Diario de la segunda visita pastoral que hizo de su Arquidiocesis el Ilustrísimo Señor Don Toribio Alfonso de Mogrovejo, Arzobispo de los Reyes. *Revista del Archivo Nacional* 1:49-81, 227-79; 2:37-79.

Morúa, Fr. M. de. [1590] 1946. *Historia del origen y genealogía real de los reyes incas del Perú*. Biblioteca "Missionalia Hispánica," vol. 2. Madrid: Consejo Superior de Investigaciones Científicas, Instituto Santo Toribio de Mogrovejo.

Netherly, P. 1977a. Organization through opposition: Dual division and quadripartition on the North Coast of Peru. Paper read at the 1977 Annual Meeting of the American Anthropological Association. Houston, Texas.

___. 1977b. Local level lords on the North Coast of Peru. Ph.D. diss. Ithaca: Cornell University.

Ramírez, S. E. 1979. Chérrepe in 1572: Un análisis de la visita general del Virrey Francisco Toledo. *Historia y Cultura*. 11:56-95.

___. 1981. La organización económica de la costa norte: Un análisis preliminar del período prehispánico tardío. In *Etnohistoria y antropología andina*, comp. Amalia Castelli et al., 281-97. Lima: Museo Nacional de Historia.

___. 1986. *Provincial patriarchs, land tenure and the economics of power in colonial Peru*. Albuquerque: University of New Mexico Press.

___. Forthcoming. Huaca looting on the Peruvian North Coast: A tale with two perspectives. In *The world upside down: Cross cultural contact and conflict in sixteenth-century Peru*. Stanford: Stanford University Press.

Ravines, R. 1978. *Tecnología Andina*. Lima: Instituto de Estudios Peruanos.

Rodríguez-Rivas, D. A. 1970. La legislación minera hispano-colonial. In *La minería Hispana e Ibero-Americana. Ponencias del I Coloquio Internacional sobre historia de la Minería*, I:657-68. León: Departamento de Publicaciones, Catedral de San Isidro.

Romero, C. A. 1937. Fundación de las ciudades de Chachapoyas y Celendín. *Revista Histórica* 11(1-2): 182-196.

Rostworowski de Diez Canseco, M. 1977. Mercaderes del Valle de Chincha en la época prehispánica: Un documento y unos comentarios. In *Etnia y Sociedad*, 97-140. Lima: Instituto de Estudios Peruanos.

Shimada, I. 1984. Preliminary report on the 1984 survey of ancient mines in the Lambayeque region, Peru. Cambridge, MA.

___. 1985. Perception, procurement and management of resources: Archaeological perspective. In *Andean ecology and civilization*, eds. Shozo Masuda, Izumi Shimada, and Craig Morris. Tokyo: University of Tokyo Press.

Shimada, I., Epstein, S. and Craig, A. K. 1983. The metallurgical process in ancient North Peru. *Archaeology* 36(5): 38-45.

Shimada, I., and Shimada, M. 1985. Prehistoric llama breeding and herding on the North Coast of Peru. *American Antiquity* 50(1): 3-26.

Salomon, F. 1978. *Ethnic lords of Quito in the age of the Incas: The political economy of north-Andean chiefdoms*. Ithaca: Cornell University, 1978.

Vega, G. de la. 1943. *Comentarios reales de los Incas*. Buenos Aires: Emece Editores, S.A.

PART 2: COLONIAL

Black Legends and Silver Mountains: Spanish Mining in Colonial Spanish America Reconsidered

Otis E. Young, Jr., Arizona State University, Tempe, Arizona, U.S.A.

Abstract

Despite an undeserved reputation for regressiveness, waste and ignorance, Spanish colonial-era mining was in fact progressive and enlightened, both technically and socially, for the age in which it flourished. It was, moreover, well adapted to its peculiar environment, which differed substantially from that of Europe, while a large number of brilliant reformers and innovators were associated with the industry. The bad name under which it has since labored was the consequence of a retrospective assessment which lacked historical insight into its evolution and difficulties.
Key words: mining, technology, colonial, Spanish America.

Carlos Prieto in *Mining in the New World* (1973) has advanced a thesis of the crucial importance of precious-metals mining to the development and economy of colonial Spanish America. Prieto suggests, and I agree, that such mining was the economic linchpin of the region. In addition, it was a major factor in the rise and maintenance of the Spanish Empire, and as such attracted talent, energy, and attention throughout that vast domain. The mercantilistic doctrine of the period held that the power of nations was proportional to the armed force it could deploy; that such armed force was mercenary, hence procurable only with specie; that therefore the nation which commanded the largest specie reserves was the leading power of the age. In the centuries between 1453 and 1775, the syllogism was universally respected in the western world. Should it sound a little quaint to

In Quest of Mineral Wealth: Aboriginal and Colonial Mining and Metallurgy in Spanish America, edited by Alan K. Craig and Robert C. West, 1994. Geoscience and Man, vol. 33, pp. 109-118. Dept. of Geography & Anthropology, Louisiana State University, Baton Rouge, LA 70893-6010, U.S.A.

modern ears, one need only substitute the word "petroleum" for "specie" to realize that the case for a sort of mercantilism is still being argued, and at the cannon's mouth.

The Spanish explorers originally sought a western sea-route to the Indies, seeking to obtain an Oriental trade ultimately translatable into gold and silver. Finding the way to the Orient barred by geography, the Spanish drive hesitated for twenty years. The conquests of Mexico and Peru then pushed interest in commerce into the background. Why pursue trade when the ultimate object in view could be had directly in exchange for a few sword strokes, a transaction quite in accord with the past eight centuries of Spanish history? As the Age of the Explorers gave way to the Age of the Conquistadors, gold by the ton flowed east to Spain, upsetting alike the diplomacy and the economy of Europe. This spate quickly diminished: the booty of Mexico and Peru was the sum of centuries of patient, grain-by-grain accumulation, and when it was gone there was not a great deal more to be had immediately. Scores, perhaps, of lesser Conquistadors attempted to duplicate the feats of Cortéz and Pizarro, but failed in their objective, for it did not exist.

Following its conquest, Mexico was divided into great semi-feudal estates whose Indian population was "commended" to the care of their new lords, much as newly-annexed Moorish principalities in Spain had been parcelled out during the Reconquest. In the early 1530s, surface silver ores were discovered at Sultepec, Taxco and neighboring towns on the southern escarpment of the Mexican central plateau, southwest of Mexico City. These discoveries set off a flurry of mining excitement, and in the following two decades the great silver lodes of Zacatecas, Guanajuato and Pachuca were prospected, all lying within a week's travel of Mexico City. These surface silver ores were cheap to excavate with impressed and unskilled labor, easily reduced with a minimum of fuel and experience, and required little refining preparatory to being minted. This initial phase produced a recorded 22 million pesos of silver, not to mention what may have escaped the notice of the royal tax-gatherers. For a time it might seem as though the brutal old Roman mining canon, based upon crude methods and the lavish expenditure of forced labor, had returned.

This proved not to be so despite the persistence of hearsay to the contrary. Within twelve years Spanish colonial mining went through a total change; factors producing this change ranged from the geological to the theological. First, mining soon went to depth in following the ore, and underground work demands more skill and far greater technical sophistication than does opencut excavation. A great part of the requisite knowledge was provided by publication in 1556 of Georgius Agricola's *De Re Metallica*, the first and greatest of mining textbooks. The Spanish were greatly impressed by this work and were well acquainted with it for three centuries before it was known, except by antiquarians, in northwestern Europe.

If textbooks alone made mines, however, all the world would go underground. Highly skilled men are required as well, and here it would seem that the domains of the Emperor Charles V had a plentiful supply from which to recruit instructors and foreman. Spain itself possessed a great, though recently somewhat decayed, tradition of mining and metallurgy reaching back to the days of the Punic hegemony. The miners of Wallachia and Croatia were famed in the Austrian sphere of his dominions. Men such as these knew how to break rock whether by fire, powder, or human effort alone. Nor should it be forgotten that the Indian laborers were heirs to the great stone-working traditions of their ancestors. The names of such technicians are seldom recorded, so we know little of them save by their works.

Another great but poorly appreciated cause of upheaval was the importation of contagious disease from Europe. Like the Black Death in an earlier period, extensive depopulation must have resulted, causing labor shortages and so cutting the ground from under forced labor systems. Allied to this was the hostility of both Crown and Church to slavery, by whatever name it was called. Perforce, the mine operators had to adopt the European mining system of comparatively small crews, composed of and supervised by well-paid professionals who substituted skill and technology for sheer mass of effort. Gunpowder played its part: the ground conditions demanded rock-blasting, but no one dared entrust powder to an unfree and resentful labor force. Chiefly Saxon in its technical approach, the new mining took on a unique flavor of its own, one largely dictated by the presence or absence of materials at hand.

The Saxon mining canon relied upon the ready availability of wood, iron, and water-power. These were scarce and expensive in the highlands of Peru or the plateau of Mexico. Therefore wood and iron were reserved only for the most essential needs, not spent on subterranean support timbering or as fuel in large-scale smelting or used as machine materials. Well-wrought masonry and arched headings provided internal support, while leather was substituted for sheet-iron or connecting bolts; to this day rawhide is the all-purpose Mexican material of expedience, just as iron hay-baling wire became proverbial at a later date in the United States. In the absence of water power, horses and mules were ingeniously pressed into service for hoisting, bailing, and milling. If to the European eye Spanish American mines appeared alien in their emphasis on stone, leather, and animal power, they were nonetheless quite effective, intelligently making the most of what they had.

The ground in most metropolitan Mexican silver mines was not prohibitively difficult. The surface strata, extending downward for 20 or so meters, consisted of hard, fine-grained volcanic andesites of the sort which miners call "kindly" rock, for it is predictable and honest. Below was several hundred meters of greatly cracked and fissured black shale in which the silver-bearing quartz veins were deposited. Not only was the shale permeated with water, but the ore leads ran through it in an unpredictable manner, now and then expanding into highly profitable lenses, and then "pinching out." To follow a thin white quartz lead hoping that it would expand into bonanza was the favorite method of subsurface prospecting: a very costly gamble which might, but probably would not, pay the lucky adventurer handsomely. Hence the proverb arose, "*una minería es una lotería*," and if fortunes were won thereby, even more fortunes were lost.

The water which permeated the shale was a perpetual problem and major item of operational cost. Unless drainage tunnels could be driven to advantage, the only solution was to bail the workings with deep drainage shafts in which great leather buckets were raised by horse-windlasses (*malacates*). Surprisingly (to modern minds) this system was quite effective. And if the mines were wet, they were at least free of gassiness, heat, clay, and most of the other hazards at which miners shudder.

While water could be hoisted by semi-mechanical means, the transport of broken ore was elementary. After being barred (by *barreteros*) or blasted down (by *pegadores*), the ore was gathered into large baskets which were carried to surface on the backs of porters, called "*tenateros*," after the *tenates* which they bore. The weight of ore carried by one man was in the range of 100 kilograms, and the equipment of the *tenatero* was traditional and exiguous. Apart from his basket with a "tump-line" running to his forehead, he wore only sandals and a loin cloth; he carried a candle fixed to the end of a stick in his left hand, and he climbed by leaning forward and steadying himself with his right hand as he ascended a deeply-notched log, fixed at a decided slant. As with bailing, the method appears almost neolithic, but it was effective, required little capital, and was quite flexible in use: the more profitable ore there was to raise, the more *tenateros* were employed, without need for much, if any, additional capital expenditure.

Rock blasting was known, if little used anywhere before 1700. The technique was dangerous and of little effect, even in the mines of Saxony where it was most highly developed. In Hapsburg Mexico, it was probably employed only when conditions were remarkably favorable or remarkably demanding. By the time of the Bourbon accession, however, the practice had grown and was being used systematically: Humboldt noted that 14,000 quintals of gunpowder were being annually imported toward the end of the period. The critical development, safety fuse, was not invented until 1831, however, too late for the introduction of blasting as a routine rock-pulling procedure. The apparent conclusion is that blasting here, as everywhere else in the world of the period, was used only as a supplement to "cold mining" by human effort: most often in the sinking of shafts through the surface andesite, which were then finished by hand labor.

Shortly before Mexico's independence, Alexander v. Humboldt visited its mines, and his *Political Essay on the Kingdom of New Spain* ([1811] 1966) contains an excellent description of the *sistema de la rata* ("rat-hole system") which had been employed for nearly 250 years. He was quite critical of some aspects of the *sistema*, particularly the mine workings themselves, which he compared to "ill-constructed buildings where, to pass from one adjoining room to another, we must go round the whole house." This was true: the condition resulted from centuries of empirical exploratory drifting, and from stoping only profitable ore deposits. Geology before the nineteenth century was poorly understood and miners everywhere used much the same methods; the only alternatives were to work empirically or to do no work at all. Nonetheless, Humboldt freely praised the ingenuity of the miners and their skill in various trades, and concluded perhaps with some exaggeration that the Mexican miner was the best-rewarded in the world. One should not forget that Humboldt's was a counsel of perfection, based upon the latest scientific discoveries of Europe, and it is possible that had he visited the contemporaneous mines of British Cornwall, he would have been appalled at their casual brutality and want of forward-looking management.

The slightly higher wages of the *barreteros* and other mine employees are difficult to compute, more especially since not all a precious-metals miner's rewards are recorded on pay sheets. So long as mining was conducted by imprisoned slaves, there was little incentive to pilfer what one could never spend. Emancipation put another face on matters, however. Humboldt noted that no sooner were the miners of Mexico freed than they began to steal, and steal expertly. Mine operators have always decried the vice, but often are privately inclined to turn the blind eye to it, since detection and punishment would quickly strip the enterprise of its most skilled and valued workmen. This generalization has stood the test of time; strong and credible reports have it that "high-grading" flourishes to the present day.

There were many points at which theft could take place, but some of the favorites were: at the working face, as when a spray of native silver was dislodged from the lode; in the cobbing and ore-sorting yards; in the patio or at the retorting house, where lumps of silver amalgam might easily vanish. Quiet arrangements to mutual benefit can see that the booty leaves the mine premises in a hundred different ways. Thus, a personal search — even so intimate a search as Humboldt suggests — would not greatly discourage the outflow. In the mining community there are always those who make it their business to purchase interesting mineral specimens — usually at a substantial discount — and cause them to enter the flow of commerce in discreet ways. Estimates (and they are only estimates) of such losses range from a tenth of the produce of high-tenor silver mines to as much as half the output of rich gold mines. From the historian's viewpoint, this ancient custom is most annoying, for although it is so widely known as to be proverbial, by its very nature it leaves no documents to prove its existence.

The silver ore, upon arrival at the surfaces was "cobbed" or up-graded by knocking away or sorting out worthless components. This work, carried on in the open air, was so elementary that completely unskilled labor could perform it. An illustration in Simonin (1869) clearly shows Apache or Yaqui slave-captives sullenly performing this task under the eye of an armed guard, but the caption calls them "miners," which obviously they were not.

The ore was next broken, a process which in Saxony was accomplished by water power in large wooden stamp mills. In Mexico, the favorite primary crusher was the Chilean wheel: one or two great stone disks set to revolve upon a common wooden axle, with their edges running in a circular masonry trough. The trough was filled with ore, horses were harnessed to the axle, and the mill was revolved, the weight of the stones crushing the ore to coarse sand. It was then removed, screened through a pierced leather hide, and passed on to the fine-crusher. It is noteworthy that the revolving wheel was adapted from a similar Mediterranean device used time out of mind for the preliminary crushing of olives or oil seeds.

The fine crusher or pulverizer was the *arrastra*, used in all sizes from great to small; large active mines had whole batteries of them. In essence it was a circular stone floor surrounded by a low coping. In the center rose a stone or wooden pillar, to which a large wooden beam was pivoted. Depending from ropes attached to its arms were stone blocks, so hung that they "nipped" the ore, then pulverized it, much like a grist mill grinding stone. The floor was filled with a layer of ore, horses or mules dragged the stones in a circular path, and an attendant added fresh ore, shoveled out and screened the produce, and supervised the animals. The *arrastra* could be run on either wetted or dry ore, and, with some small refinements, persisted in use until the time of the American Civil War in both silver and gold developments of the North.

The last noteworthy achievement of the early period which made the *sistema de la rata* what it was, was incited by a change in the nature of the silver ore itself. As the workings deepened, the silver values became "thinner," and were found combined with sulfur and base metals into a very refractory ore. The only method then known of reducing such ores was large-scale smelting. In itself a recondite art, smelting required quantities of high-grade fuel (hardwood charcoal) which were expensive in Mexico and Peru. In 1555, a Spanish-born merchant, Bartolomé de Medina, devised at Pachuca a completely novel method for "wet" chemical reduction of this ore. His "patio process" required no fuel and no great amount of mechanical power, consuming only common salt, mercury, and cheap copper-iron sulfate ("greenstone" or "bluestone"), the latter obtainable from local sources. The process was carried on in the open upon large stone-paved patios, hence the name.

The upgraded, finely-crushed and moistened ore was spread thinly about the patio, and the quantities of salt, mercury, and "*magistral*" (copper and iron sulfate) were sprinkled over the ore by the *azoguero,* an experienced leading workman. The mix was then trampled and stirred by driving horses about the patio for days on end. From time to time the mix was shoveled into piles for a complicated chemical reaction which, at the expense of some mercury consumed, amalgamated the reduced silver with the rest of the mercury. At the end of some days, or weeks, the reaction was complete. The gray mix was washed free of stony constituents and the amalgam was recovered as a paste to be retorted or distilled. The heat drove off the mercury, leaving the silver as a spongy metallic mass which needed only to be melted and cast into ingots.

The patio process was so unique that virtually every traveler to Mexico remarked upon it, and was so effective that nothing better for the circumstances was devised until modern times, it being used without great modification in the silver mines of American Nevada until 1880. The Spanish Crown profited greatly from its monopolies on salt and mercury, to be sure, and collected the royal *quinto* on the silver bullion produced, but the mines were now so profitable that these exactions could be easily borne when they were not quietly evaded. In sum, Spanish ingenuity and enlightened self-interest launched in the Mexican and Peruvian mines a technological revolution upon whose basis Spain as a world power would remain preeminent until the end of the eighteenth century.

Matters went well until about 1700, when it became apparent that the days of profitability seemed to be coming to an end. The mines had grown ever deeper, hence costlier to work; the ore scarcer and of lower tenor, with the highest-grade ore bodies worked out; and flooding had become a serious problem. It appeared to many that the industry was dying. Scientific rationalization, implying major changes in planning and operational economy, was required. This in turn subtended political and educational reform. Fortunately, the Age of Reason saw a flowering of talent — even genius — in Mexico itself, as well as the coming to the Spanish throne of Carlos III, possibly the most enlightened monarch of the century of enlightened despotism. The combination of reform-minded monarch, brilliant and devoted subjects, and pressing need would have remarkable consequences.

Such men as the mathematics professor of the University of Mexico, Carlos de Sigüenza y Góngora, the self-taught mining engineer José Antonio Alzate, "the Pliny of Mexico," and Juan Ordóñez de Montalvo attempted as individuals to improve the efficiency of the mines. About 1750 the mine operators themselves organized the Cuerpo de Minería, a producers' association, which initiated reform across a broad front. The guild petitioned the Crown for a revolving credit system to finance exploration and development, termination of the royal monopolies of salt and mercury, establishment of a general tribunal to put an end to the costly, unending litigation that is the curse of mining everywhere, and, lastly, royal patronage of a college of mining engineering patterned after the preeminent establishment at Freiberg, Saxony.

Carlos III was pleased to grant these requests, and in 1792 the Royal College of Mining was established at Mexico City. Its rector, Fausto de Elhuyer, was a world-renowned chemist, the director-general of the tribunal of the Cuerpo de Minería, and a sound educational administrator. The college enjoyed swift and sturdy growth, and it appeared for a time as though a new era for Hispanic American mining was at hand. Between the institutional reforms, the brilliance of its scholars, and the importance of mining to Spanish America, a great upsurge of science and technology seemed probable.

This suggestion may be best illustrated by a comparable later development in the United States. Before 1865, there was very little higher education west of the Appalachians, and what there was was largely dominated by religious interests seeking to train their clergy. In the main, however, learning was derided as elitist, and so anathema to the prevailing Jacksonian democracy. In a word, higher education in the West faced not only the handicaps of poverty and clerical domination that it encountered in Mexico, but also it enjoyed none of the prestige which Spanish culture traditionally associates with learning and the arts.

The requirements of the American mineral frontier soon altered this dismal picture. A great many of its leaders (Agassiz, Douglas, Silliman and Wharton to name but a few) were educated men who appreciated the value of their training. They in turn helped inspire the foundation of state-financed schools of mining whose graduates soon convinced parsimonious legislators by their works that higher education paid well.

Within a generation, state colleges and universities rose upon the foundations of public acceptance established by the schools of mines, western pragmatism having overcome its suspicion of learning. By the turn of the twentieth century, engineers, and mining engineers in particular, were virtually public heroes in the United States. Even to this day, the favorite argument of western college presidents remains the argument of economic betterment. On balance, therefore, it seems quite possible that a similar efflorescence could have occurred in Mexico, a century earlier than it did in the United States.

Sad to relate, the dawn proved false. Carlos III was succeeded by an unpromising heir, whose chief minister was distinguished only by cowardice and duplicity. In 1789 began the French Revolution and with it a generation of war that would leave metropolitan Spain in ruins. The palsy of Spain encouraged revolutionary sentiment in Spanish America to such effect that by 1821 the region was in turmoil, quietly abetted by European intrigue. A smooth transition to independence was prevented by the fact that Spanish policy, ever fearful of local autonomy, was to do nothing and permit nothing that might prepare the region for eventual self-government. In Mexico there followed a half-century of tumult which proved disastrous for education and development. Among its other casualties was the College of Mines, which withered under the demands of unrest.

One might summarize by suggesting that the Hispanic mining industry had coped successfully with a long series of technical and social problems but, through no fault of its own, was unable to evolve any solution to its political quandary. In self-defense against virtual anarchy, the mine owners adopted a medieval expedient: they built stone walls about their headworks, armed their followers to defend them, and held what they had while working in the traditional manner. This technical retreat ensured that mining would not be a scientific venture, but a lottery in which luck and preservation of the status quo were everything.

When in 1877 Porfirio Díaz restored internal order and external integrity of a sort, he invited foreign mining corporations to revive the old source of national wealth. The foreign technicians who came were themselves the product of fifty years of ever-accelerating science, and were quite scornful of what they saw in the *antiguas*, the mines of the previous age. They attributed this backwardness to flaws of national character, for they were largely ignorant of the history of the nation. Whether English or American, these engineers had moreover been exposed in common school to the "Black Legend" of Spanish perfidy, cruelty and superstition, nor was either group disposed to much sympathy for the lower orders of workers, they having little enough for their own. For the most part they were prepared to see no good in classical Spanish mining, which they derisively called "the rat-hole system," and so they reported on the basis of their misconceptions.

Now and again these foreigners saw something which did not square with their projections. They admittedly were awed by some of the old engineering works, such as the great master drainage shaft of the Valenciana mine at Guanajuato (a work which is still in use). It was 10 meters in diameter, over 500 meters deep, sunk through hard rock and finished to the most exacting standards. They scarcely mentioned the fact that this marvel was constructed before Spain lost Mexico. Instead, they offered a new Black Legend of their own making: that the Spaniards as mining men were ignorant, greedy, and indisposed to improvement. They did not even consider that what they saw had been frozen in time, then fought over for more than fifty years.

Apart from a modest attempt to rehabilitate the reputations of an unjustly-maligned group of men, are there general lessons to be drawn from this instance? In my opinion, the answer is a decided "yes." We can learn, or rather relearn, the historiographical principle that it is vain to judge the ethics and actions of one era by the professed standards of a later period. For to do so not only perverts history into polemic, but hastens the slide from polemic to downright distortion for ideological purposes. It is disingenuous to suggest that a little historically based polemic may support a Higher Good, for Higher Good, like beauty, resides only in the eye of the beholder. In the final analysis, history consists only of an honest effort to ascertain what happened, and why.

References

Agricola, G. [1556] 1950. *De Re Metallica*. Basileae: APVD Hieron Frobenivm et Nicolaven Episcopivm. Herbert C. Hoover and Louise H. Hoover, eds. and translators. New York: Dover Publications.

Bakewell, P. J. 1971. *Silver mining and society in colonial Mexico: Zacatecas, 1546-1700*. London and Cambridge: University Press.

Brading, D. A. 1971. *Miners and merchants in Bourbon Mexico, 1763-1810*. London and Cambridge: University Press.

Howe, W. 1969. *The mining guild of New Spain and its tribunal general, 1770-1821*. New York: Harvard University Press.

Humboldt, A. v. [1811] 1966. *Political essay on the kingdom of New Spain*. Facsimile of 1811 edition. New York: AMS Press.

Manje, J. M. 1954. *Luz de tierra incógnita [Unknown Arizona and Sonora, 1693-1721]*, trans. Harry J. Karnes. Tucson: Arizona Silhouettes.

Prieto, C. 1973. *Mining in the New World*. New York: McGraw-Hill.

Probert, A. 1969. Bartolomé de Medina: The patio process and the sixteenth century silver crisis. *Journal of the West*, 7:90-124.

Rickard, T. A. 1907. *Journeys of observation*. San Francisco: Dewey Publ. Co.

Simonin, L. L. 1869. *Mines and miners, or underground life*, trans. and ed., H. W. Bristow. New York: D. Appleton & Co.

Ward, Sir H. G. 1828. *Mexico in 1827*. 2 vols. London: H. Colburn.

Young, O. E., Jr. 1970. The Spanish tradition. In *Western mining: An informal account of precious-metals prospecting, placering, lode mining and milling from Spanish times to 1893*, O. E. Young, Jr., ch. 3. Norman, OK: University of Oklahoma Press.

Early Silver Mining in New Spain, 1531–1555

Robert C. West, Louisiana State University
Baton Rouge, Louisiana, U.S.A.

Abstract

Henry Wagner published one of the first essays on early silver mining in New Spain in 1942, and some 27 years later Alan Probert's 1969 essay dealt, in part, with the so–called "mining crisis" in Mexico during the late 1540s and early 1550s, a period just prior to the introduction of the amalgamation process. Some of the data presented by these two scholars are reviewed, and additional information concerning the first two decades of silver mining in Mexico — the 1530s and 1540s — is presented, pointing out: problems connected with the extraction and refining of silver ores; the systems of labor employed; the supply of equipment and food for the mines; and quantity of silver production.
Key words: Mexico, silver mining, metallurgy, labor.

Introduction

Spaniards began exploitation of precious metals in New Spain immediately after the conquest of the Aztec and Tarascan states,first by looting native gold and silver artifacts followed by mining of gold placer deposits and lodes of silver ores. Figure 1 illustrates the location of the major Spanish mining activities to about 1555: (1) the exploitation of gold placers (1522–1550) with Indian slave labor, most of the sites located on streams that drain the ancient igneous and metamorphic rocks of southwestern Mexico, and (2) the mining of silver lodes found within the more recent volcanics and metamorphics within the central plateau and its southern escarpment, as well as in the Sierra Madre del Sur.

Spanish silver mining on a commercial scale began in the early 1530s, the lodes of Zumpango del Río south of the Balsas River and those of Sultepec discovered in 1531 on the escarpment southwest of Mexico City. Perhaps a year later, silver deposits around Taxco, also on the escarpment, were being exploited. In western Mexico (Gobierno of Nueva Galicia) the

In Quest of Mineral Wealth: Aboriginal and Colonial Mining and Metallurgy in Spanish America, edited by Alan K. Craig and Robert C. West, 1994. Geoscience and Man, vol. 33, pp. 119-135. Dept. of Geography & Anthropology, Louisiana State University, Baton Rouge, LA 70893-6010, U.S.A.

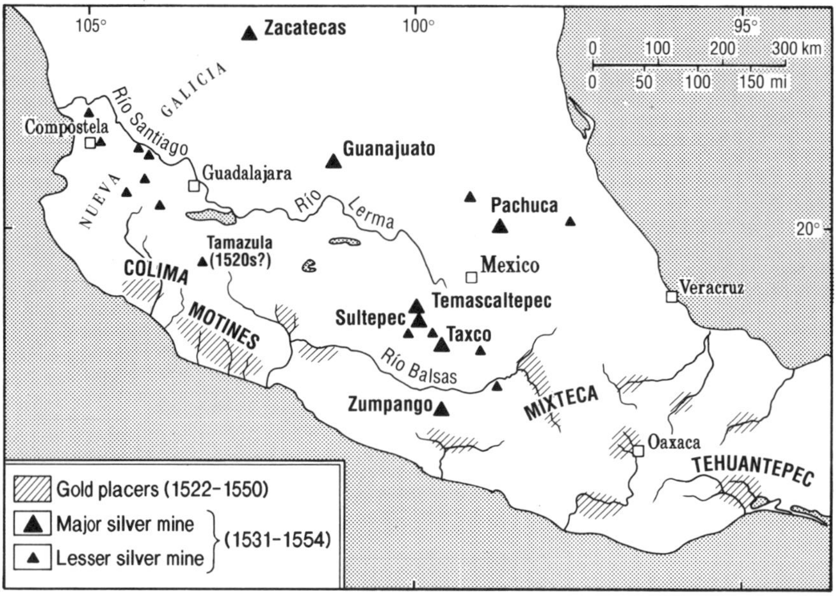

Fig. 1. Spanish mining activity in New Spain, 1522–1554.

great mines of Zacatecas were found in 1546, and many lesser ones opened south of the Río Grande de Santiago in the 1540s, especially that of Guachinango west of Guadalajara. Mining at Pachuca, north of Mexico City, began about 1552; at Guanajuato perhaps a year later.

Spanish Looting

It is well known that the first gold and silver that the Spaniards obtained in Mexico came not from mining but in the form of native Indian artifacts looted from palaces, temples, graves and other sacred places, where the Aztec and Tarascan nobility had accumulated such treasures for generations. These metal objects were used by native priests in ritual and status functions, rather than serving as symbols of economic wealth (Pollard 1987). Made of copper–silver or copper–gold alloys, such artifacts were silvered or gilded so that their surfaces resembled pure silver or gold (Hosler 1988). The first substantial amounts of native "silver" artifacts obtained by the Spaniards were taken from Tarascan territory in west–central Mexico, roughly the area covered by the present state of Michoacán. The Tarascans and their neighbors near the Pacific coast were the foremost metallurgists of pre–Conquest Mexico. Figure 2 indicates the source of Tarascan copper, silver and gold before the Conquest: silver from the western part of the political state in the vicinity of Tamazula in present–day Jalisco; copper from

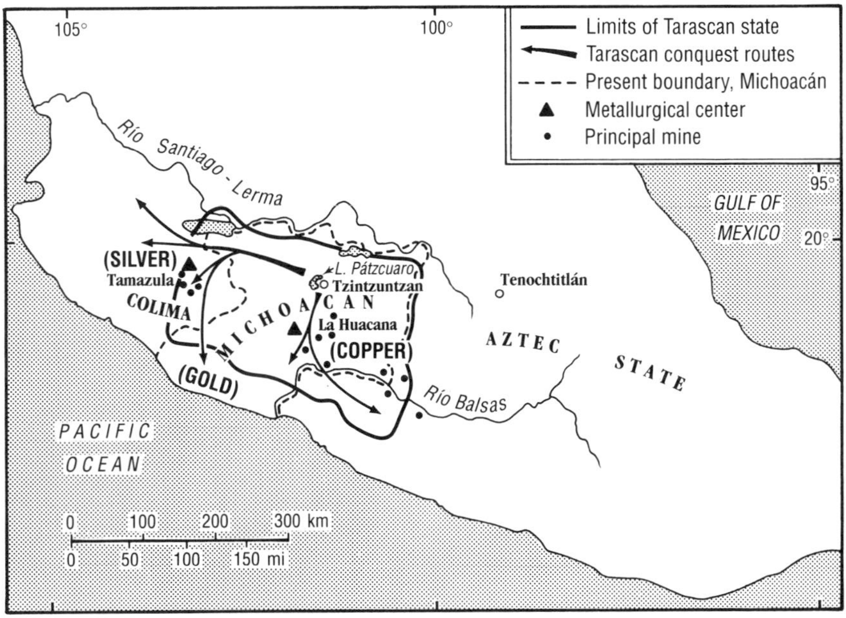

Fig. 2. Tarascan mining activity before the Spanish Conquest.

the La Huacana area and the western portion of the Balsas River basin; and gold from the placers along streams within the coastal ranges of Michoacán and Colima. Artifacts of copper–silver alloys were in the form of *rodelas*, or small replicas of round shields used in ceremonies, armlets, bracelets, cups and other forms. The Tarascan nobles had long accumulated such artifacts in palaces and graves around the political capital and ceremonial center of Tzintzuntzan on the shore of Lake Pátzcuaro and on islands within the lake — all sources of rich booty that the Spaniards, including Hernán Cortés, Nuño de Guzmán and their cohorts, forced from Tarascan leaders (Warren 1985; Craine and Reindorp 1970). The loot was usually melted down into larger pieces, which formed the first shipment of "silver" (which the Spaniards called *plata baja* because of the high copper content) from the New World to Spain in the early 1520s.[1] Significantly, the underlying religious motive for the acquisition of the precious metals on the part of the New World aborigines was fundamentally different from that of the Europeans, who coveted gold and silver as the basis for wealth within their mercantile economy.

Cortés quickly claimed as his own the Tamazula area as a main source of silver (Warren 1985, 160). This area may have been the locale of the legendary mine of "Morcillo," supposedly exploited by Spaniards with Tarascan labor in the 1520s, but was lost through a cave–in and subsequently abandoned.[2] During the early 1530s natives of Tamazula and neighboring towns of Tuxpan and Zapotitlán, as well as the inhabitants of Tzintzuntzan, continued to supply small amounts of low–grade silver as tribute to Spanish *encomenderos* (AGI Contaduría 657). As late as 1550, villages in the Lake Pátzcuaro area were still being assessed tribute that included "*plata baja*," indicating the persistence of a mining and metallurgical tradition among the natives of Michoacán (Paso y Troncoso 1905–06, [Suma de Visitas] I:123, 181, *passim*).

Discovery of New Silver Mines, 1530s

Their experience in Michoacán during the 1520s alerted the Spaniards to the occurrence of other silver lodes in New Spain and to the possibility of quick wealth once the deposits were found. Perhaps delayed by the exploitation of gold placers, search for silver outside Michoacán did not begin in earnest until the early 1530s. Thereafter reports of new discoveries of silver deposits increased rapidly as the following translation attests: "... this year [1533] many rich silver mines are being discovered and every day more are being found." (AGI Patronato 184, ramo 21, Real Audiencia de México to the Crown, 3 Nov. 1533).

Documents of that period indicate that the Spanish conquerors and early settlers in New Spain, being mostly soldiers, adventurers and hidalgos, were abysmally ignorant of mining and ore reduction. This despite the fact that many came from Andalucia in southern Spain, where in the Sierra Morena mining of lead and silver had been practiced sporadically for centuries. Once the silver lodes of Zumpango del Río, Sultepec and Taxco were discovered, throughout the 1530s royal officials in Mexico dispatched urgent requests to the Crown to send over experts in ore reduction because of the current lack of metallurgical knowledge on the part of the Spanish conquerors.[3]

Zumpango del Río

Located in the rugged Sierra Madre del Sur south of the Balsas river basin, the mines of Zumpango del Río were the most isolated of the big early discoveries; great distance and difficult trails separated them from central authority in Mexico City and needed supplies. Two Spaniards claimed to have discovered the rich deposits in late 1531: Francisco de Hoyos and Juan Jaramillo, who were returning with a military expedition sent to quell an uprising of Indian workers (the Yopelcingo) forced to work in the gold placers of present–day Guerrero state (Icaza 1923, II:143, 299; Paso y Troncoso 1905–06, VI:317, 317 n.2). In view of their mining inexperience, just how the Spaniards themselves could ferret out silver deposits is in question; likely, Indian auxiliaries (Aztecs or Tarascans) who usually accompanied such expeditions, some of whom may have had knowledge of metallic ores, probably were the actual prospectors, encouraged by their Spanish overseers.

The mines were found in a mountain range some five leagues southwest of the village of Zumpango (Gerhard 1972, 317). The deposits attracted a fair number of Spanish "miners," who used Indian slave labor to extract and crush the ore and to reduce it in smelters. Among the more prominent Spaniards who owned mines in Zumpango was Juan de Burgos, royal accountant in Mexico City, active in buying and selling mines, slaves, and livestock in several parts of New Spain; in 1539 the overseer of his mines in Zumpango reported an output of 2000 marcos (1000 pounds) of silver to the treasury in Mexico City (AGI Justicia 131, exped. 2). Hernán Cortés, leader of the Conquest, himself owned several mines in Zumpango, acquired in 1534 (AGN Hospital de Jesús 264, exped. 7).[4] A single productive mine together with its gang of Indian slave labor and refining equipment was often held in partnership among two or more Spaniards, a common custom of the day, called *compañía*. After 1550, when enforcement of the ban on Indian slavery was finally achieved, many of the mines were abandoned for lack of cheap labor, but as late as 1582 some silver was still being produced at Zumpango (Paso y Troncoso 1905–06, VI:317). During the later colonial period, mining was never revived in the area and today traces of the early sixteenth century bonanza may have disappeared.

Provincia de la Plata

The most productive early silver mines in New Spain were worked in a relatively small mineralized area on the southern escarpment of the central Mexican plateau (fig. 3). This rugged area has been dissected by streams that in places have exposed outcrops of silver–lead–bearing veins in mainly metamorphic rocks, such as schist and slate (figs. 4 and 5). Probably with the help of Indian prospectors, Spaniards first discovered silver deposits in this area, possibly in 1531 near the native villages of Amatepec and Sultepec (Gerhard 1972, 269). Perhaps a year later silver ores were being worked in the vicinity of Taxco, some 12 leagues (50 kilometers) southeast of Sultepec. Early on, the area of Amatepec and Sultepec was known as the "Provincia de la Plata," but that term might well include also the mines of Taxco, for in the late 1530s and early 1540s the entire mineralized zone became New Spain's major center of silver production. Around 1540 Spaniards opened other mines near the town of Zacualpan, a few leagues northwest of Taxco, and finally, perhaps in 1545, the mines of Temascaltepec were discovered on the northwestern margin of the Provincia de la Plata, bringing to five the number of major *asientos de minas*, or mining centers, within the area.

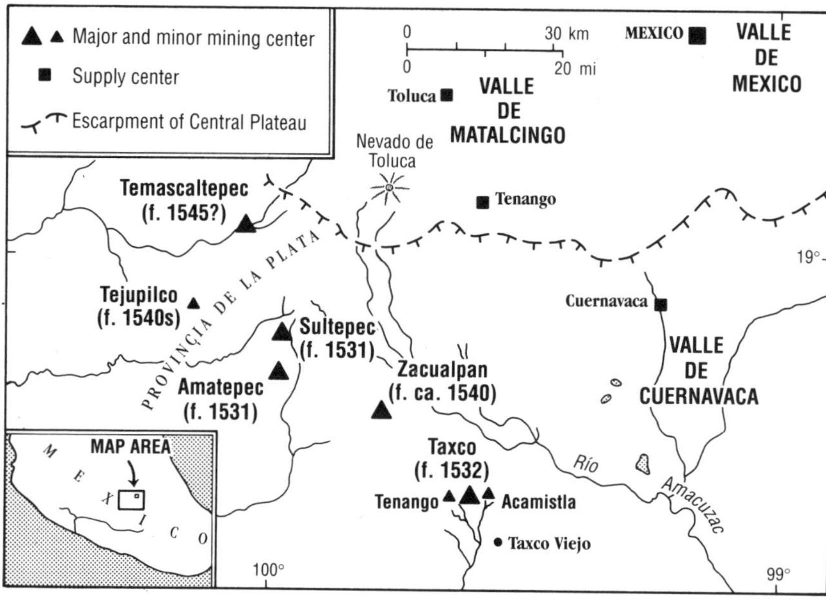

Fig. 3. Spanish mining activity in the "Provincia de la Plata," southern escarpment of the central plateau of Mexico, 1531–1554.

Early documents relating to these centers rarely give the names of the actual Spanish discoverer of the ore deposits, but, once opened, the mines attracted numerous high officials who purchased mines or registered claims in the hope of quick returns. By the early 1540s, Hernán Cortés owned all or parts of 20 mines in Sultepec, 12 in Taxco, 3 in Zacualpan (AGN Hospital de Jesús 266, exped. 14). The royal treasurer of New Spain, Juan Alonso de la Sosa and the royal accountant, Juan de Burgos, were active mine owners in the area, usually holding claims in *compañía* (partnership) with others. The practice often led to rancorous civil law suits, such as one dated 1543 over joint ownership of a mine in Sultepec between Cortés and Juan Alonso de la Sosa, which details mine and labor practices (AGN Hospital de Jesús 276, exped. 87). Most of the early Spanish miners, however, were small operators.

In contrast to the isolation of the Zumpango mines in the Sierra Madre del Sur, those of the Provincia de la Plata were relatively close to Mexico City, source of mining equipment and luxury food imported from Spain. Also nearby were the valleys of Matalcingo (Toluca) and Cuernavaca, both producers of food and livestock. Moreover, unlike the relatively short life of the Zumpango mines, those of Sultepec, Taxco, Zacualpa, and Temascaltepec, although subject to periods of "boom and bust,"[5] produced silver throughout the colonial and independence period of Mexico. Today

Fig. 4. Canyon of the Río Sultepec below the mining center of Sultepec, showing steep slopes and traces of mining activity. The grass–covered slopes were once covered with pine–oak forest, destroyed by overcutting to obtain wood for making charcoal.

Fig. 5. The present–day town of Sultepec, which served as a major *asiento de minas* of New Spain during the colonial period. Photo taken in December 1949, by the author.

the mines in the vicinity of Taxco are still active, producing industrial metals of lead and zinc, rather than silver.

Mining Methods, 1530s–1540s

Prospecting

Above I have suggested that the search for ore deposits during the early years may have been done by Indians, slaves or freemen, working under the direction of Spaniards. At Taxco in 1539 during a court hearing regarding a lawsuit over the ownership of a mine, witnesses testified that "The Indians are the ones who discover the mines, and they do so following orders of their Spanish masters...." Again, an Indian witness, slave of one Rodríguez, a Spanish miner in Taxco, claimed at the same hearing that he and three other slaves in 1539 discovered an ore outcrop and made the first *cata*, or exploratory dig, at the site (AGI Justicia 132A f. 234, 244, 294). The practice of using Indians as prospectors was still common in the mid–sixteenth century, when in 1553 early lode discoveries near Pachuca were made by free Indians hired specifically to find ore deposits and to make initial assays to determine ore quality (AGI Justicia 167).

Excavation

Early Spanish mining of silver deposits in New Spain, as elsewhere in the Americas, was confined mainly to the oxidized portion of outcrops near the surface, a zone containing small amounts of native silver and larger deposits of grey–black cerargyrite (silver chloride mixed with native silver); cerussite (lead carbonate); and argentite (silver sulfide), which was especially abundant near the bottom of the oxidized zone above the water table. Such ores were usually easily smelted. Initially mines were no more than "gopher holes" a few meters in depth, and, as digging progressed, vertical shafts were constructed. At Taxco in 1548, some 16 years after discovery, the deepest shaft was reported at 30 *estados* (some 65 meters), approaching the water table. Most shafts, however, measured less than 10 *estados* (25 meters) in depth (AGN Hospital de Jesús 266, exped. 14, folio 218 ff.). By the 1540s, adits, called *socabones*, either horizontal or slightly sloping tunnels, were being dug into hillsides at both Taxco and Sultepec in order to reach productive veins. Neither shafts nor adits appear to have been reinforced with cribbing, and by the 1540s cave–ins had occurred in Sultepec, killing Indian workers and causing the closure of several mines (AGN Hospital de Jesús 276, exped. 87, folio 298). At that time, excavations had not yet reached the water table (at 70 meters), and drainage problems did not become serious until the next century.

Ore Reduction

Carried to the surface in fiber sacks by Indian workers, the ore was separated from the gangue in the *lavadero*, or "washer" and then crushed either by hand with hammers of wood reinforced with iron ribbing or in the few stamp mills owned by the more prosperous miners. In 1543 the royal treasurer of New Spain, Juan Alonzo de la Soza, owned a water–powered mill along a stream near Sultepec, but it could be operated only during the wet season (AGN Hospital de Jesús 276, exped. 87, folio 296^{v}). Probably most mills in Taxco and Sultepec were powered by horses or mules.

The crushed ore was then smelted in small crude cylindrical furnaces of clay and stone, to each of which were attached two hand–operated bellows of sheep or goat skin for air blast. Once the easily reduced surface ores had been exhausted, the deeper and more difficult sulfide ores required large amounts of either metallic lead or of lead oxide (litharge) as flux for the smelting and refining of silver. According to one report of 1539, a miner at Taxco needed around 25 hundredweight of litharge to refine only 75 to 126 ounces of silver (AGI Justicia 1029).

Some of the processes used to reduce silver ores probably were not of Spanish origin, but may have been introduced by Germans who settled in Sultepec in 1536. Among them was one Johann Engel who is said to have acquainted Spanish miners with the practice of using litharge and other lead–bearing substances to mix with silver ores as flux in their small furnaces. The stamp mill, either animal– or water–powered, may also have been a German introduction.[6] Sultepec appears to have been a center of mining and metallurgical innovations from which new techniques spread to other mining centers in Mexico. Even the amalgamation process which revolutionized the silver industry in Mexico, although perfected at Pachuca, was first employed on a large scale at Sultepec in 1556.

Labor in the Mines

Indian Slavery

From the 1530s through the 1540s, the main labor force in the mines of Sultepec, Taxco and neighboring *asientos de minas* consisted of Indian slaves, many of whom had been transferred from the declining gold placers in southern Mexico. During the Conquest, Spaniards legally enslaved large numbers of natives — men, women and children — as booty of warfare, branding each individual on the cheek with the symbol of the Spanish Crown as proof of owners' payment of the royal fifth on acquisition of each. Most of these were apportioned among the high ranking officers; Cortés owned several hundred, used mainly in gold placering (Berthe 1958). As slavery was a well established institution among the Aztecs and their neighbors, after the Conquest Spaniards also purchased slaves from native caciques, and later many miners bought slaves from other Spaniards (Zavala 1968, xi, 1). Although the New Laws of 1542 formally abolished Indian slavery in New Spain, the regulations could not be enforced, and the institution persisted as late as the early 1550s (Simpson 1940).

Each of the more prominent miners in Taxco and Sultepec owned from 100 to 150 slaves, who were assigned the most arduous work, such as ore extraction and work in the smelters (Zavala 1968, 54–55). As most of the miners' capital was invested in the labor force, it behooved them to feed and clothe their chattels adequately, adding to the high cost of silver production. Even so, slaveholders were continually plagued by death of slaves through disease, and by runaways, who were rarely caught and returned. After 1550, blacks from Africa and Indian freemen gradually replaced native slaves in the mines; by 1569 over 800 black slaves worked in the Taxco mines (García Pimentel 1897, 170, *passim*).

Early on there developed among the Indian slave workers the practice of stealing pieces of rich silver–lead ore during their work shift to barter with Spanish traders or to extract silver in their small, one–room family huts by using aboriginal blowpipes (*soplillos*) for air blast or a small cupelling furnace (*cendradilla*) equipped with a single bellows. The fumes generated in the huts by this practice led to serious cases of lead poisoning, especially among the family children, many of whom died. One of the provisions of mining ordinances formulated by the Audiencia of Mexico in 1542 alerted mine owners of Taxco to this illicit practice and urged its control (Zavala 1982, 179).[7] The theft of ore from the mines by Indian workers may have been the beginning of the custom of the *pepena* or *partido*, the legalized daily take of a portion of the ore mined by each worker — a time–honored practice of a method of ore–sharing payment permitted as a labor inducement in most Mexican mines during the colonial and independence periods. Moreover, later in the sixteenth century the use of small furnaces by workers to smelt the *pepena* ore spread to many silver mines, including Zacatecas, where it became a common practice (Bakewell 1971, 125).

Encomienda *Labor*

Indians living in villages under an *encomienda* in the vicinity of the mines also afforded a significant source of labor for the early silver industry. Through various arrangements, *encomenderos* were allowed to commute tribute required of their subjects, thus permitting a certain number of Indians to work in the mines, usually in two–week relays, the miners paying the *encomenderos* for the use of their subjects (Zavala 1984–87, 1:218–220; Millares and Mantecón 1945–56, II:30, 135). These Indians performed the less arduous labor, such as cutting wood in adjacent forests and hauling it to the smelters for making charcoal, used as fuel; building huts in the mining center for the slaves; and often carrying food and other necessities, sometimes from distant parts (AGI Justicia 123). During the late 1540s and early 1550s most of the *encomienda* workers used in the Sultepec, Taxco and Amatepec mines came from the Tarascan pueblos on the central plateau, necessitating a journey of 200 to 250 kilometers on foot (fig. 6) (Paso y Troncoso 1905–06, I:34, *passim*). Other areas supplying *encomienda* labor were pueblos in the Balsas basin and around Toluca in the valley of Matalcingo to the north. The heavy reliance on Tarascan labor is puzzling, but might be accounted for by the relatively large native population in the Tarascan area at that time, and the miners' appreciation of Tarascan metallurgical knowledge which they often utilized in the smelters.

One of the duties of *encomienda* labor in the mines — the cutting of wood used to make charcoal for fueling the smelters —soon led to depletion of woody vegetation surrounding the mining centers. This was especially true in the vicinity of Sultepec and Taxco where oak–pine forests originally covered the mountain slopes. The mining ordinances issued in 1542 for Taxco specified that the surrounding forests be guarded to prevent cutting of new growth (Zavala 1982, 180). Later, ca. 1550, Viceroy Mendoza warned his successor, Luis de Velasco, regarding depletion of

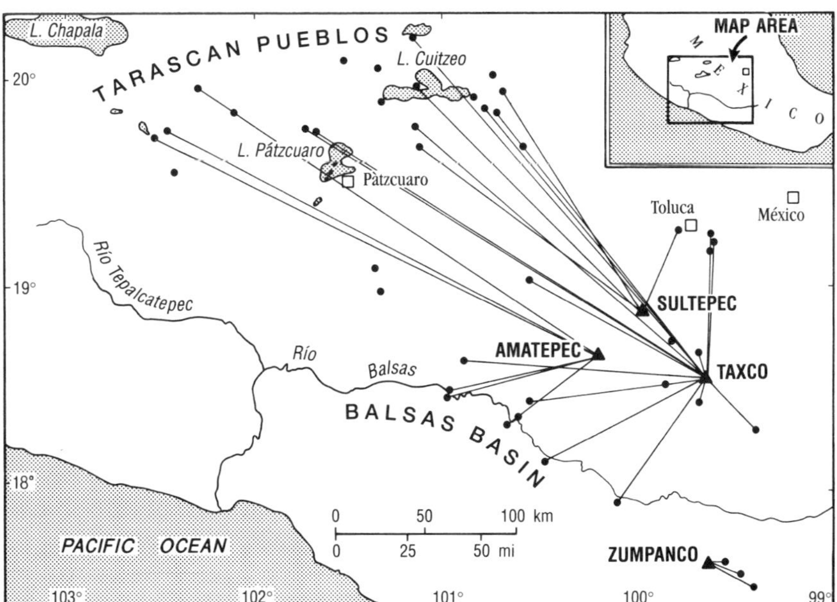

Fig. 6. Source of *encomienda* labor for the silver mines of Amatepec, Sultepec, Taxco, and Zumpango del Río, ca. 1550. Data from the *Suma de Visitas* in Paso y Troncoso (1905–06, I.).

the forests: "In just a few years a large area of forest has been destroyed [near the mines], and it appears that the wood supply will be depleted sooner than the ore. Ordinances have been made regarding the conservation of the forests, and likewise regarding the paths that the Indian workers use for making charcoal, cutting wood, and on the maximum loads they may carry" (*Colección de documentos inéditos para la historia de España* [Mendoza to Velasco, n.d., ca. 1550] 26:288).[8] The problem of vegetation depletion around mines dependent on charcoal for the smelters continued throughout the colonial period.

Supply of Food and Equipment

The discovery of silver deposits in Sultepec and Taxco in the early 1530s soon led to a rush of not only would–be Spanish miners of many social classes seeking quick wealth, but also of merchants and *rescatadores*, or traders, hoping for a lucrative market for food, clothing and mining equipment. Many of the miners were residents of Mexico City, some of whom were also *encomenderos* who controlled Indian villages near Toluca or in Cuernavaca. As an *encomendero*, Hernán Cortés exacted tribute of food and cloth from his villages in both areas to supply slaves in his mines in Sultepec and Taxco, and to sell to other miners (Wagner 1969, 455, 457). The main diet of Indian slaves consisted of maize tortillas, beans and chile peppers imported from distant and nearby villages. Luxury foods for Spaniards (olive oil, wine, vinegar) and mining tools (iron picks, crowbars, sledge hammers), as well as bellows equipped with copper or bronze nozzles (*tuyères*) were sold by Mexico City merchants, who imported them from Spain (AGN Hospital de Jesus 432, exped. 1). Reagents used in smelting ores, such as metallic lead and litharge came from Zumpango del Río, Tehuacán and the Mixteca Alta to the south (Zavala 1982, 186).

Merchants and traders, as well as jobless vagabonds, thieves and other riffraff became so numerous and obnoxious in Taxco that in 1542 the Audiencia in Mexico City issued special ordinances, to control commerce in the mining centers. Stores selling equipment, cloth and luxury goods were to be limited to no more than two in each center; only one tavern dispensing wine and brandy was to be permitted, and sale of intoxicants to Indians and slaves was forbidden; a general open market was to be held on Sundays at a specified place. All other commercial transactions dealing with the sale of food and equipment was to be permitted only outside a radius of four leagues from the mining town (Zavala 1982, 177 ff.). How effective these rules were in controlling commerce in Taxco and other centers in the general area is unknown, but their issuance is indicative of the hectic manner of life in the early mining communities.

The "Mining Crisis," 1539–1554

During the late 1530s, particularly in 1539, the miners of Taxco, Sultepec and neighboring *asientos de minas* began to complain loud and long of the high costs of mining and refining silver ores: lead and lead compounds were costly; within eight years the price of food and clothing for mine workers had doubled; the number of Indian slaves in the mines was decreasing through deaths and runaways. A lengthy document in the Archivo General de Indias in Sevilla records a hearing in which various witnesses, mainly miners, testified as to the increasing costs (AGI Justicia 1029; Patronato 180, ramo 72 [1539]). Several indicated that many miners had been jailed for failure to pay their debts owed to merchants for food and equipment and others had been forced to sell their mines at great financial loss. Most witnesses favored petitioning the viceroy and royal officials in Spain to decrease the severance tax from the fifth (20 percent) to the tenth (10 percent) so that miners might realize a profit.

Nonetheless, despite such financial difficulties the amount of refined silver registered at the *fundición* in Mexico City steadily increased from 1534 to 1548 according to statistics examined in documents as graphed in figure 7 (AGI Patronato 199–202). Unfortunately, statistics beyond 1549 were not found, and even for that year the data were available for only seven months. If we extrapolate for the missing five months, the curve would continue upward. The increase in silver produced from 1534 to 1548 or 1549 would indicate that although the less affluent miners in Taxco and Sultepec were in financial difficulty, the more affluent ones, such as Hernán Cortés, Juan de Burgos and others, continued to produce and register substantial amounts of silver. Moreover, the start of production at Zacatecas and the smaller mines of Nueva Galicia probably helped to maintain the increase as shown in the graph.

However, several reports of the royal officials in New Spain directed to the King, dated between 1552 and 1556 indicate that since 1549 a precipitous decline of silver production had occurred. According to one report the King's share of production had declined by two–thirds from 45,500 marks in 1549 to 16,000 marks in 1553 (Zavala 1984–1987, 2:163; AGI Mexico 323 [1552]). If correct, these data indicate a definite depression in the mining industry — the so–called "mining crisis" of New Spain — which was probably brought on in part by the decrease in the availability of rich surficial ore deposits and the difficulty and expense of smelting the less productive sulfide ores (e.g. galena) obtained at depth. But according to most reports, the main cause of the crisis revolved around labor: the final freeing of Indian slaves; the commuting of tribute exacted from *encomienda* Indians from agricultural products to payment in money (gold); and the abolishing of personal services in *encomiendas* so that Indians were no longer available to work in mines (AGI Mexico 323 [1552]).[9]

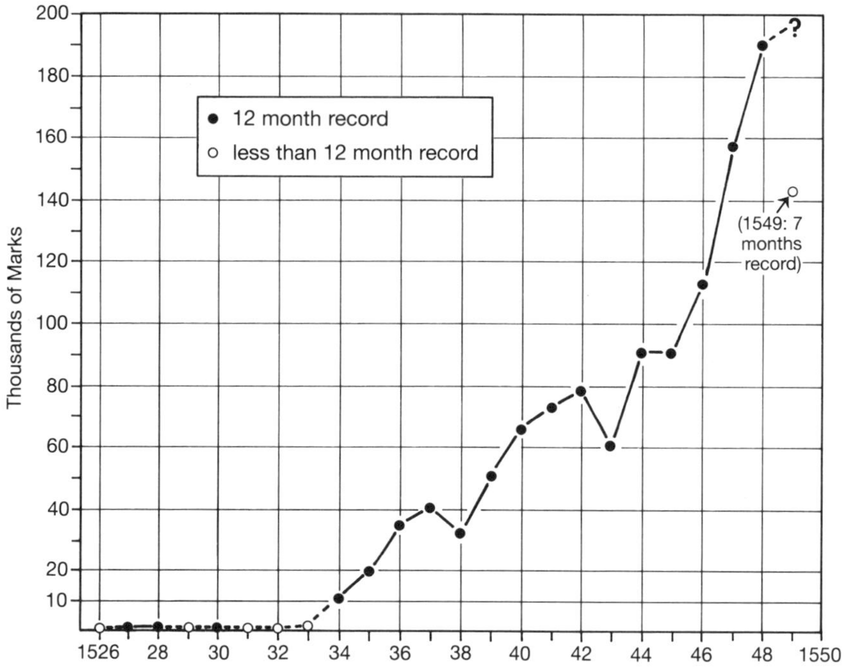

Fig. 7. Silver, in marks, registered at the *Caja de Fundición*, Mexico City, 1526–1548. Open dots indicate data covering less than 12 months of a given year. Data from AGI Patronato 199–202.

The introduction in 1555 of the amalgamation process perfected by Bartolomé de Medina and improved by the German, Gaspar Lohmann, permitted the refining at comparative low cost large quantities of the poor–grade silvers ores mined at depth. This process helped arrest a deepening of the "mining crisis" of the mid–sixteenth century in New Spain.

Acknowledgment

For much of the information on early mining in Sultepec, Taxco and adjacent areas the author is indebted to Dr. Elinore Barrett, professor of geography, University of New Mexico, who kindly loaned microfilm and transcripts of documents that she obtained from the Archivo General de la Nación, Mexico City.

Endnotes

1. In November 1521, a list of aboriginal metal artifacts and other items was registered before royal officials in the house of Hernán Cortés at Coyoacán, Valley of Mexico, for the purpose of taking the royal fifth. The list enumerated a portion of a gift sent to Cortés by the

native ruler of Michoacán. Included were artifacts of low-grade silver weighing some 86 marks (688 ounces) in the form of *rodelas,* or discs, bracelets and earplugs. When melted down into bars, such may have among the first shipments of Michoacán silver to Spain (AGI Contaduría 657).

2. The geographer Carl Sauer (1948, 89–92) suggested that "Michoacán may contain the oldest record of Spanish mining in the New World," citing early Spanish accounts of silver in the Tamazula area. Today, however, traces of mines or processing of ore that might be assigned to the lost Morcillo mine in that vicinity have not been found. Accounts of the 1580s do indicate the existence of silver mines near the town. The *Relación Geográfica* of 1582 for Tamazula states "*En los términos deste pueblo [Tamazula] ay minas de plata y las a habido ricas antiguamente, y ay minas de cobre, y ninguna se labra de presente*" (Rea Vargas, 1946, 127). Again, according to the account of Padre Alonzo Ponce's visit to New Spain made in the 1580s (written by Antonio de Ciudad Real, one of his two companions)(Ponce [1584] 1875, II:114–115), at that time the inhabitants of Tamazula knew about the silver mine of Morcillo, located nearby, and that mining there had been revived. The name of the mine may refer to one Francisco Morcillo (Icaza 1923, I:207) listed as an early resident of Michoacán who took part in the conquest of the province.

3. For example, a letter from the royal officials in New Spain to the Crown, dated 8 Feb. 1533, reads in rough translation: "... since in many parts of New Spain for the last half year a particular ore has been found that yields from two to three marks (16 to 24 ounces) of silver per *quintal,* and if there were in this land experts of this profession [metallurgy] there would be such a wealth available that even the cripples would go out in search of it...; but lacking are those who know how to refine the ore successfully...; if your majesty would order people versed in this occupation to come to these parts, there would result more wealth than that obtained from gold [placering] and less work for the Indians" (AGI Mexico, leg. 323; *Epistolario de Nueva España* 2:25). Many more such requests are found in similar documents written in the early 1530s.

4. Other prominent miners at Zumpango included Martin Dircio (de Yrzio) and Pedro de Castilleja, both of whom were members of the expedition sent to quell the rebelling Yopelcingo Indians and probably took part in discovery of the mines (Icaza 1923 I:30, II:305; AGN Hospital de Jesús, leg. 264, exped. 6, 7).

5. After their opening in the early 1530s the mines of Taxco produced modest but steady amounts of silver, but in 1542 with the discovery of a rich vein of ore in a mine belonging to one Luis de Castilla the first big bonanza of the area took place, lasting about 15 years. In 1552 the mines along this vein (Cerro Rico) were in full production. By 1570, however, the rich ores had played out and a long period of decline lasted much of the seventeenth century. Production was revived by the bonanza of 1748–1757 in mines belonging to the Borda family (Carrera Stampa 1956; Bargalló 1955, 57–58; Gerhard 1972, 254).

6. Engel and his German companions apparently were sent to New Spain by colleagues Lazarus Martin Berger and Cristoph Kaizer, who were living in Sevilla. Later, the latter two petitioned the Crown for remuneration, citing Engel's contributions to the silver industry in Sultepec: "... [he] came with equipment and knowledge for smelting ores in the silver mines, [information] theretofore unknown [in New Spain], and he constructed mills for grinding ores [stamp mills?] and smelting them [type of furnace?], from which the country received much benefit and a great service to his majesty, for thereby the royal fifth was greatly increased, and for all [these services] the said applicants spent a large sum of money, more than 50,000 *maravedis,* and for which the applicants have not been remunerated" (Icaza 1923, II:258–259). Activities of Germans in New Spain during the early sixteenth century are discussed in Pferdekamp (1938). See also Bargalló (1955, 91, 94).

7. A copy of the original ordinances of 1542 pertaining to Taxco and neighboring mining centers is housed in the Library of Congress, Washington, D.C. (Krause Collection 140) and is published in Zavala 1982, 176–180, followed by several pages of commentary dated

1551–1552 written by Viceroy Velasco. The document is one of the best extant sources of information on problems and organization of the early silver mining complex.

8. Mendoza's ordinances regarding cutting of forests around Taxco, issued n Feb. 1550, are published in O'Gorman 1939, 260–263.

9. In 1552 royal officials in Mexico reported to the Crown: "All of the silver mines that sustained New Spain are in great decline; if it were not for new mines discovered at Zacatecas about two or three years ago ... neither your Majesty nor the miners would profit. Also, mines that previously produced four or more marks of silver per hundredweight [of ore] without the use of lead or litharge [as flux] are now very deep and do not produce one or two per hundredweight, and so much lead is needed [for ore reduction]. Add to this the sudden freeing of the Indian slaves; the shortage of food due to the commutation of tribute [from products of the land] to money, which the Indians do not have; wood and charcoal are distant [from the mines]; negro slaves are expensive — 180 to 200 pesos per head..." (RAH, Colección Muñoz, vol. 86, ff. 139–140. Officials to the Crown, 8 March 1552).

References

Archival Sources

AGI — Archivo General de Indias [Sevilla]
Contaduria 657
Justicia 123, 131, 132A, 167, 1029
Mexico 323
Patronato 180, ramo 72; 184, ramo 21; 199–202

AGN — Archivo General de la Nacíon [Mexico City]
Hospital de Jesús 276, 264, 276, 432

RAH — Real Academia de Historia [Madrid]
Colección Muñoz, vol. 86.

Published and Other Materials

Bakewell, P. 1971. *Silver mining and society in colonial Mexico: Zacatecas, 1546–1700.* London: Cambridge University Press.

Bargalló, M. 1955. *La minería y la metalurgia en la América española durante la época colonial.* Mexico City: Fondo de Cultural Económica.

Berthe, J.-P. 1958. Las minas de oro del Marqués del Valle en Tehuatepec, 1540–1547. *Historia Mexicana* 8:122–131.

Carrera Stampa, M. 1956. El mineral de Taxco y Don José de Borda. *Gacetas históricas,* International Geological Congress, 20th session. Mexico City: Editorial Jus.

Colección de documentos inéditos para la historia de España 1842–1895. 112 vols. Madrid: Imprenta de la viuda de Calero.

Craine, E. R. and Reindorp, R. C. 1970. *The chronicles of Michoacán.* Norman: University of Oklahoma Press.

Epistolario de Nueva España. 1939–1942. 16 vols. Recopilado por Francisco del Paso y Troncoso. Mexico City: Antiguo Librería Robredo.

García Pimentel, L., ed. 1897. *Descripción del arzobispado de México en 1570 y otros documentos.* Mexico City: José Joaquín Terrazas e hijas, Imps.

Gerhard, P. 1972. *A guide to the historical geography of New Spain.* Cambridge: University Press.

Hosler, D. 1988. Ancient west Mexican metallurgy: South and Central American origin and west Mexican transformations. *American Anthropologist* 90:832–855.

Icaza, Francisco A. de 1923. *Diccionario autobiográfico de conquistadores y pobladores de Nueva España.* 2 vols. Madrid: El Adelantado de Segovia.

Millares, C. A. and Mantecón, J. I. 1945–46. *Indice y extractos de los protocolos del Archivo de Notarías de México, D.F.* 2 vols. Mexico City: El Colegio de México.

O'Gorman, E. 1939. Mandamientos del virrey D. Antonio de Mendoza. *Boletín del Archivo General de la Nación* 10:213–311.

Paso y Troncoso, F. del. 1905–06. *Papeles de Nueva España.* Segunda serie, Geografía y Estadística. 7 vols. Madrid: Establecimiento Tipográfica "Sucesores de Rivadebera."

Pferdekamp, W. 1938. *Deutsche im frühen Mexiko.* Berlin: Deutsche Verlags–Anstalt.

Pollard, H. 1987. The political economy of prehispanic Tarascan metallurgy. *American Antiquity* 52:741–752.

[Ponce, A.] [1584] 1873. *Relación breve y verdadera de algunas cosas de las muchas que sucedieron al padre fray Alonso Ponce en las provincias de la Nueva España ... escrito por dos religiosos, sus compañeros....* 2 vols. Madrid: Imprenta de la Viuda de Calero.

Probert, A. 1969. Bartolomé de Medina: the Patio Process and the Sixteenth Century Silver Crisis. *Journal of the West* 8:90–124.

Rea Vargas, ed. 1946. *Papeles de Nueva España.* (Coleccionados por Francisco Paso y Troncoso) Segunda Serie, Geografía y Estadística, vol. 7 (supplement) (Relaciones Geográficas de Michoacán). Mexico City: Biblioteca Aportación Histórica.

Sauer, C. O. 1948. *Colima of New Spain in the sixteenth century.* Ibero–Americana 29. Berkeley and Los Angeles: University of California Press.

Simpson, L. B. 1940. *Studies in the administration of the Indians of New Spain. IV. The emancipation of the Indian slaves and the resettlement of the freedmen, 1546–1553.* Ibero–Americana 16. Berkeley and Los Angeles: University of California Press.

Wagner, H. 1942. Early silver mining in New Spain. *Revista de Historia de América* 14:49–71. Mexico City: Instituto Panamericano de Geografía e Historia.

——. 1969. *The rise of Fernando Cortés.* New York: Kraus Reprint.

Warren, F. 1985. *The conquest of Michoacán: The Spanish domination of the Tarascan kingdom in western Mexico.* Norman: University of Oklahoma Press.

Zavala, S. 1968. *Los esclavos indios en Nueva España.* Mexico City: El Colegio Nacional.

——. 1982. *Libros de asientos de la gobernación de la Nueva España.* Mexico City: Archivo General de la Nacíon.

——. 1984–87. *El servicio personal de los indios de la Nueva España.* 3 vols. Mexico City.

Technology of Mining in Colonial Mexico: Installations, Tools, Artifacts and Machines Used in the Patio Process, Sixteenth to Eighteenth Centuries

Ramón Sánchez Flores, Sociedad Méxicana de Historia de la Ciencia y de la Tecnología, Mexico City

Abstract

A detailed investigation of instruments used in the amalgamation process, or "patio process," has been largely ignored among historians of mining in colonial Mexico. This process became the most efficient for refining gold and silver ores from the sixteenth until the late nineteenth centuries, when cyanidation was introduced. Miners invented and adapted tools and instruments for using mercury, salt, *magistral* (copper–iron sulfate) and other ingredients in the amalgamation process, such as tubs, or "*canoas*," used in washing ores, and tanks employed in the preheating or roasting of the ores. The method of refining the amalgam as well as its reduction to disks and bars, the classification of the instruments and processes, and their comparison with those used in Europe all contribute to an understanding of the advances made in mining technology in the American colonies.
Key words: mining technology, amalgamation, Mexico.

Resumen

El estudio del proceso de amalgamación de los metales de oro y plata por el llamado "método de patio," ha sido muy fecundo entre los historiadores de la minería del México colonial. No obstante es practicamente desconocida una relación pormenorizada de las herramientas, artefactos y máquinas empleados en este método, que fue el mas eficiente desde el siglo XVI hasta la invención del método de cianurización en el siglo XIX. En nuestro ensayo se refiere como los mineros que usaron el método de azogue, sal, magistral, y otros ingredientes, inventaron y adoptaron sus instrumentos, así como las artezas o canoas para el lavado de las menas y los cajones para el precocimiento o tostado de las mismas. Se incluye una descripción de los métodos de afinación de los metales, así como su conversión a tejos o barras. La tipificación de todo este instrumental y su comparación con el ya utilizado en Europa, per-

---

In Quest of Mineral Wealth: Aboriginal and Colonial Mining and Metallurgy in Spanish America, edited by Alan K. Craig and Robert C. West, 1994. Geoscience and Man, vol. 33, pp. 137-153. Dept. of Geography & Anthropology, Louisiana State University, Baton Rouge, LA 70893-6010, U.S.A.

mite conocer los adelantos e invenciones logrados en una tecnología de origen americano, ignorada en las historias convencionales.
Palabras claves: Tecnología minera, amalgamación, México.

As a main source of productivity, the mining industry in colonial Mexico has been the subject of study by outstanding historians of economy and science. Such studies have centered around modes of ore extraction and production, as well as refining methods, especially that of cold amalgamation, also known as the patio process. Few, if any works, however, have treated the historical development of the particular tools and instruments that made the patio process viable.

The present study proposes to consider, within the context of the labor practices of the day, the tools, artifacts, machines, and other installations that were developed in New Spain in relation to the evolution of the patio process.

Knowing full well that the task of depicting the artifacts used in a given process bears the danger of diminishing the description of the technique, we accept the challenge of enlightening this chapter in the history of Mexican mining. From another approach we endeavor to emphasize the inventive creativity and labor of those who championed this important aspect of mining history.

Introduction of the Patio Process and Related Machinery, Sixteenth Century

The first successful endeavors in New Spain to solve the problem of substituting the practice of smelting for that of amalgamation occurred during the 1550s with the interesting solution developed by the Sevillano, Bartolomé de Medina. His experiments, carried out in the Real de Purísima Grande at Pachuca, involved the use of mercury and large quantities of salt and *magistral* (a copper–iron sulfate) which were mixed with finely ground silver ore to produce an amalgam without the application of heat (Menes Llaguno 1989). In the course of his experiments between 1554 and 1555 he also developed tools and other mechanisms needed for this particular system.

To be sure, since 1530 and even earlier, other miners in New Spain had introduced equipment and concepts related to the development of the amalgamation process. We know the names of individuals such as Alvaro López, who was the first to use aquafortis, a nitrous acid for separating gold and silver (1520–1526); Antonio García Saldaña, who was noted as "the first to devise methods for making a type of smelter and a stamp mill"; Rodrigo de Alarcón, an expert smelterman (1535); as well as Juan de Palencia, who introduced "horse–powered machinery for milling and smelting ore" (1535). Generally, one can affirm that these miners used equipment already known in Europe, especially that used by German miners (Bargalló 1955). For example, the excellent equipment used by Martin Verger and Christoph Keiser, residents of Sevilla and former employees of the Fugger family, exploiter of precious metals in Spain since the beginning of the sixteenth century, through contract with the Catholic kings (Sánchez Gómez 1989). Associates of Verger and Keiser arrived in New Spain in 1536 with "machinery and knowledge for reducing silver ores, which, according to them, was theretofore unknown" in the colony (Sánchez Flores 1980).

Although the introduction of Western mining technology into New Spain has been well studied and discussed in textbooks, it is necessary here to remember the differences in use and nature between those employed by aboriginal miners in Mexico since time immemorial and those introduced from Europe; in this way we will avoid the exaggerations which in one sense or another have distorted influences on mining in New Spain.

Although the tools, machines and other instruments used were numerous and complex, they should be classified in relation to their importance and origin as (1) *domestic* and (2) *industrial*. The devices of domestic origin arose through actual improvisation, inspired by the necessity of adapting them to whatever kind of work at hand, including the tools of aboriginal origin. Evident in the design of these domestic tools are the ingenuity and skill of the miner and the foundryman who took part in their manufacture, often without any mechanical and, much less, any scientific or mathematical system.

In contrast, the design and fabrication of tools and machines destined for use in the extraction, grinding and smelting of ore and the draining of mines, among other tasks, required the joint use of various factors, such as the introduction of European prototypes and adaptation of equipment for installation in New Spain, mainly to ensure proper operation in terms of regional requirements. In the design and construction of such artifacts and machines one sees the work of carpenters, mortisers, machinists and even artisans skilled in the building of masonry structures, containers, and complicated mills (Sánchez Flores 1980, 91–108). In fact, in design, form and dimension, the mechanical variety used in New Spain is notorious and, quite apart from the prototypes devised by German miners, commonly used in Europe. Not until the end of the seventeenth and during the eighteenth centuries did mechanics, ironsmiths, and even well educated university professors devote themselves to the design and construction of instruments and machines that approached those that could be called "mathematically derived" (Moreno de los Arcos 1986).

Domestic Artifacts

In the process of mixing and washing the mineral–bearing ores with mercury and salts, some types of domestic artifacts served as containers. Among the more common were *canoas*,[1] or wooden tubs, and *bateas*, or wooden bowls; other receptacles consisted of bowls made of fired clay, as well as *jícaras*, or hard rinds of gourds from the calabash tree (fig. 1). For transporting ground ore, miners used rawhide sacks, wicker baskets, *tenates*, or cylindrical baskets of woven palm fronds, as well as pieces of coarse cloth and *tilmas*, or cape–sized cloths woven of agave fiber, such as that of *lechuguilla* and *henequén*. For storing mercury and for sprinkling it on ground ore, sacks of rawhide and chamois skin were employed; also, cloths of padded cotton were used to squeeze out unused mercury from portions of the amalgam paste. For sieves or strainers, pieces of coarsely woven agave fiber and woven stands of wire were used. All cordage of aboriginal origin consisted of *maguey* (agave) fiber.

Within the aggregate of domestic tools and machines is the *malacate* (from the Nahuatl *malacatl*), a capstan or vertical winch of Mexican origin. Although the European capstan, or whim, was introduced together with other machinery to drain mines, experience showed that the native winch was more useful because of its simple construction. It consisted of a vertical axis having a radius of from 4 to 8 times smaller than the crossbars, to which rope could be wound more easily and the whim turned more rapidly than with the more complex European whim (fig. 2). Finally, the horse–operated *malacate* was a new adaptation that was slowly perfected. It was not until the late eighteenth century that this "mestizo" apparatus received notable improvement from learned engineers, such as Alzate, Velázquez de León, and Elhuyar (Sánchez Flores 1980, 167–169).

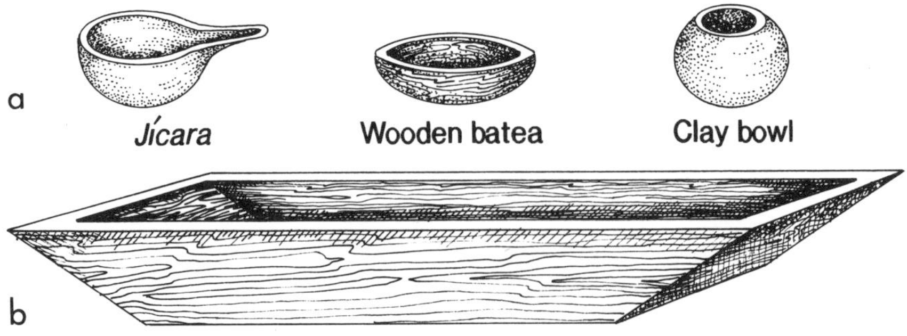

Fig. 1. Aboriginal–type artifacts used in the patio process. a) Bowls used in measuring ingredients. b) *Canoa*, or vat, used in mixing ore and mercury.

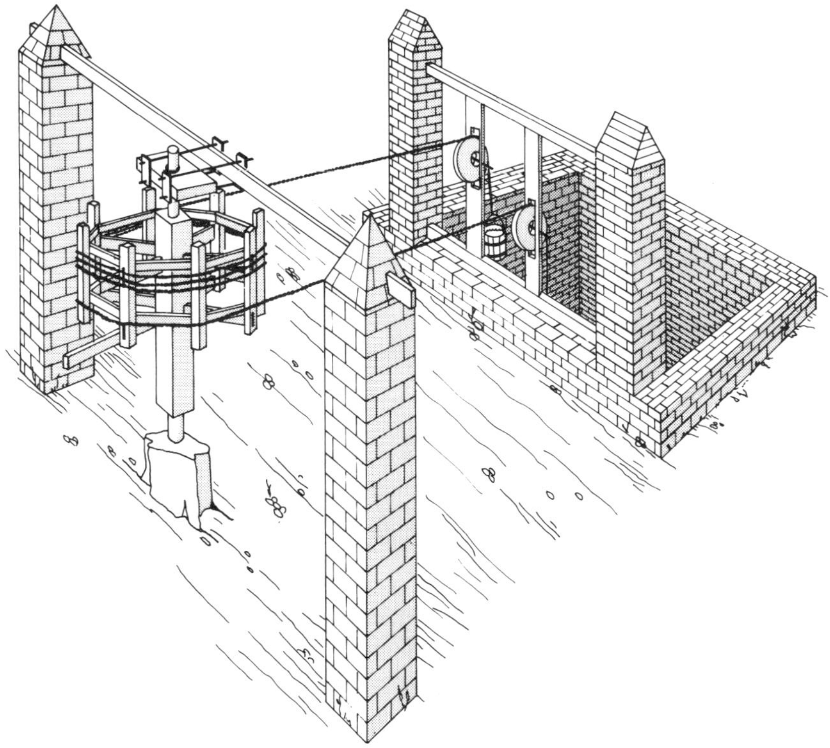

Fig. 2. A small *malacate*, or hand–powered water–lift, ex–hacienda los Lebrillos, southeastern Guanajuato state. Used to draw water from a well.

Domestic and Industrial Artifacts, Sixteenth–Eighteenth Centuries

The industrial operation of the patio method, whereby the processing of several tons of mixed ores took place at one time, required composite tools and installations. In this way the domestic artifacts attained technical levels. This is the case with furnaces, the construction of which required the knowledge of expert masons and carpenters. Furnaces were of diverse forms: square, conical, pyramidal, and so forth, according to the need for initial heating of the ores for maximum yield of silver. The technical significance of the furnace lay in the design of its vents, its doors, and its chimney draught. The size of these features was usually calculated in measurements no greater than four square meters.

The nature of the receptacles in which mixing of ore with mercury, salt and copper sulfate was carried out, and where loads of ore containing gold or silver were washed, had equal importance. Such receptacles were called *canoas, cajones*, and *buitrones*.[2] As is well known, the term *canoa* has an Antillean origin, but the receptacle was similar to the Mexican *acalli*, or dugout canoe, fashioned from a single log of special wood, hollowed out in one piece in the form of a boat. The size of *canoas* varied with time. Gonzalo Gómez de Cervantes (1969), who in 1599 described the patio process, mentioned *canoas* 6 to 7 meters long, although some of 10 meters were found. In these, the mud of finely ground silver ore, mercury, and other ingredients were mixed by trampling. Gómez de Cervantes (1969, 151–152) describes the process: "... then workmen get into the *canoas* and with their feet knead and mix the ore until it is joined with the mercury and salt, so that the mercury is no longer visible...."

During the time span from the sixteenth through the eighteenth centuries, the *canoa* type of receptacle was altered according to the changes that the amalgamation system went through, in both the cold and roasting processes. In 1580, Bernardino de Santa Cruz solicited a license from the viceroy Martín Enríquez Almansa to patent a method for reducing the amount of mercury used in amalgamation. The invention involved the use of a large chest made of wooden planks with a stone or brick base, in which ore, after being mixed with mercury and other ingredients, was heated in order to accelerate the amalgamation process (Bargalló 1955, 36–40). This method gave rise to the famous "fire chests," or *buitrones* (fig. 3), receptacles that could be used as furnaces. These attained their greatest use in Peru along with other local innovations (Bargalló 1969).

Although the *canoas*, or wooden tubs, and the *buitrones*, or fire chests, never disappeared, as the refining of silver ores by means of amalgamation with mercury increased, the mixing operation was carried out on a large scale on open–air bricked surfaces. Thus the method came to be called "the patio process," because the work was done in the large patios of the mining haciendas.

A type of mill, called the *molinete* and a kind of blender known as the *rastra*, both used to separate the mercury and wash the mixture of gangue and silver, constituted a combination of technologies. The same system of movement with terminal gears used in the stamp mill was applied to the *molinete* to operate a pair of forked reels. It was placed in *canoas*, tubs and water chests to produce vortices which diluted and separated the amalgam from the rest of the mixture. These mills were constantly being improved; for example, in 1582 Pedro Díaz de Baeza obtained a license to use his "*lavadero de metales*," or ore washer, in the mines of Sultepec, Mexico. His model placed the forked reels not in the center of the *canoa*, but in the canals where the stream of mixed ore entered the tub; he also placed the

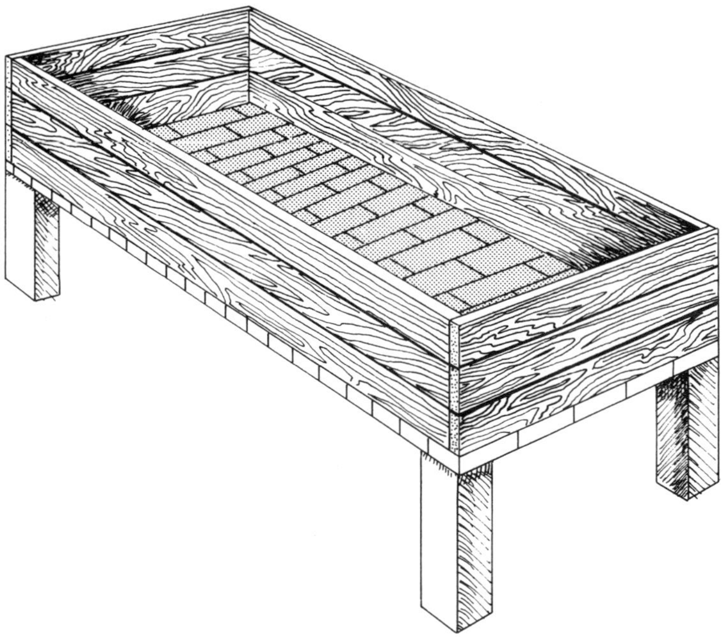

Fig. 3. A *buitrón*, or wooden fire chest with brick base, used for mixing ore and mercury with the aid of heat, sixteenth century.

reels near the bottom of the tub, so that the rich mixtures would be deposited there (fig. 4) (Sánchez Flores 1980, 99).

The *rastra*, or blender, employed the same type of terminal gears as the *noria*, or water lift. It was used to mix the ore slime by tamping, and was operated by animal power. The blender attained improvements within the patio process only in the seventeenth and eighteenth centuries (Othón de Mendizabal 1980).

Auxiliary Tools

The system of amalgamation with mercury required few minor tools. But through time, especially after the development of heated amalgamation, the system increased in productivity, which is well described by Padre Acosta and Gómez de Cervantes. (See appendices A and B.)

As indicated previously, the auxiliary tools were fabricated with native materials. The types of shovels, rakes, hammers, spoons, ladles, wheelbarrows, and so forth, although the designs were taken from European models, acquired the stamp of their native makers. In the first place,

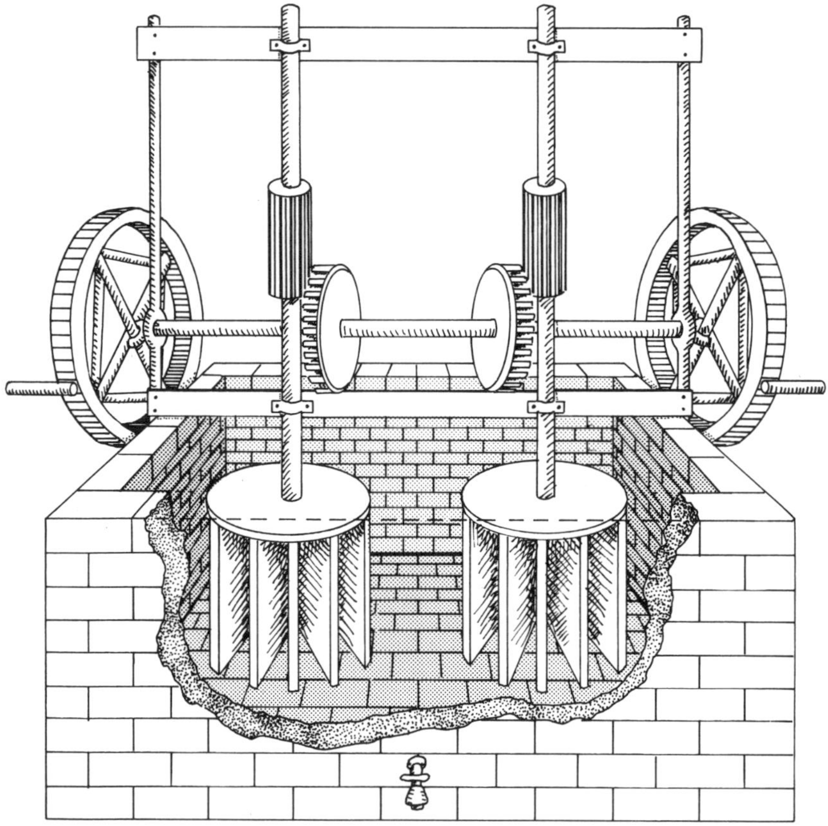

Fig. 4. Hand–powered *molinete* for washing the ore mix, a type invented by Pedro Baeza, sixteenth century.

the dimensions of these tools were adjusted to fit the nature and skill of the indigenous mine workers. For example, the large Carolingian sledge hammer, used by the strong German miners, in New Spain was converted to a lighter and more portable tool (fig. 5). A comparison of European tools with those of Mexico indicates substantial modifications, since iron was often replaced by hard wood as the main material. This change resulted in important advantages, because wooden tools could be replaced easily, they were cheap in price, and above all, certain pieces lasted longer than iron ones, since wood did not contaminate the chemical process of amalgamation, the ore mixture being corrosive.

In general the artifacts and tools of iron, and of iron and wood, were employed in the last phases of the patio process, when the silver amalgam

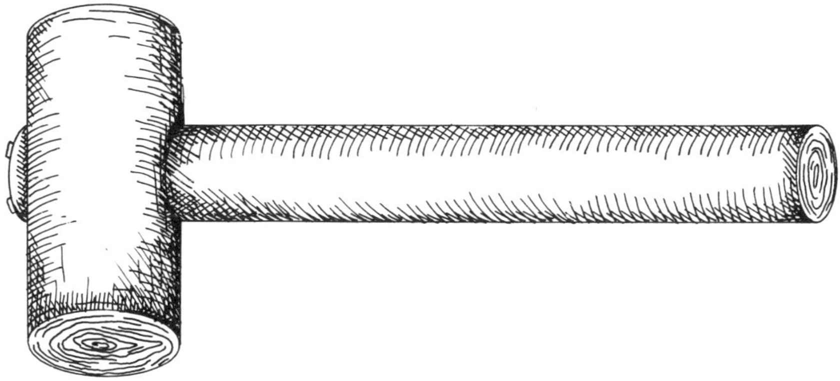

Fig. 5. *Triturdor,* or wooden mallet used for crushing ore.

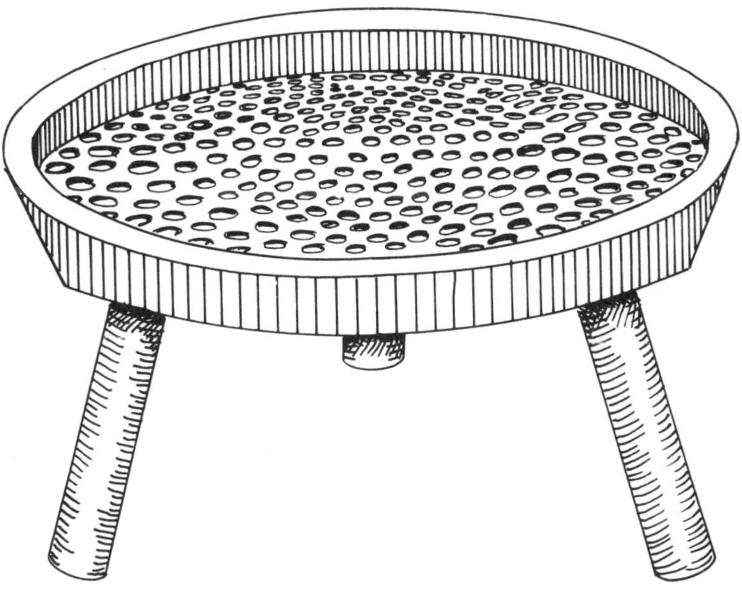

Fig. 6. An example of a *trébade,* or iron tripod oven.

rich in mercury was subjected to heat. This was done by placing a ball-shaped portion of amalgam in an instrument called a *trébade* which consisted of an iron plate or platter perforated like a sieve and elevated on three legs in form of a tripod, under which heat was applied (fig. 6). Mercury vapors released from the amalgam were passed through a container of water, where they condensed and were recovered as liquid mercury.

Improvements of mercury retorts and condensers were continuous, the most famous ones having been invented by the miner Juan Capellín, for which he obtained a license in 1576. The invention consisted of a new design of the condenser, improved by means of coiled tubes similar to those used in a still, in order to recover a greater quantity of mercury (Sánchez Flores 1980, 101–102).

Finally, one should mention the *alcribices*, or *tuyères*, devices of potter's clay or of metal that were placed in openings or taps of smelters in which amalgamated silver was subjected to fire in order to complete its purification (fig. 7). The ladles and tongs of many forms and sizes used in

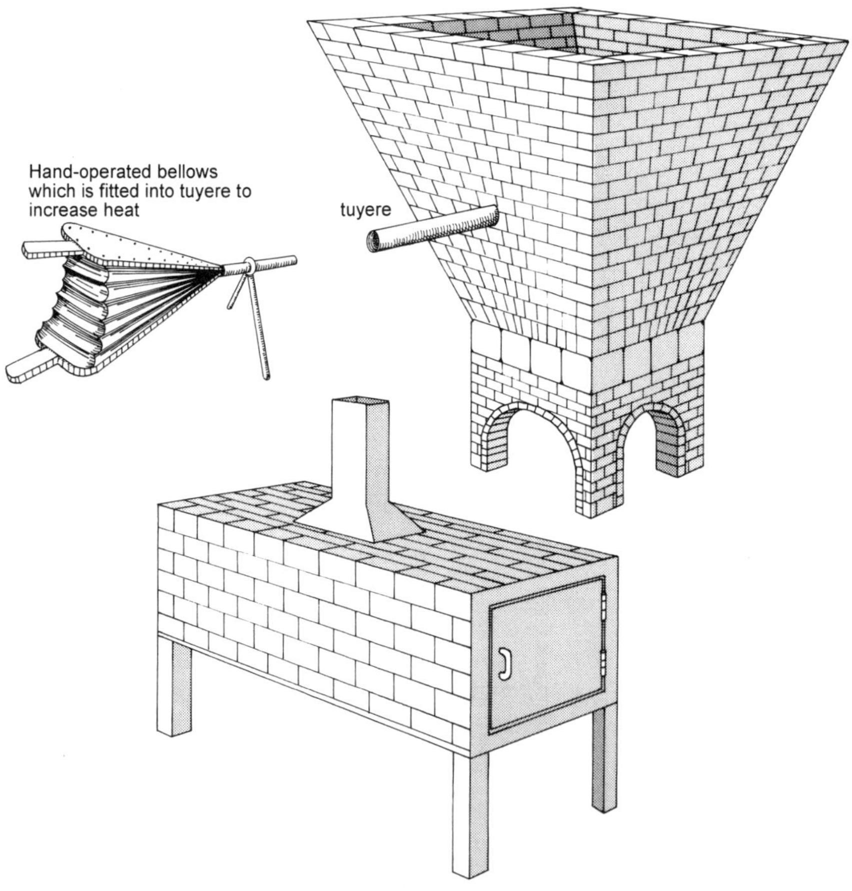

Fig. 7. Furnaces for refining amalgamated lumps of silver (*pellas de plata*), the final operation of the patio process.

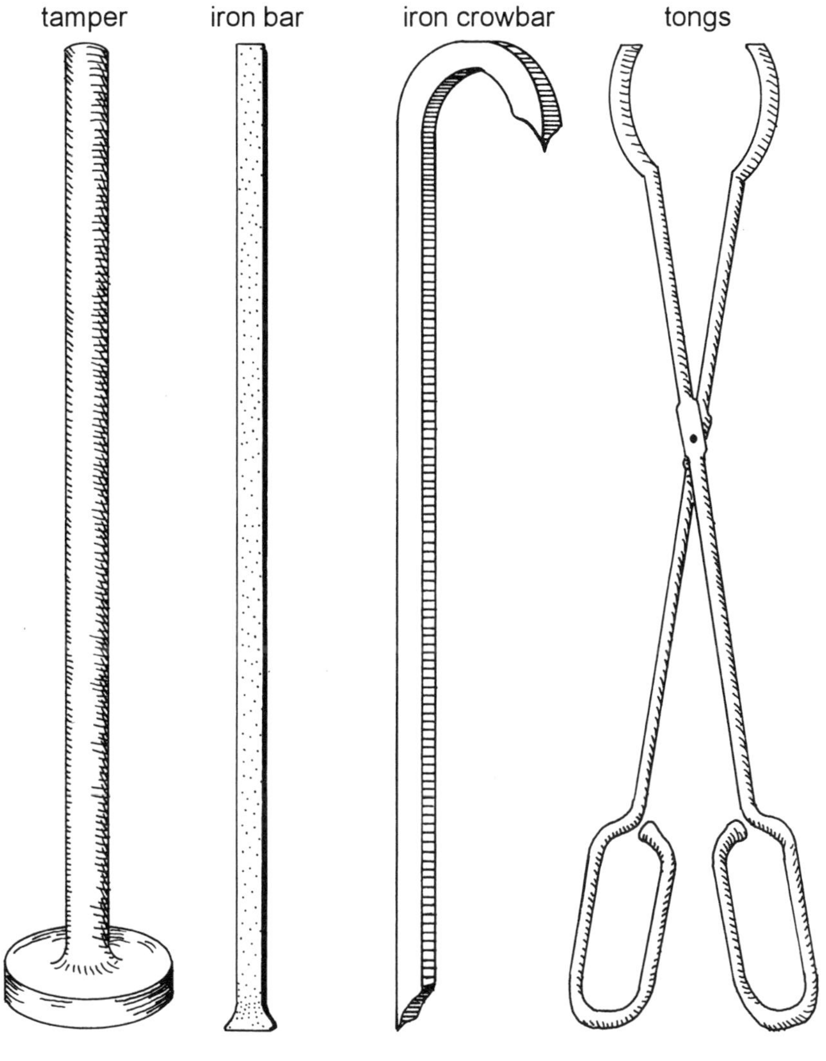

Fig. 8. Examples of metal instruments used to manipulate lumps (*pellas*) of amalgamated silver during the refining process.

this operation complete the picture of this final phase of silver production by amalgamation (fig. 8).

During its time of utilization, the cold amalgamation process basically increased the production of silver; but in doing so the system itself underwent comparatively few improvements. This is manifested by the fact that during the course of three centuries there occurred few substantial changes in process, save in design and quality of instruments and installations. All of this utility, truly of medieval origin, disappeared when in the mid–nineteenth century the system of cyanidation surged forth to revolutionize mining in Mexico.

Appendix A

The following is a translation of a report on how ores are reduced with mercury, written in 1599 in Mexico by Gonzalo Gómez de Cervantes (1969, 150–155).

The ores that are taken from the mines are refined in one of two ways: either with mercury or by smelting. Those that are refined by smelting must yield one mark [8 oz.] or more per hundredweight of ore; and those that are refined with mercury should yield at least one ounce or more per hundredweight; and those who refine ore of that low yield will rarely, if ever, emerge from misery and hard work because the costs of mercury, salt, labor and tools, and other necessities are so great that little profit will remain. And in discussing the two ways of refining, as to that with mercury, ore is refined in the following order.

First, ore is taken from the mines; it has a consistency of clay, because it is mined at such depths that it appears wet; and once taken from the mines, it is placed in sunlight to dry, and once dried, is taken to a water–powered stamp mill, and where there is no water, to a mill powered by mules; there the ore is finely crushed with oaken stamps, the heads of which are plated with iron; and, after being milled, the ore is carried in hoppers to be screened through wire sieves to obtain the smallest particles. Once screened, the ore is taken to another place called the *encorporador*, which consists of several large *canoas*, or wooden vats. Ore is placed in the vats by hundredweight portions, to be mixed with mercury. But before mixing occurs, each hundredweight of ore is assayed as to its silver content — usually from 2 to 4 ounces of silver per hundredweight — in order to determine the amount of mercury to put in the mix, as a rule 6 ounces for each ounce of silver assayed. Mercury is sprinkled evenly over the ore so that it will join with the silver in a thorough manner; then for each hundredweight of ore a fourth as much salt is added [25 pounds], regardless of the silver content of the ore. Once the mercury and salt have been thrown into the mix, water is then added, for it is necessary that the finely milled ore, having been reduced to a powder, be turned into a mud or slime; and the mud should not be very thin, but solid enough so that it can be well mixed; and then workers get into the *canoas*, or vats, and with their feet tread and mix the ore until it is joined with the mercury and salt, so that the mercury is no longer visible. And after the mixing is done the Indians who have been treading the mixture make balls of it the size of a man's head, and then they put these into ovens, making piles of 20 to 25 hundredweight, more or less, as they wish, and closing the door of the oven, they apply fire so that the piles are heated; in that way the ore gives off its silver content more quickly.

After being in the oven for six days, the heads of ore are removed and again placed in the *canoas*, or wooden vats, where they are once more treaded with the feet as before; water is applied as needed, for the heat of the oven has caused the balls to become dry; and having made this remixing, the ore is returned to the ovens as was done at first. And in this way the ore is remixed 4 times, and if it is seen that the ore is not yielding according to the assay, another remixing, or even 3 more, can be undertaken, and these remixings are longer in duration than the first four; during the intervals between the mixings no longer is it necessary to heat the heads for 6 days at a time, but for only 3 days, in which time the mercury becomes more amendable and is able to combine with the silver within the ore, however intractable and strong it may be.

And if refining is done without the ovens, the time of amalgamation takes longer, because the mixings must be done every 10 days, and after each remixing the mud is left to dry in piles in a large room that has many barred windows on all sides, so that sunlight and air may enter and that the mixture may dry quickly; for the ovens only serve to speed amalgamation because their heat dries the ore sooner than in the piles, since the cold of the latter prevents the silver from incorporating quickly with the mercury.

After all these mixings have been completed, a small amount (about a pound) of the mud is taken from one or two piles and is washed in a *batea*, or wooden bowl, and assayed to see if the proper fineness of silver has been attained, and if the mercury appears either very curdled or quite thin. If it appears curdled, it is an indication that amalgamation has occurred, and if thin, that it has not taken up the silver, and more remixings will be necessary; and if the mercury and silver have been incorporated, as attested by an assay, the mud is thrown into a large tub in which there is a certain machine, or mill, equipped with forked reels that are powered by a mule or by water [to agitate the water within the tub]. The whirling action of the water within the tub separates the amalgam from the rest of the ore mud; and once separated, a stopple is opened at the bottom of the tub, the water draining through narrow channels lined with riffles which catch any mercury that may drain through the stopple. The rest of the mud that drains out goes to an outside stream, and because it is heavy most of the amalgam settles at the bottom of the tub, where it is gathered with some sand and fine sediment; this is put in average–sized barrels and is washed in smaller vessels to purify it; and the sediment that remains is kept, because it is rich in mercury that can be retrieved by heating; because the mercury in the sediment has not combined with silver, it is very clear and pure, for there was insufficient amount of silver in the ore to amalgamate; in such a case this material is placed in a canvas through which the free mercury can be filtered, and that which remains in the canvas is amalgam incorporated with the silver; and if the mixture appears to be very curdled, a small amount of mercury is put in so that the amalgam will not be floured.

Then, by hand, the amalgam is fashioned into small balls the size of an egg; and these are placed in an iron or copper plate, which is perforated like a sieve, and the balls are placed one upon the other until they make a pile about a span [8 in.] high, and then the plate is put over a *trébade*, or iron tripod oven near a jug of water that is placed within a copper or bronze retort; and this retort is placed in a small oven that is over the water; and a fire is made close by the above covering, which causes the mercury in the amalgam balls on the perforated plate to vaporize; and this vapor rises into the retort, but unable to escape below, it rises; and making contact with the water, it condenses to liquid form. This process lasts 4 hours or more with a good fire, so that the covering of the retort becomes quite hot; by that time the mercury that has been removed from the silver is deposited in the water, and the silver remains in pure form in the plate; and in case the retort might have an abrasion or crack (caused by long use and having been subjected to fire many times) through which heat and smoke may be emitted, the mercury is lost without recourse; this is the major loss that can happen and does happen in the refining process; and when it does so that value of the silver refined is insufficient to cover the cost of the mercury lost.

Another considerable loss results when the mercury is not completely removed from the amalgam when heated and all of the mercury contained therein is not vaporized, causing considerable loss to the miner; and this occurs because of lack of care in heating the retort or by removing the heat prematurely.

After the retort has cooled it is lifted out and the plate containing the pellets of amalgamated silver is taken off the *trébade*; then the pellets are placed, one by one, in a clay bowl, which is a kind of cupel of wood ash and ground brick; and with a fire of dry wood the pellets are melted down to form pieces of silver of suitable size to pay creditors and for other necessities; in order to cupel the silver and maintain its fineness, it is well to add an ounce of lead for each mark of silver, although some miners, in order to increase the weight of the silver, apply more lead than necessary and extinguish the fire prematurely, so that lead is retained in the silver over and above the amount permitted; and to continue the fire until the silver is purified, always some lead remains; and to avoid criminal charges, it is convenient to find an assayer to test correctly the fineness of the silver.

After these losses and costs have been assessed, the miner weighs his silver to see how much has been refined, and weighs the mercury recovered to see how much has been lost; and the most moderate loss of mercury is about a pound for every mark [8 oz.] of silver refined, a loss that can never be recovered, for part of the mercury is carried off by water during the mixing and part lost by heat; and although some miners lose more, others less, this is the most usual of the losses, on average, a pound of mercury per mark of silver.

. . . Once the silver is refined and cleansed of all impurities, it is taken to the magistrate [*alcalde mayor*] and the scribe for the mines, before whom the miner swears that the metal is from his mine and that there has been no attempt to circumvent payment of his Majesty's royal fifth [severance tax]; and this done, the silver bars are marked with the tithing [*diezmo*] stamp, which is kept in the safe locked with three keys, one of which is kept by the magistrate, another by a deputy, and the other by the scribe.

Appendix B

Translation of brief extracts from Padre José de Acosta's *Historia natural y moral de las Indias* (first edition published 1590 in Sevilla), Libro IV, Cap. XII.

. . . The ore is finely milled, first by means of stamps that strike the stones like the hammers used in a fullering mill, and after the ore is well ground it is screened through sieves made of woven wire, resulting in a cribble as fine as hog bristles; with these sieves, if well assembled, 30 hundredweight of powdered ore can be screened in a day. The finely ground ore is then put into heated water chests called *buitrones*, in which it is mixed with salt; for each 50 hundredweight of ore, 5 hundredweight of salt is mixed in; this is done so that the salt may disintegrate the fine clay and silt in the ore mix, and thus permit the mercury to combine more readily with the silver. Next, mercury, contained in coarse linen sacks, is sprinkled over the ore on which it settles like dewdrops, and then the ore is stirred so that all of it comes in contact with the drops of mercury. . . .

When it is thought that the mercury has done its duty, that is, to join with the silver, whether the latter be much or little, and without leaving any behind, and soaking it up like a sponge does with water, incorporating it and separating it from the tailings and the lead and copper minerals with which the silver is associated; then attempts are made to obtain the silver by separating it from the mercury, which is done as follows: The ore mix is put into tubs or water, where with *molinetes*, or revolving reels that stir the water and the ore, as one might dilute or make mustard; the clay and fine sediments within the ore are expelled from the water, which runs out, and the silver and mercury amalgam, being heavy, sinks to the bottom of the tub. . . .

Since it is clean and glitters, because all of the fine silt has been expelled, the silver–mercury amalgam is put in a canvas container, which is strongly squeezed, ejecting some of the free mercury . . . leaving a *pella*, or lump of amalgam, one–sixth of its contents being silver, the rest, or five–sixths, mercury. . . .

And to separate the silver from the mercury, the amalgam is subjected to strong heat, being covered with a *cuperuzón*, or clay receptacle shaped like a conical mold used to make cakes of sugar; and this is covered with burning charcoal; by this process the mercury in the amalgam is vaporized, and on contact with the overlying conical covering it condenses and precipitates like steam on a covered pot; and by means of a cylindrical tube, as in a still, all of the distilled mercury is recovered, leaving a lump of pure silver, which in weight is 5 times less than the original lump of amalgam, and is crisp and spongy — which is something to see. From two of these *piñas*, or lumps of silver, is made a silver bar that weighs 65 or 66 marks, and in this form is taken for paying the fifth and for marking.

Endnotes

1. [Ed. note.] The Spanish words employed in this paragraph are colloquial Mexican terms not in vogue elsewhere.

2. [Ed. note.] The word *buitrón* is a term used in Alto Perú (Bolivia).

References

Acosta, José de. 1590. *Historia natural y moral de las Indias*. Sevilla: Impr. en casa de Juan de León.

Bargalló, M. 1955. *La minería y la metalurgia en la America española durante la época colonial*. Mexico City: Fondo de Cultura Económica.

—. 1969. *La amalgamción de los minerales de plata en Hispanoamérica colonial*. Mexico City: Cía. Fundidora de Fierro y Acero de Monterrey, S.A.

Gómez de Cervantes, G. 1969. *La vida económica y social de Nueva España al finalizar el siglo XVI*. Mexico City: Antigua Librería Robredo.

Menes Llaguno, J. M. 1989. *Bartolomé de Medina: Un sevillano pachuqueño*. Pachuca: Universidad Autónoma del Estado de Hidalgo.

Moreno de los Arcos, R. 1986. *Ensayos de historia de la ciencia y la tecnología*. Mexico City: Universidad Nacional Autónoma de México.

Othón de Mendizabal, M. 1980. *La minería y la metalurgia mexicana, 1520–1943*. Mexico City: Editorial CEHSMO.

Sánchez Flores, R. 1980. *Historia de la tecnología y de la invención en México*. Mexico City: Fomento Cultural Banamex.

Sánchez Gómez, J. 1989. *De minería, metalúrgica y comercio de metales. 1450–1610*. 2 vols. Salamanca: Instituto Tecnología Geominero de España.

The Territorial Organization of Metallic Mining in New Spain

Alvaro Sánchez-Crispín,
Universidad Nacional Autónoma de México, Mexico City

Abstract

The distribution of metallic mining in New Spain, from 1521 to 1810 shows that territorial occupation was based on mineral exploitation in the central and northern regions. Various cartographic sources showing the expansion of mining reveal the linkages to development of social and economic structures. For over four centuries, the same mines have been responsible for the largest part of the national ouput.
Key words: mining, metallic mining, New Spain.

Colonial Mining in Mexico and Historical Geography

The literature review of papers and books dealing with the topic of the territorial organization of colonial mining in New Spain reveals that, whereas at a national scale there is a lack of serious studies, some regions have been thoroughly examined. For instance, areas such as Pachuca–Real del Monte (Herrera 1989),the Sierra Gorda in Querétaro–Hidalgo (Langenscheidt 1988), Zacatecas (Bakewell 1976a and 1976b; Brading 1970), Santa Eulalia (Hadley 1979), and Hidalgo del Parral (Alatriste 1983; West 1949) have been studied in detail. At a national level, Duhau (1988) examines the relationship between the location of mine sites in New Spain and the consolidation of internal market places, which later became urban localities. However, this is only one of the many aspects of the territorial organization of mining in New Spain.

Regardless of the scale of analysis for colonial mining in New Spain, according to Coatsworth (1990) there are two major explanations in relation to the role of mining as an articulator of economic space. One inter-

---

In Quest of Mineral Wealth: Aboriginal and Colonial Mining and Metallurgy in Spanish America, edited by Alan K. Craig and Robert C. West, 1994. Geoscience and Man, vol. 33, pp. 155-170. Dept. of Geography & Anthropology, Louisiana State University, Baton Rouge, LA 70893-6010, U.S.A.

pretation argues that over–exploitation in the mineral sites distorted the economy of New Spain. The other explanation refers to the role of mining as the leading sector of the colonial economy; under this interpretation it is assumed that the backward linkages of the mining sector were responsible for the territorial organization of the economy in New Spain. For the purposes of this paper, we take for granted that mining created backward linkages that shaped particular economic regions of New Spain and strongly influenced many other areas within the colony. This is shown in the works by Herrera (1989) on Parral, Fresnillo, Sombrerete, and Guanajuato, and by Commons (1989a) in her study on the consolidation of agricultural and livestock–raising areas centered at San Bartolomé and Casas Grandes, which depended on the mineral sites of central Chihuahua. Therefore, we assume that mineral exploitation became the dominant economic activity during the colonial period. The development of mining promoted the creation of a strong internal market whose imprint is still seen today in Mexico (Duhau 1988).

The *National Atlas of Mexico* and the History of Mining in Mexico Through Its Maps

An important source of information on the territorial organization of mining in New Spain is the *National Atlas of Mexico*.[1] The maps contained therein dealing with the history of mining in Mexico form a valuable cartographic source to help understand how mineral exploitation became the dominant economic activity and the basis of the territorial organization of New Spain.

There are three volumes in the National Atlas of Mexico: Society, Nature and Economy. Four maps included in volumes one and three are related to the topic of this paper. Three of these maps are part of the volume on Society: two show the expansion of the Spanish interests in New Spain during the period 1521–1810 (Camelo and Fernández 1990); another map is concerned with the distribution of population in Mexico throughout the colonial period (Juárez et al. 1990). The fourth map belongs to the volume on Economy and contains information about the territorial organization of mining in Mexico, from pre–Hispanic times to the beginning of the twentieth century (Coll–Hurtado and Sánchez–Salazar 1989). Some of the information contained in these maps was used to prepare the present paper.

New Spain: Basic Features of the Territorial Organization of Metallic Mining

Prior to the Spanish Conquest, metallic mining in Mexico was mainly concerned with the exploitation of native metals such as silver and gold and, much less, with that of copper, tin and lead (Bernstein 1964). Gold was exploited in placers along streams, particularly in the lower basin of the river Balsas (Guerrero and Michoacán states) and in northern Oaxaca. When the Spaniards arrived in Mexico, silver deposits were already being exploited at Taxco and Zumpango. Non–metallic mining was related to the exploitation of salt on the Pacific coastal plains in southern Mexico. There was also some production of semi–precious stones, such as turquoise and opals (Ortiz et al. 1987).

After the arrival of the Spaniards, metallic mining, basically concerned with the exploitation of gold and silver, constituted the core of mineral production. A systematic exploitation of metallic deposits in New Spain started soon after the conquest of Mexico–Tenochtitlán in 1521. The Spanish Crown was eager to support the discovery and exploitation of metallic deposits, which at that time seemed to be endless (Semo 1985). The interest in gold exploitation, on the part of the Spaniards, gave way to a new dynamics in the economic space of New Spain; this period lasted until 1540 (Vázquez 1965). As early as the 1530s, silver production had become the dominant type of production, after the first major mine discoveries occurred in southern New Spain, especially in the Taxco area.

In the sixteenth century, the first bonanza period in New Spain was related, basically, to exploitations near the surface, in the oxidized layers of the ore bodies. Another factor that contributed significantly to the bonanza period was the method used for metal separation known as *beneficio de patio*, or the patio system, based on the amalgamation process. This method was useful to separate metals coming from medium–to–low metal content deposits and also permitted the exploitation of ore bodies where silver was associated with lead and zinc, a mixture commonly found in the mineral structures of central and northern Mexico (Bernstein 1964).

The variety of silver deposits found in Mexico demanded a number of engineering works for their exploitation. Some of the mineral deposits were superficial, such as those found in Durango, and needed relatively few works. Other mineral bodies were found down deep in the sulphide layers and were more difficult to exploit. The succession of bonanza periods and abandonment of the mines of New Spain is closely related to the nature of the ore bodies and to the mining technology available at that time.[2]

The metallic deposits of Mexico are mainly located in young mountains, folded and raised in the Tertiary geological period: the western Sierra Madre, the southern Sierra Madre, the transverse sierras of central Mexico (Durango–Puebla) and the Mexican Volcanic Belt (De la Peña 1959; González–Reyna 1944). The variety of genetic agents intervening in the mineralization of these geologic structures resulted in many types of silver ore bodies related to epithermal, mesothermal or hypothermal processes. Thus, Mexican silver is commonly associated with the non–ferrous metals. The only gold–silver ore bodies found in the country are located at Pachuca, Guanajuato, Batopilas, and Tayoltita.

From the sixteenth century onwards, metallic mining shaped the territorial structure of New Spain's economy. For the mining sector, the sixteenth century is a century of bonanza; however, it is in this period that two of the traditional problems of the mining sector became apparent: first, ore bodies were gradually found at deeper layers and, second, mining entrepreneurs had difficulties in attracting a labor force to the mines (Velasco et al. 1988). The first area of New Spain to receive the impact of the presence of mining was the so–called "Silver Province," southwest of Mexico City. In it the rich deposits of Taxco, Sultepec and Temascaltepec were exploited (Commons 1989b; Sánchez–Salazar 1990). When the Spaniards explored the western part of New Spain, different expeditions led to the discovery of silver ore bodies in the vicinity of Compostela (in what today is the State of Nayarit) and the finding of copper deposits in the Tarascan plateau (Michoacán). Guadalajara was also the departure point of various journeys whose aim was to explore the mining zones of northern Nayarit and of Sinaloa.

Not until the 1540s did mining spread into the arid and semi–arid regions of central and northern Mexico. The first mine site in that area was Zacatecas (founded 1546); then mineral exploitation started in Guanajuato (1548); and also Pachuca and Real del Monte, the mining towns nearest to Mexico City, started to produce precious metals. When mining was established at Zacatecas and Guanajuato, the northern limit of Mesoamerica was crossed, and for the first time systematic mineral exploitation took place in the arid regions of Mexico. This is important because the large majority of the mines of New Spain were discovered outside of Mesoamerica.

In this context, new settlements were founded in the arid and semi–arid regions of New Spain, particularly in the mining districts of Zacatecas, Durango and Chihuahua. The founding of these settlements is related to the exploitation of silver deposits nearby. At the same time, central New Spain reinforced its strategic position as the site of political power and the area across which metal was transported from the mines in the North to Veracruz, the port from which silver and gold were shipped to Spain (Duhau 1988). The settlement of the arid and semi–arid regions of northern and north–central New Spain was strategic and totally different from the one achieved in southern New Spain, Yucatan or Baja California (fig. 1). Thus, the sixteenth century in New Spain can be characterized as a period of frontier expansion: for the mining sector, for the agricultural lands, and for the political control of the new lands.

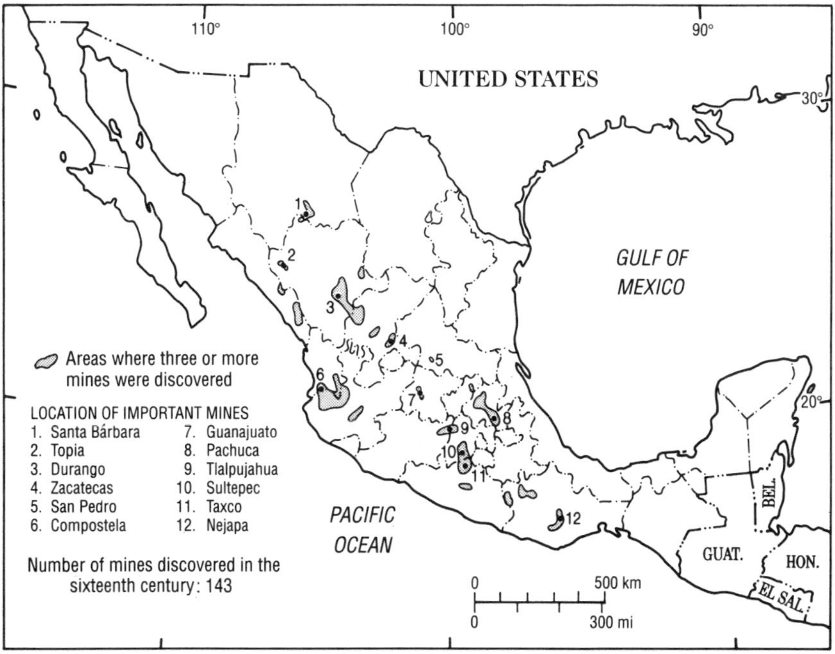

Fig. 1. New Spain: Discovery of mines in the sixteenth century. ***Source*****: Comelo and Fernández 1990.**

The communities founded in the arid and semi–arid regions of New Spain needed economic and military support, particularly because of climatic conditions. As mines were open in the new mining districts, known as *reales de minas*, other kinds of settlements appeared in the barren plateaus of central and northern New Spain: forts, *presidios* and centers dedicated to the diffusion of the Roman Catholic faith (Sánchez–Salazar 1990). The construction of roads, known as *caminos de arriería* (mule trails), that linked the distant mining areas with central Mexico, was fundamental to sustain metallic production throughout the sixteenth century. These roads facilitated migration to the mining settlements, particularly that of Indian groups from central New Spain (Tlaxcala, Michoacán, México) to the North. Salaries in the mining towns were relatively higher than the average in northern New Spain and also some sort of *gambusinaje*[3] was permitted, which attracted labor forces from distant places. Thus, there was a considerable flow of labor force to and within the mining regions of northern and north–central New Spain. The presence of roads linking these mining areas with the economic and political hub of the colony contributed, as well, to the development of interregional commerce, land transport, agriculture and livestock raising.

This is the setting for the opening of the "Silver Route," linking Mexico City with Guanajuato and Zacatecas; many communities were founded, as posts and forts, along this route. Examples of such places can be found in Querétaro and Guanajuato whose names reveal the reason of their founding. In this context, the city of Zacatecas was the base for the exploration of arid northern New Spain and also became the most important marketplace in the area. Different expeditions departing Zacatecas reached the northernmost settlement of New Spain at that time: Santa Bárbara (Sánchez–Salazar 1990).

Thus, it is possible to argue that the mining towns, particularly those located in the northern and central areas of the colony, were the basis of the articulation of the economic space of New Spain. The settlement of northern and central New Spain resulted in a diffused pattern of occupation, where empty spaces were combined with agricultural lands (haciendas), ranches, *presidios*, and missions, and some points where economic activity was more dynamic: the mining communities (Sánchez–Salazar 1990). Those points were islands of incipient economic activity, within the limits of the "Silver Frontier". Therefore, it is possible to say that in the sixteenth century the basics of the territorial organization of metallic mining in New Spain (and in Mexico, indeed) were established.

A steady decline in the mining sector was experienced in New Spain throughout the seventeenth century. This situation resulted in the need for land exploration in search of new mines. Consequently, mining districts such as Parral, Urique, Batopilas and Cushuiriachic were opened in Chihuahua (fig. 2). The total occupation of this part of New Spain was completed when the *real de minas* of Santa Eulalia was founded. This *real de minas* preceded the founding of Villa de San Felipe el Real, which later would be known as the city of Chihuahua, the most important town of northern New Spain in this period (Sánchez–Salazar 1990). Meanwhile, in the Northwest, the first mines were established in Sonora, especially the *real* of Alamos, and that of Rosario in Sinaloa.

In the eighteenth century, due to external factors, the Spanish Crown supported and encouraged mineral production in New Spain. The mining sector entered a new era: new mines were opened and the territory explored in search of metals. Therefore, the occupation of the rugged and barren lands of the Northwest and North was completed and the number of mining districts grew, particularly in areas of the western Sierra Madre in Sonora (tables 1 and 2; fig. 3). Nearly 70 percent of the mines in Sonora being exploited at the end of the eighteenth century were discovered precisely in that century. The remoteness and difficult location in the rugged terrain in the Sierra Madre, made most of these ore deposits inaccessible.

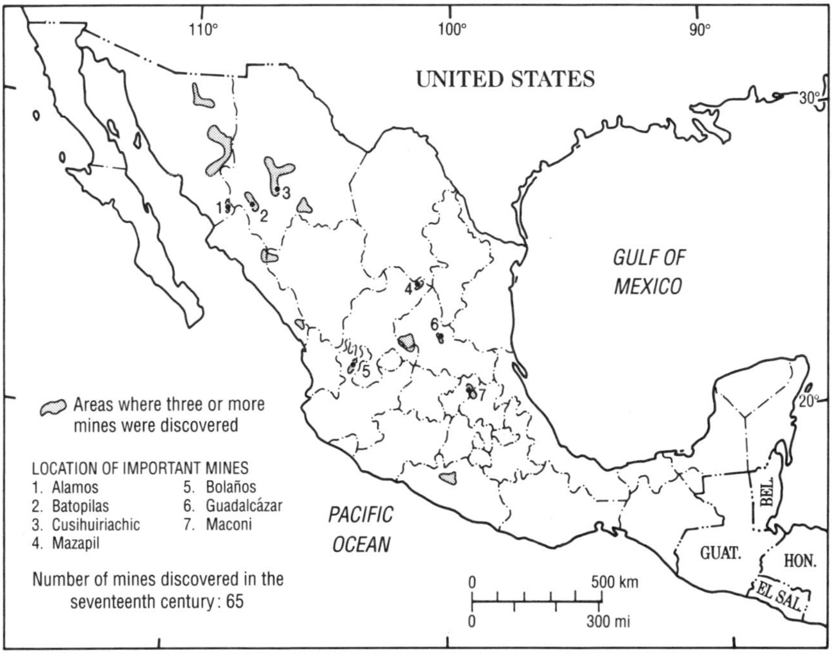

Fig. 2. New Spain: Discovery of mines in the seventeenth century. ***Source:*** **Comelo and Fernández 1989.**

Consequently, these mining districts were the last to come into existence in New Spain. Between the mines in the Northwest and Mexico City there is a distance of over 2,000 kilometers. The presence of a greater number of active mines and the favorable policy of the Spanish Crown regarding mineral exploitation resulted in an increasingly larger metal production between 1720 and 1810. As a consequence, silver production coming from the mines of New Spain amounted to over one–half of the total world silver output in this period.

The most common technique used in New Spain to separate silver from other metals was the method of amalgamation. However, in the eighteenth century smelting continued as a suitable technique to recover silver from gangue materials. The exploration of many areas of northern New Spain (Sinaloa, Sonora and Chihuahua) led to the discovery of a large number of mining districts where ore bodies had a high metal content, which in turn resulted in bonanza times for these regions (Velasco et al. 1988). As pointed out by Bernstein (1964), up to the beginning of the eighteenth century there were no important foundries in New Spain. However, in that century the first of these plants were constructed at Sombrerete, the city of Zacatecas and Santa Eulalia, in Chihuahua.

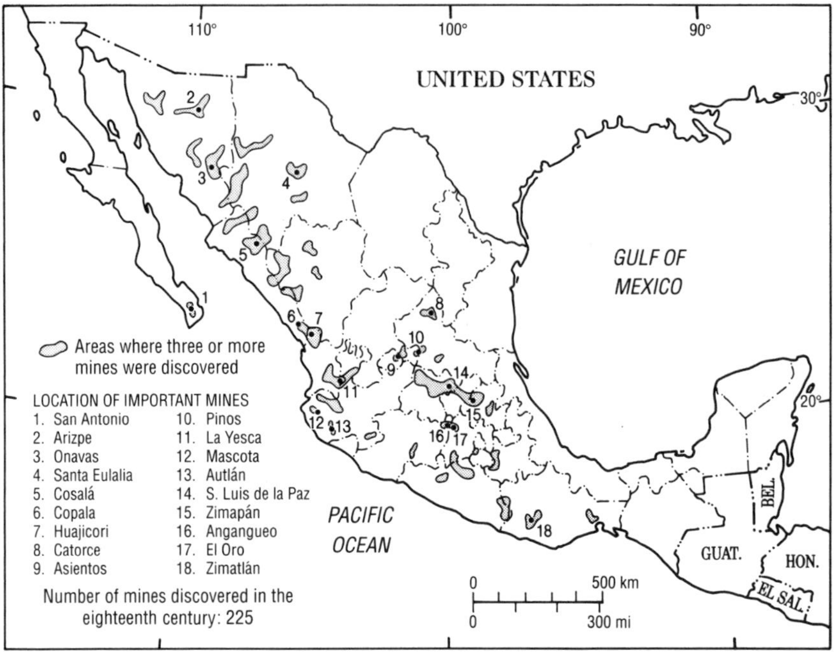

Fig. 3. New Spain: Discovery of mines in the eighteenth century. ***Source*****: Comelo and Fernández 1989.**

According to Poinsett (1973), at the beginning of the last century, the largest mines of New Spain, with relation to their volume of production, were: Guanajuato, Catorce, Zacatecas, Real del Monte, Bolaños, Guarisamey, Sombrerete, Taxco, Batopilas, Zimapán, Fresnillo, Ramos and Parral. Therefore, it is safe to say that metallic mining in New Spain, at this time, was heavily concentrated in a small number of places. For example, the first three mines (Guanajuato, Catorce and Zacatecas) produced over half of the silver output in New Spain at that time.

The Distribution of Population in the Mining Areas

The pattern of distribution of population in the early years of the colonial period (1521–1640) was characterized by a dispersed occupation of the northern and north–central areas of New Spain. Consequently, the provinces containing the largest mines were practically uninhabited as in the case of Chihuahua, Coahuila, Durango and Zacatecas (Juárez et al. 1990). In this period, central and southern New Spain were densely populated areas; within them the key mining districts were located in northern Michoacán, northern Guerrero, northern Oaxaca and southern Hidalgo.

Table 1. New Spain: Number of mines discovered in the sixteenth, seventeenth and eighteenth centuries, by state

	16th Cent.	17th Cent.	18th Cent.	Total
Aguascalientes	—	—	2	2
Baja California Sur	—	1	2	3
Coahuila	2	1	1	4
Chihuahua	3	13	39	55
Durango	17	6	19	42
Guanajuato	4	1	15	20
Guerrero	13	5	10	28
Hidalgo	10	1	9	20
Jalisco	28	4	17	49
México	6	0	4	10
Michoacán	5	3	6	14
Morelos	—	—	3	3
Nayarit	8	1	11	20
Nuevo León	3	1	2	6
Oaxaca	13	0	12	25
Puebla	2	1	5	8
Querétaro	2	1	3	6
San Luis Potosí	3	3	7	13
Sinaloa	12	4	19	35
Sonora	—	13	31	44
Zacatecas	12	6	8	26
Total	143	65	225	433

Source: Camelo and Fernández 1990.

Table 2. First mine sites of New Spain, sixteenth to eighteenth centuries, by state

State	Century	First mine site
Aguascalientes	1700s:	Asientos-Tepezalá
Baja California S.	1600s:	North of Santa Rosalía
	1700s:	Todos Santos-Santiago, 1748
Coahuila	1500s:	Monclova, 1578
	1600s:	Torreón vicinity
	1700s:	Cuatro Ciénegas
Chihuahua	1500s:	Santa Bárbara, 1567
	1600s:	Cusihuiriachi and Batopilas
	1700s:	West Sierra Madre
Durango	1500s:	Topia, 1563 and Durango
	1600s:	Canatlán and Cuencamé
	1700s:	West Sierra Madre
Guanajuato	1500s:	Guanajuto, 1557
		San Luis de la Paz, 1590
	1600s:	Dolores Hidalgo
	1700s:	North Guanajuato
Guerrero	1550s:	Taxco, 1532
	1600s:	Ajuchitlán-Totalapan
	1700s:	Mixteca-North Guerrero
Hidalgo	1500s:	Pachuca, 1552
	1600s:	Zimapán
	1700s:	Zimapán area
Jalisco	1500s:	South Sierra Madre
	1600s:	Bolaños
	1700s:	Mascota-Autlán
México	1500s:	Sultepec, 1531
	1700s:	El Oro
Michoacán	1500s:	Tlalpujahua, 1522
	1600s:	Coalcomán
	1700s:	Anguangueo and Ario
Morelos	1700s:	Northwest edge

(Table 2 cont'd.)

Nayarit	1500s:	Compostela, 1531
	1600s:	Ahuacatlán
	1700s:	Huajicori and La Yesca
Nuevo León	1700s:	Cerralvo and Cadereyta
Oaxaca	1550s:	Nejapa, 1550 and Totontepec
	1700s:	Mixteca-South Sierra Madre
Puebla	1700s:	West Puebla
Querétaro	1600s:	Maconí
	1700s:	Sierra of Japlan
San Luis Potosí	1500s:	San Luis Potosí, 1592
	1600s:	Ramos, 1608
		Guadalcázar, 1616
	1700s:	Matehuala
Sinaloa	1500s:	South Sinaloa
	1600s:	Rosario
	1700s:	Cosalá, Copala, Northwest Sinaloa
Sonora	1600s:	Ostimuri (San Miguel), 1666
		Alamos and Arizpe
	1700s:	West Sierra Madre
		Onavas, Arizpe, Cananea
Zacatecas	1500s:	Zacatecas, 1546
	1600s:	Concepción del Oro, Pinos
	1700s:	Pinos and Noria de Angeles

Sources: Camelo and Ferández 1990; Chance 1989; Commons 1989b; Langenscheidt 1988; Sánchez-Salazar 1990.

This pattern of population distribution changed in the seventeenth and eighteenth centuries, especially due to the discovery of the mining districts in the far North and in the Northwest. There is a close relationship between the discovery of these mines and the settlement of northern Mexico. The largest mining towns of New Spain in this period were: Pachuca, Guanajuato, Zacatecas, Durango and Chihuahua, all of which had over 50 thousand inhabitants. Other mining towns, whose population ranked between 15 and 50 thousand inhabitants, included Huajicori in Nayarit, the city of San Luis Potosí and Monclova in Coahuila (Juárez et al. 1990). Like other communities making a living from agriculture, forestry or commerce, the majority of the mining communities of New Spain were small, with less than five thousand inhabitants.

The Mining Areas and Their Economic Articulation with Other Regions of New Spain

The main economic linkages generated by the mining sector of New Spain were related to the provision of foodstuffs, livestock, forest products, and salt. In this context, the following linkages are worth mentioning:

(1) On one hand, the purchase of *ixtle* (a hard fibre) and hides necessary for the production of bags for minerals and, on the other, the provision of fats, wax and wick for the preparation of candles used to illuminate the interior of the mines. *Ixtle* came from the arid regions of northern and central New Spain and the hides and fats were bought at ranches located near the mines. Forest products were also important for the economic activity of the mines; so important was wood that the exploitation of former wooded areas next to the mines of Guanajuato, Zacatecas, Charcas, Fresnillo and Sombrerete brought about the depletion of the forest. Today, areas near those mines, at an altitude over 2,000 meters, are covered by shrubs and xerophytic plants or worse, with little vegetation at all. In another context, Puebla and Querétaro were textile towns which had important links with the mining communities of central and northern New Spain.

(2) Many interpretations have been given to the role of mining as the promoter of the development of other economic activities in New Spain, particularly in relation to the agricultural sector (see, for instance, Baroni 1990; Coatsworth 1990; Herrera 1989). Some authors argue that the mines of New Spain established backward linkages with the agricultural regions located within the hinterland of the mining community; this is clearly shown in the case of the city of Zacatecas (the mining town) and Juchipila (the agricultural community) or in the case of Sombrerete–Poanas and Fresnillo–Río Grande (Sánchez–Salazar 1990).

At the same time, agricultural regions such as El Bajío and others located in Michoacán and Jalisco, not far from the mines of Guanajuato and Zacatecas, were open due to the presence of mining in the latter areas. Livestock raising was another sector that benefitted from the presence of the mining communities, especially in the case of cattle and sheep ranches. In this manner, livestock raising spread over the arid and semi–arid regions of central and northern New Spain, where climatic conditions and the type of natural vegetation favored the development of cattle ranching. So important was cattle and sheep raising in these ranches that even southern New Spain benefitted from their production.

(3) It is difficult to demonstrate that there was a correlation between salt production in New Spain and the presence of metallic mines (Ewald 1985). However, since most of them used the amalgamation process for the separation of silver, salt was a basic element to keep production in the mines steady. In this respect, salt was produced at specific places whose umland–hinterland was clearly defined. In the eighteenth century the coastal plains of Colima provided salt to the nearby mining districts of Angangueo (Michoacán) and Autlán (Jalisco). The salt pit of Iztapalapa was important for the mineral production of Pachuca–Real del Monte and Zimapán. The salt deposits at Peñón Blanco (San Luis Potosí) were second to none due to their favorable central location which allowed easy communications with the rest of New Spain. Consequently, salt from Peñón Blanco was sent to the major mines in north–central and northern New Spain. Thus, the hinterland of Peñón Blanco included areas as far away as Avino (Durango) in the northwest, Bolaños (Jalisco) in the west, Charcas (San Luis Potosí) and Fresnillo–Zacatecas (Zacatecas) in the north, and Guanajuato to the south (Ewald 1985).

Final Remarks

This paper has dealt with the territorial organization of metallic mining in New Spain. The pattern of territorial occupation was based on the presence of mineral exploitation in the central and northern regions of New Spain, which generated many social and economic structures. After more than four centuries of existence, the old mining towns and regions of New Spain continue as key places for mineral production in contemporary Mexico.

The territorial organization of mining in New Spain had three basic aspects: first, a pattern of geographical dispersion, since the mining towns were widely separated and far from Mexico City. Only in the case of Taxco–Sultepec–Temascaltepec and of Zacatecas–Guanajuato–San Luis Potosí would it be safe to speak of mining "regions," because of relative geographical proximity of the centers. Second, mineral production was concentrated in a handful of districts; so strong was this concentration that at certain times almost two–fifths of the colony's total silver production came from the Zacatecas mines. Finally, two axes of economic linkage came into existence in New Spain, one associated the mineral districts of the North and Northwest with Mexico City, the other connected the latter with the mines in the South (Taxco, northern Oaxaca).

For over four hundred years the same mines and the same mining towns of Mexico have been responsible for the largest part of the national mineral output. It is due to the production of these mines that Mexico is an outstanding mineral producer in the international scene. The contemporary territorial organization of metallic mining in Mexico is almost identical to that that prevailed during the colonial period. The first imprint given to the space by the mining sector, some four hundred years ago, is still present and valid.

Endnotes

1. In 1985, the Institute of Geography decided to undertake a new research project whose final product was the National Atlas of Mexico. Its complete version was published at the beginning of 1992. Over 200 specialists from all fields and a variety of institutions collaborated in the making of nearly 600 maps, covering a wide range of topics, from the geomorphology of the country to the international relationships of Mexico.

2. Bonanzas in the major mines of New Spain were experienced as follows:

Sixteenth Century: Taxco, Zacatecas and Real del Monte

Seventeenth Century: Zacatecas, Sombrerete, Parral

Eighteenth Century: Taxco, Guanajuato, Bolaños, Catorce, Sombrerete, Parral, Santa Eulalia, Pachuca–Real del Monte

In the eighteenth century, a larger number of bonanzas occurred in the major old mines of New Spain, perhaps as a result of the policies of the Spanish Crown aiming to increase mineral production in the colony at that time. Such was the amount of silver obtained in the bonanzas of the eighteenth century that La Valenciana mine, in Guanajuato, produced over one–fifth of the total silver output of New Spain at the end of the colonial period (Sánchez–Salazar 1990).

3. *Gambusinaje* is an old form of mineral exploitation in Mexico which is based on the effort of independent individuals, working as a group or as one producer, whose techniques for the separation of metals are rather traditional. At present, *gambusinaje* in Mexico is part of the informal sector of the mining economy.

References

Alatriste, O. 1983. *Desarrollo de la industria y la comunidad minera de Hidalgo del Parral durante la segunda mitad del siglo XVIII (1765–1910).* Thesis, Facultad de Filosofía y Letras, Universidad Nacional Autónoma de México. Mexico City.

Bakewell, P. 1976a. *Minería y sociedad en el México colonial: Zacatecas (1546–1700).* Mexico City: Fondo de Cultura Económica.

——. 1976b. Zacatecas: An economic and social outline of a silver mining district, 1547–1700. In *Provinces of early Mexico*, eds., Altman and Lockhart. Los Angeles: University of California at Los Angeles.

Baroni, A. 1990. *La formación de la estructura agraria en el Bajío colonial, siglos XVI y XVII*. Mexico City: CIESAS–Cuadernos de la Casa Chata.

Bernstein, M. D. 1964. *The Mexican mining industry, 1890–1950*. Albany: State University of New York.

Brading, D. 1970. Mexican silver mining in the eighteenth century: The revival of Zacatecas. *Hispanic American Historical Review* 50:665–681.

Cameo, R. and Fernández–Aguila, A. 1990. Nueva España. Siglo XVI y Nueva España. Siglos XVII y XVIII. *Atlas Nacional de Mexíco (Sección Sociedad)*. Hojas II.2.1 y II.2.2. (Escala 1:4,000,000. Mexico City: Instituto de Geografía, Universidad Nacional Autónoma de México.

Chance, J. 1989. *Conquest of the Sierra: Spaniards and Indians in colonial Oaxaca*. Norman, Okla.: University of Oklahoma Press.

Coatsworth, J. 1990. *Los orígenes del atraso: Nueve ensayos de historia económica de México en los siglos XVlll y XlX*. Mexico City: Alianza Editorial Mexicana.

Coll–Hurtado, A., and Sánchez–Salazar, M. T. 1989. Minería historíca. *Atlas Nacional de México (Seccion Economía)*. Hoja VI.8.2. (Escala 1:8,000,000). Mexico City: Instituto de Geografía, Universidad Nacional Autónoma de México.

Commons, A. 1989a. La minería en Nueva Espana en el siglo XVIII. *Boletín del Instituto de Geografía* 19:89–103.

——. 1989b. Principales zonas mineras en la segunda mitad del siglo XVlll. *Boletín del Instituto de Geografía* 20:105–120.

De la Peña, L. 1959. Recursos naturales no renovables de México. *Minería y Metalurgia* 11:13–119.

Duhau, E. 1988. *Mercado interno y urbanización en el México colonial*. Mexico City: Universidad Autónoma Metropolitana.

Ewald, U. 1985. *The Mexican salt industry, 1560–1980: A study in change*. Stuttgart: Gustav Fischer Verlag.

Gerhard, P. 1988. *Geografía Histórica de la Nueva España*. Mexico City: Universidad Nacional Autónoma de México.

González–Reyna, J. 1944. *Minería y riqueza minera de México*. Mexico City: Monografías Industriales, Banco de México.

Hadley, P. 1979. *Minería y sociedad en el centro minero de Santa Eulalia, Chihuahua (1709–1750)*. Mexico City: Fondo de Cultura Económica.

Herrera, I. 1989. Empresa minera y región en México: La Compañia de Minas de Real del Monte y Pachuca (1824–1906). *Siglo XlX. Revista de Historia* 4:103–123. (UniversidadAutónoma de Nuevo León, Monterrey, México).

Juárez, M. C., et al. 1990. Distribución histórica de la población en México. *Atlas Nacional de México (Sección Sociedad)*. Hoja III.1.1. (Escala 1:8,000,000). Mexico City: Instituto de Geografía, Universidad Nacional Autónoma de México.

Langenscheidt, A. 1988. *Historia mínima de la minería en la Sierra Gorda*. Windsor (Canada) and Mexico City: Rolston–Bain.

Ortiz, F. et al. 1987. *Tierra profanada: Historia ambiental de México*. Mexico City: Instituto Nacional de Antropología e Historia y Secretaría de Desarrollo Urbano y Ecología.

Poinsett, J. 1973. *Notas sobre México [1822]*. Mexico City: Editorial Jus.

Sánchez–Salazar, M. T. 1990. *Análisis de la organización territorial de la actividad minera en México*. Doctoral thesis, Facultad de Filosofía y Letras, Universidad Nacional Autónoma de México, Mexico City.

Semo, E. 1985. *Historia del capitalismo en México: Los orígenes, 1521–1763* (13th ed.). Mexico City: Era.

Vázquez, E. 1965. *Distribución geográfica y organización de las órdenes religiosas en la Nueva España (Siglo XVI)*. Mexico City: Instituto de Geografía, Universidad Nacional Autónoma de México.

Velasco, C., et al. 1988. *Estado y minería en México (1767–1910)*. Mexico City: Fondo de Cultura Económica.

West, R. 1949. *The mining community of northern New Spain: The Parral mining district*. Iberoamericana, 30. Berkeley and Los Angeles: University of California Press.

Agricultural Colonization and Mining Colonization: The Area of Chihuahua During the First Half of the Eighteenth Century

Salvador Alvarez, Universidad Autónoma de Ciudad Juárez, Cd. Juárez, Chihuahua, Mexico

Abstract

The traditional historiography regarding the development of mining in colonial Spanish America considers the search for metals as the instrument for occupation of territory, especially in frontier areas. This, however, is an oversimplified viewpoint which only describes the short term results of mining bonanzas without analyzing the origins of these opulent periods or the economic organization within the mining districts. An analysis of the process of populating the mining and agricultural district of Chihuahua indicates how the slow development of farming and the organization of the supply network forms a necessary antecedent for mining exploitation on a large scale. In Chihuahua, where mines had been discovered since the mid-seventeenth century, more than 70 years of previous agricultural colonization were necessary to make possible a mining boom during the first third of the eighteenth century. The agricultural haciendas, together with the reduction, or resettlement, of the Indians, led to the Spanish occupation of the frontier: mining was a factor that accelerated the process, but was not its cause. In Chihuahua, during the eighteenth century, most of the population was concentrated in the haciendas, while the urban sector — the town of San Felipe el Real — became a place essentially reserved for the powerful elite. Thanks to their access to essential resources, the big hacienda owners controlled both the mining and refining of ores, and monopolized a good part of the area's regional commerce.
Key words: mining, agriculture, Chihuahua, colonization, Indians.

Resumen

La historiografía tradicional sobre cuestiones mineras en América Española colonial hace de la búsqueda de metales el motor de la ocupación del territorio, en especial en regiones de frontera. Sin embargo, esta es una versión demasiado sumeria de la cuestión que sólo describe las consecuencias a mediano plazo de las grandes bonanzas mineras, sin analizar ni los

In Quest of Mineral Wealth: Aboriginal and Colonial Mining and Metallurgy in Spanish America, edited by Alan K. Craig and Robert C. West, 1994. Geoscience and Man, vol. 33, pp. 171-204. Dept. of Geography & Anthropology, Louisiana State University, Baton Rouge, LA 70893-6010, U.S.A.

orígines de esos auges, ni la organización económica de los distritos mineros. Un análisis del proceso de poblamiento del distrito minero y agrícola de Chihuahua, muestra como una lenta colonización agrícola y la organización de redes de abasto son el antecedente necesario para la explotación minera a gran escala. En Chihuahua, donde las minas eran conocidas desde mediados del siglo XVII, más de 70 años de colonización agrícola fueron necesarios para hacer posible la bonanza minera durante el primer tercio del siglo XVIII. Fueron las haciendas agrícolas junto con las reducciones de indios el motor de la ocupación del territorio: la minería fue un factor que aceleró el proceso mas no fue su causa. En la Chihuahua durante del siglo XVIII eran las haciendas que reunían la mayor parte de la población, mientras que la mancha urbana de la villa de San Felipe el Real se convirtió en espacio reservado esencialmente para los poderosos. Gracias a su acceso a los recursos esenciales, los grandes hacendados dominaban tanto la explotación como la refinación de los minerales, e incluso acaparaban buena parte del comercio regional.
Palabras claves: minería, agriculture, Chihuahua, colonización, indios.

Introduction

The emergence of the mining and agricultural region of Chihuahua at the beginning of the eighteenth century was one of the most important events in the history of Spanish colonization of northern New Spain. The development of the area was a symbol of the maturity of the Latin American colonial world; the area was likewise the seat of the last urban agglomeration formed during Spanish imperialism in America: the town of San Felipe el Real de Chihuahua. Founded in 1718, San Felipe, in less than two decades, came to be the center of one of the most populous areas of northern New Spain at that time, with a population of 25,000 inhabitants and a mining district of great fame (Cramaussel and Alvarez, 1992). In fact, the history of the Chihuahua region of the eighteenth century was intimately linked to its mineral wealth. The event that marked the beginning of the mining boom was the establishment of the *reales* of Santa Eulalia de Mérida and San Francisco de Cuéllar, in 1707 and 1709 respectively. Nonetheless, as it frequently happens in time, behind the brilliant mining successes was hidden a much slower and complex history: in that frontier zone when the first mines were opened, agricultural settlement was already approximately 70 years old.

Agricultural Haciendas and Missions: The Origin of the Chihuahua Mining Region

Spanish and Indian Relations

Frequently in the literature on mining history, the idea is presented that the discovery of a rich ore vein was the key event that led to the founding of a new *real de minas*. According to this oversimplified image, once the newly found deposit was made known, large numbers of prospectors and fortune seekers rushed to the discovery to settle there and develop all kinds of activities. Entire regions, for long inhospitable, unpopulated and inaccessible, far from large centers of pre-Hispanic civilization, were overnight converted to emporiums of stable economic activity and, above all, to permanent settlements. Peter J. Bakewell (1990, 51) sums up this process thusly:

> The search for ore deposits carried the Spaniards far and wide across the Americas, contributing much to the amazing rapidity with which they explored and settled their portion of the continent....As the rich districts begin to disgorge ores, towns sprang up in various inhospitable regions — coastal New Granada, highland Charcas and the north Mexican plateau, for example — where only sparse and primitive populations had lived before. Roads and commerce spread rapidly as new economic circuits, energized by mining, developed.

According to this interpretation, the discovery of mines was antecedent to settlement of territory. Settlers came in as the production of silver increased; the economy was organized and the landscape transformed. The great semi-arid plateaus of northern New Spain offered a perfect scenario for this kind of mining colonization. First Zacatecas, then Parral, and later Chihuahua, have been described as classic examples of this process:

> Following the route opened by the silver mines, the officer-administrator, the missionaries and the farmers made their entrance....Along all of the "silver belt," which leads from Zacatecas to Chihuahua, the same phenomenon is repeated: after the discovery and exploitation of the mines, it was necessary to feed the "newly arrived immigrants" who settled the *reales* and to supply the essential animals that drove the ore refineries...soon farming centers and stock ranches multiplied in the nearby valleys and along streams. At the end of the colonial period the sad, semi-arid North had been radically changed by the colonizing efforts induced by the mines...(Florescano 1973, 48, 56-57).

Nevertheless, the history of the mining centers of New Spain indicates that the above is a model to which numerous shades of meaning must be added. The colonial society of New Spain was such that available human labor could never undergo displacements of population comparable in size (neither in numbers nor in distances) that were suffered, for example, by Indians in Potosí. The bonanza which that large mountain of silver experienced could never have been possible without the prodigious organization of the Peruvian *mita*: in no mining district of New Spain was there anything similar.

We are dealing here with a discourse on a process of colonization that emphasizes much more than the impulse which, under certain conditions, mining had directly on the occupation of a territory, and much less the process of settlement itself. Proponents of the former thesis thus favor an almost automatic sequence of events: the discovery of a productive ore deposit results in a great mobilization of human and natural resources, and this in turn leads to the formation of a new mining settlement (Bakewell 1976, 31, 305). However, as we have demonstrated previously in the cases of Zacatecas and Parral, developments were never as simple as that (Alvarez 1989). Chihuahua is one more example which shows us that to discover an ore deposit is not the same as to populate.

The region that came to be called Chihuahua had already been well known by Spaniards since the early seventeenth century. At that time Chihuahua was within the dominion of the *encomenderos* of the province of Santa Bárbara, who frequently entered in search of Indians who could be forcibly transferred as workers on the haciendas in the Valley of San Bartolomé (Cramaussel 1991).

Moreover, the *camino real* to New Mexico crossed this region. Already by 1619, for example, there is mention for the first time of a locality called Tabaloapa (Naylor and Palzer, 272-293), only some 20 kilometers from where, a century later, the mining center of Santa Eulalia was founded. Although the place was attractive, for decades it was no more than a rest stop with trees and good spring water on the long road to New Mexico. Among the *encomenderos* in the Valley of San Bartolomé, many of whom had taken their Indian workers from that area, there were experienced individuals for whom it was difficult to disregard the favorable aspects of the place. It was precisely they who made the first attempts to colonize Chihuahua.

In 1646, a wealthy hacendado of Parral, named Ambrosio Sáenz de Chávez, was the first to denounce claims to land and mines in the Chihuahua region, at a place that he found 40 leagues from Parral, along the New Mexico road, on the Sacramento River (Almada 1984, 10). Since 1631, Sáenz de Chávez also had been a miner in Parral, where he possessed a refinery and charcoal kilns; later, he was also a miner in San Diego, nearby (AHP, ramo Minas, Solares y Tierras, 1631, 1641). It is not known if Sáenz de Chávez ever worked his mines in Chihuahua, but his claim could not have been ignored by experienced miners in Parral. During the following years the region continued to be explored. In 1652 another *encomendero* of San Bartolomé, Pedro del Castillo, denounced a new mining claim in the "Discovery of Chihuahua" and he also included the spring of Tabaloapa as well as some parcels of land. Since that had been an official expedition in which other miners from Parral participated (whose names are not extant), del Castillo was named chief magistrate (*justicia mayor*) of this new jurisdiction (AHP, 1652, Nueva Sección; Almada 1984, 11-15). Clearly, during this time the region was being thoroughly explored; in the same year Diego Rodríguez de Amaya denounced mines in Bachimba, where ore deposits had been found (Almada 1984, 14). Possibly, this discovery may have been in the same area later called "Old Chihuahua," but there is insufficient documentation to affirm that (Cramaussel and Alvarez, 1992). In any case, it is important to emphasize that mines in Chihuahua had already been found, explored and claims denounced more than a half century before the opening of Santa Eulalia.

Although the initial discoveries were considered important, and other miners of Parral made repeated attempts to settle and exploit ore deposits in the area, an avalanche of settlers never arrived. Because of almost continuous warfare with local Indians, the miners could only extract a little ore and return to Parral before being surprised by a party of wild nomads. Filing away the papers that confirmed their new mining claims, the del Castillo family decided it was more prudent to continue operations in Parral. Diego del Castillo had been a miner there at least since 1645, and we know that after his death his brother Pedro continued to exploit mines in nearby San José at least until 1671 (AHP 1641, 1643; AGI, Contaduría 926, 927).[1] Nothing changed in Chihuahua with the discovery of ore deposits there; during the next two decades the region remained under the exclusive domination of the *encomenderos* of San Bartolomé, who continued to fight the Indians of Chihuahua, forcing captives to work on the haciendas farther south, a process that had been going on for a half century.

A Period of Slow Settlement

The antithesis that we have presented regarding the discovery of ore deposits and their actual exploitation is much more than a simple peculiarity of the mining centers in northern New Spain: it is a problem associated with the very nature of mining settlements in general. Most of the writings on the subject to date have stressed that the impetus which mining activity exerts on colonization of an area is essentially different from that which is developed through agricultural settlement. However, no kind of mining activity was possible without the assurance of a constant and stable supply of imports, most of which came from agricultural activity. Another necessary condition was the availability of human resources. The latter condition, that of the origin of settlers of the large mining centers of New Spain, until now has been little considered in the historiography of the subject, despite its importance in the study of mining economy.

The province of Santa Bárbara, whence came the discoverers of the first mines of Chihuahua, was one of the most prosperous regions of northern New Spain. Beside the *real* of Parral, the province contained the *reales* of San Diego, San Francisco del Oro, and Santa Bárbara; the agricultural town of San Bartolomé (made up of Indians and Spaniards); as well as the Jesuit and Franciscan missions along the San Pedro, Conchos and Florida rivers, without counting the numerous haciendas present throughout the region. Its total population can be estimated as more than 20,000 inhabitants, making it the most densely settled area of Nueva Vizcaya. The 40 leagues that del Castillo claimed to have traversed from Parral to Chihuahua did not comprise a long distance; much of the Indian labor used in the mines and on the farms came from areas even more distant, such as Sinaloa, Sonora and New Mexico. Chihuahua proper, as indicated previously, was one of the areas from which the hacendados obtained Indian labor. But as the crow flies, the 40 leagues was a distance that could be traversed in a few days. From a practical viewpoint, that was hostile territory that had remained obstinate to Spanish penetration. Despite the great mining bonanza that had occurred in Parral since the mid-seventeenth century, no one at that time would venture to exploit the mines of Chihuahua, however rich they might have been: the problem was not "economic motivation," but nature of human resources.

From this viewpoint, an examination of the history of the founding of large mining districts in New Spain suggests that it is possible to invert this model: it was not the mines that governed the rhythm of settlement, but, on the contrary, it was settlement that determined the rate of mining development. To understand the conditions that gave rise to mining districts in Chihuahua, it is necessary to place the area within the historical context of northern settlement, which can be divided into various phases, each with well defined characteristics. One might say that the first phase involved the rapid expansion during the sixteenth century, a period of extensive exploration and important population movements, which in the case of northern New Spain occurred from 1550 to 1570. During that time Spaniards founded Zacatecas, Durango, Culiacán, and Chiametla, among other outstanding agricultural and mining centers (Alvarez 1989, 1992). A new phase occurred between 1570 and the 1630s. Although the rapid advance of the sixteenth century had abated, a period of intense colonization began in the northern part of Nueva Vizcaya, which culminated in the founding of the *real* of San José de Parral in 1631. The colonization of northern New Spain was slow. In the extreme north, in the territory of the Pueblo Indians, the colony of New Mexico was formed — a fragile Spanish enclave which temporarily ended with the Pueblo rebellion of 1680. Meanwhile, farther south, population growth prospered and occupation of the land stabilized. The province of Santa Bárbara and surrounding area enjoyed a period of growth in all phases of development. The mining bonanza of Parral made this center a serious competitor of Zacatecas, while the agricultural haciendas transformed the Valley of San Bartolomé into the most important granary in the north: during this period its local population also grew at an increasing rate. But neither the mining bonanza nor the population or agricultural growth resulted in occupation of new land. From the 1630s and for a half century thereafter, the Conchos River marked the virtual limit of colonization. No Spanish settlement of any kind was able to prosper north of this river basin. Aside from hunters of Indian slaves and *encomenderos*, no one dared stay for long in that area.

For example, in 1648 the town of Aguilar was founded on the Papigochi River, an enclave of Spaniards within the dominion of Jesuit missions among the Tarahumar Indians. This settlement lasted only a short time, for two years later a violent uprising of the Tarahumar left it in ruins (Naylor and Polzer, 1986, 394). Even during this era of rebellion Spaniards continued trying to establish themselves in that area, and the first attempts at colonizing the region of Chihuahua date from this period. The presence of Spaniards was not accidental; besides ore deposits the area contained a relatively dense Indian population; in a radius of some 50 kilometers the Spaniards encountered a series of well populated Indian towns and hamlets, and together with the first miners Franciscan missionaries arrived. Although located within the domain of the Conchos Indians in the upland plains well away from the Sierra Madre, all indications were that the Chuvíscar River basin was home to both Tarahumar and Conchos who lived side by side, the former being the most numerous. At the site of Chihuahua, and near the place where the first mines were exploited, two hamlets of Indians were encountered: Tabaloapa and Chuvíscar, both identified in 1653 as Tarahumar settlements. To the west there were three other Tarahumar pueblos; Santa Isabel (present-day General Trías), San Bernabé (near Cusihuiriachi) and San Andrés (Riva Palacio). About the same distance away to the southeast were Bachimba, a Conchos hamlet (where mines were also discovered in 1652) and Babonoyaba which between 1648 and 1656 was a small Franciscan mission for Conchos and Tarahumar. The larger pueblo was Santa Isabel, location of the main Franciscan mission, and where, in 1653, a Tarahumar, Don Bernabé, was named Indian governor (AHP 1653b, Griffen 1979, 76). But that colonization effort lasted for only a short time. The Tarahumar rebellion, active since 1648, was followed by that of the Conchos, and a series of outbreaks in both the Sierra and the plains (1648-1656) caused the few colonists who had settled north of the Conchos River to flee southward; the incipient missions, both Jesuit and Franciscan, also collapsed.

Two decades of effort were necessary to restore the loss. Indian warfare continued sporadically, and finally armed violence and epidemics diminished the number of Conchos and Tarahumar. Many of the fertile valleys at the foot of the Sierra Madre, formerly dominated by Indians, were left unpopulated and at last were occupied by Spanish colonists.[2] Most of the new settlers were soldiers who had been fighting the rebellious Indians. The most important was Juan de Portillo y Lisalde, a veteran of the so-called "*compañía de campaña*" (forerunner of the cavalry troops of the seventeenth century) that had been fighting the Tarahumar and Conchos since 1661 (AGI Contaduría 927, 1661). In that year Portillo "purchased" from the Indian governor of San Pedro de Babonoyaba a *caballería* of land some six leagues from the Indian pueblo, where he formed a ranch called San Onofre. With him was Juan de Solís who also "purchased" another ranch at a place nearby and later sold it to Portillo. Solís was also a veteran of the campaigns against the Conchos and Tarahumar. Both men settled on their holdings with servants and armed guards. Seven years later (1668) the governor of Nueva Vizcaya, Lope de Sierra Osorio, named Portillo the "protector" of the Conchos Indians, which made him a virtual *encomendero* in the area (Almada 1928, 363; Griffen 1979, 77). The role of "protector of Indians" involved armed protection for pacified Indians as well as for Spanish settlers; because the region was being rapidly settled, and to avoid political conflicts, it was decided to incorporate the region into the jurisdiction of the *juez de mesta y cañadas* of Parral (the magistrate for regulation of sheep drives) (AHP 1671a, ff. 368-369).

The organization of armed guards was an important step for the pacification of the area. Thanks to the decrease of Indian population, the number of Spaniards slowly became sufficient to resist further uprisings. The improved environment encouraged the arrival of missionaries: the pueblos of Santa Isabel, Babonoyaba, Chuvíscar, San Bernabé, San Andrés, and Bachimba once again came under the jurisdiction of the Franciscans. Thereafter a slow consolidation of missions began by means of transferring Indians to reinforce a Spanish population that was still weak in numbers. This was not a new procedure, for even before the great rebellion of 1648, the position of the missions had been precarious: Indians could escape mission authority with ease and both religious orders competed to attract the largest number of neophytes within their respective jurisdictions, for on that depended the missions' existence.[3]

In the area of Chihuahua the largest of the Franciscan reductions was Santa Isabel, which was made the head mission for the pueblos of Chuvíscar, San Andrés and San Bernabé. Its convent began to operate officially September 1, 1673. At that time a program called the "*nueva conversión*," under Fray Nicolás de Chavarría, was begun, with an annual gift of 400 pesos from the *caja real* (treasury) of Durango (AGI Contaduría 927).[4] The annual gift helped to sustain the padres for an entire year. The same kind of aid began for Babonoyaba, which was also named as a "*nueva conversión*" under the name of Santiago Babonoyaba. Apparently Bachimba remained within its jurisdiction: the first director of Santiago Babonoyaba was Fray Juan de Cervantes, who was granted an annual gift of 400 pesos for progress of his mission (AGI Contaduría 927).

Chihuahua presented a sample of what was happening in all parts of northern Nueva Vizcaya. The retreat of the warring Indians and the expansion of local Spanish settlement into aboriginal territory was occurring almost daily. Although by law the presence of lay civilians within areas under mission jurisdiction was prohibited, such a restraint could not be enforced. During this period the missionaries, including the strict Jesuits, accepted the presence of civilians, especially the military within their jurisdiction. The cases of Miguel del Iriate and Nicolás Caro, who obtained land in Jesuit territory, was only the beginning of a long series of such acquisitions. In 1671, Felipe Rentería settled in San Gerónimo Huejotitlán (AGI Contaduría 927) and two years previously Gregorio Joseph de Mendoza did likewise near the pueblo of Papigochi (AIPG, Tierras y Aguas 40, (9), 1769). Although such cases involved small properties, their occurrence illustrates a practice that would increase in time (Alvarez 1991).

Similar cases occurred in the region of Casas Grandes: the arrival of the military on the border of Indian territory was a prelude to missionary activity.[5] Spanish settlement in this area, however, was still unstable; the local Suma Indians were easily disturbed and the presence of the military was always necessary. During the same time the province of Sonora was also undergoing a period of growth, and we have found mention of old Sonoran settlers established in Casas Grandes. One is outstanding: Francisco Ramírez, formerly an officer under General Juan Fernández de la Fuente, the most influential military leader of Sonora. Ramírez lived in Casas Grandes (apparently shortly after López de Garcia had arrived); he became the *alcalde mayor* and military leader of the jurisdiction. Ramírez, together with Fernández de la Fuente, would later be engaged in the great rebellion of the Conchos in 1684. But for the time, the presence of this kind of colonist served to pacify the area and permit the arrival of other Spaniards, such as Pedro Núnez Falcón and Luis Fernández de Andrade, among others (AGI Contaduría 927 [1676-1677]). Under these conditions the Franciscan mission of Casas Grandes was established and installed September 1, 1673 under the custody of Fray Juan de Balboa (AGI Contaduría 927 [1675-1676]).

Consolidation of the Jesuit and Franciscan Missions

The scene of northern Nueva Vizcaya was changing radically. The developments began in the 1670s and continued for the last third of the century; this was the beginning of what might be called the "golden age" of the mission system. Within the surroundings of Chihuahua, while the older missions were being consolidated, new ones were being established. Westward in the old domain of the Tarahumar, Coyachi was founded in 1673, and Julimes just before 1677 (Porras Muñoz 1980, 202 and Gerhard 1982, 197-198). In 1685 a new Franciscan mission called San Pedro de Alcantará de Namiquipa was established, with an annual gift of 100 pesos (AGI Contaduría 928 [1685-1686]). Somewhat later, at an undetermined date, the missions of Janos and Bachíniva were started; the latter place was listed as a Tarahumar pueblo during the Concho rebellions of 1684 and 1698 (Griffen 1979, 80).[6] Westward, not far from present-day San Buenaventura, the new Franciscan mission of Santa Ana de Torreón, supported by an annual gift of 100 pesos, was founded in 1685 (AGI Contaduría 928 [1685-1686]), while the old smaller mission of Nuestra Señora de Carretas was made a convent under the custody of Fray Antonio de Aguilar (AGI Escribanía de Cámara 400-C; Contaduría 298 [1685-1686]). The violent uprising of the Sumas, Janos and Conchos which erupted during this time did not succeed in dispelling the colonists, in large part due to the strong offensive undertaken by the miners and soldiers of Sonora.

Another event, also linked with the arrival of Spaniards of Sonora, was the colonization of the Papigochi River basin. We have already mentioned the presence of white settlers in the neighborhood of the Jesuit missions of the Baja Tarahumara. The case of Papigochi was quite similar. The pacification that occurred during 1670s and made possible the reestablishment of the old mission of Papigochi, which was soon converted as the center for numerous *pueblos de visita* (Indian settlements occasionally visited by missionaries).[7] From the start, missionary and civil colonization in Papigochi went hand in hand. A few years after the refounding of Papigochi mission, Spanish colonists there increased in number and already by the start of the 1680s, in addition to the ranches and farms, the initial exploitation of mines began. Soon, in 1687, the mining center of Santa Rosa de Cusihuiriachic was established (Almada 1928, 130), and years thereafter gained importance as a producer of silver. Sufficient to say that if the mines of Cusihuiriachic did not cause the beginning of colonization in the area, they indeed were an important factor in its final stabilization; starting from this period its jurisdiction became a confluence of two important movements in territorial expansion: one came from the province of Santa Bárbara, the other from Sonora and the region around Casas Grandes, Janos and Carretas. Soon some of the most important hacendados of Parral and the province of Santa Bárbara settled within the jurisdiction of Cusihuiriachi, where they started important haciendas. Among the oldest hacendados were Juan Cortés del Rey, Marcos Fernández de Castañeda, and Alonso de Leyva Isasi (APC, *registros matrimoniales*).

While a new and important *real de minas* was being opened at Cusihuiriachi, in Chihuahua the old mining sites continued to be forgotten. Although the existence of ore deposits in the area was well known, during these years life in Chihuahua took on a different orientation. The production of grains became the most important economic activity, as much in the Spanish haciendas as in the Franciscan missions. This is one of the least known aspects in the history of Chihuahua. Notwithstanding, it deals with a key period in the development of the future mining region, for it was then that the forthcoming network of food supply was created. We emphasize that the key to this process lay in the consolidation of Indian reductions and the formation of agricultural haciendas. Contrary to what had been the general previous tendency during the last third of the seventeenth century, Indian pueblos in Chihuahua, instead of diminishing, increased in population. But it was not because of natural growth of population; the increase should be attributed fundamentally to the transfer of Indians that was accomplished during this period. Not only was it intended to concentrate the Indians in already established reductions, but also to create new ones. The most important in the Chihuahua region was that of Nombre de Dios, a settlement of Conchos Indians established in 1690 near the flourishing hacienda of Tabaloapa. In 1694 and 1696, Nombre de Dios was then a *visita* of the mission Santa Isabel, and in it 44 Conchos families were gathered (Griffen 1979, 74).[8]

To be sure, all of the reductions underwent difficult times during the last third of the seventeenth century, but in the end demonstrated their efficacy. Indian uprisings were constant during all of this period. The best studied to date are those of the Tarahumar of 1689 and 1694 and those of the Sumas, Jacomes and Janos of 1684 and 1694 (Galaviz de Capdeville 1967, 133-140). However, in truth, little or no peace occurred in Nueva Vizcaya at the end of the seventeenth century. As expected, the Conchos, forcibly herded into reductions, ended up by rebelling; from 1680 on, their traditional territory was converted into a war zone. During the entire decade these fugitives from the reductions extended their attacks from San Francisco de Conchos as far as Casas Grandes, including parts of Sonora (Hackett 1926, 248; AGI Escribanía de Cámara 400-C, 1684). Meanwhile their nearby neighbors, the Tobosos, also continuously pillaged most of the towns located from the Conchos River and its tributaries as far south as Parral (Hackett 1926, 218-226). During the same period, farther north, attacks by the Apaches and Cíbolos reoccurred, and in time the Conchos also participated, together with the Julimes and Sumas in attacks in all of northwestern Nueva Vizcaya, including Sonora. During the 1690s warfare extended toward the Río Grande del Norte (today the Río Bravo) (Hackett 1926, 342), and the Conchos joined the Acoclames and other warring northern tribes (Hackett 1926, 342-344). Also the Tarahumar again rebelled in 1691 in the Papigochi area (Hackett 1926, 380) and hostilities continued for the rest of the century. But after a fashion, the Indians within the reductions learned to resist the attacks and were induced to shield the Spaniards. In 1697, for example, the Tarahumar neophytes of the Franciscan missions of Babonoyaba, Guadalupe, San Bernardino, Santa Isabel and La Concepción, accompanied Indians from the Jesuit missions to war against their rebelling relatives (AGI Escribanía de Cámara 390-B, 1697).

Currents of Migration and Occupation of Territory

As the reductions grew, Spanish settlement gained strength: the one cannot be explained without the other. Traditionally it is said that the founding of the mines of Cusihuiriachi was the source of mining development in the Chihuahua region, and that most of the latter's miners came from the former *real*. However, as we shall see, that interpretation has little foundation and is quite biased. All that we know about Chihuahua prior to 1705 comes from sources outside the region. Although the extant documentation does not permit us to reconstruct step by step the arrival of new colonists, we do have sufficient data to delineate the process during the last third of the seventeenth century and the first years of the eighteenth. The first conclusion obtainable from this analysis is that the peopling of Chihuahua was not a consequence of Cusihuiriachi, but rather, that both developed simultaneously. The difference between the two is that during the last third of the seventeenth century Chihuahua became a region eminently agricultural, not one devoted to mining.

During the last three decades of the century, the landscape of Chihuahua was transformed completely: the Indians were no longer the only permanent inhabitants of the region. As indicated earlier, the haciendas of Juan de Portillo y Lisalde and Juan de Solís were only two in the long list of the first registered land grants. More and more Spanish colonists were arriving and acquiring land during this period. An especially interesting case is that of Benito Pérez de Rivera, who in 1678 acquired a grant of four *sitios* for livestock and 12 *caballerías* of land for farming from the governor of Nueva Vizcaya, Lope de Sierra Osorio. Apparently this grant led to the origin of the hacienda San Juan de Encinillas, north of Chihuahua, and one that would become one of the most important in the North (Porras Muñoz 1989, 4). Although his original grant was modest, by the beginning of the eighteenth century properties of Pérez de Rivera had increased substantially, for in 1709 he requested lands "without measurements" for 220 pesos, which generally was equivalent at the time of a request for more than 10 *sitios* (AGI Contaduría 873, Composición de Tierras 1709).[9] Pérez de Rivera had been living in Parral, not in Cusihuriachi, at least since 1685 and was a hacendado in Chihuahua until 1710 (AGI Contaduría 875 [1699]; 928 [1688-1689]; 929, Cuentas y Fenecimientos de 1708-1710).

According to Phillip Hadley (1979, 69), at the beginning of the eighteenth century "the only permanent inhabitants of the area of Santa Eulalia were a few cattlemen who lived in the settlements they had formed near the missions...." This statement is inexact; at that time already there existed in Chihuahua a series of important agricultural haciendas, some of them of considerable size. That of Pérez de Rivera was only one of them: another important land owner who arrived before 1700 was Tomás de Chávez who in 1706 owned a property of 21 1/2 *sitios* in size (AGI Contaduría 873 Composiciones de Tierras de 1706). It is also known that before 1676 Domingo de Apresa Falcón acquired the land of Tabaloapa spring; this man was one of the richest miners and hacendados of Parral, where he possessed farms and refineries. (AGI Contaduría 926, 1652-1653); he was also a merchant and owner of a cart train that transported silver to Mexico City. Thirty-three years after his death, in 1709 his daughter, María de Apresa Ibarra, inherited the estate that probably consisted of over 20 *sitios*. In Parral she made two requests for land of 22 3/4 and 29 *sitios*, respectively. One of the requests may have included the lands of Tabaloapa, although payment is not indicated (AGI Contaduría 873 Composición de Tierras, 1709). Land acquisition also occurred near Indian reductions. In Babonoyaba, for example, Joseph de Flores acquired a grant in 1685, where he settled with his people (AGI Contaduría 928, 1685-1786).

Though not yet complete, the data give us a clear idea of the evolution of land holdings in the Chihuahua region. Since the end of the seventeenth century a stream of immigrants mainly from the province of Santa Bárbara settled in Chihuahua. But this was not a totally spontaneous immigration, open to any class of people. It consisted fundamentally of big hacendados who either settled with their families, servants, and Indian workers in the region, or established holdings there to be managed by others. A clear example of what we might call a "typical immigrant" is illustrated by Diego Cano de los Ríos, a rich hacendado whose large family had been successful in accumulating a network of properties in the Valley of San Bartolomé. These holdings were acquired as much through purchase and grants as through marriage.[10] At the end of the century the Cano de los Ríos clan could consider itself as one of the great hacendado families. In addition to the members already mentioned, there were Francisco, Miguel and Antonio, all of whom carried the Cano de los Ríos family name. All owned haciendas in San Bartolomé, San Diego, and Parral (APMN). But aside from genealogical history, the family connections of the clan represent well established hacendados, owners of old properties in Nueva Vizcaya that were stable and well populated.

Between the end of the seventeenth century and the beginning of the eighteenth, the haciendas in the Valley of San Bartolomé experienced a period of expansion. The demographic situation was favorable and the haciendas had grown as much inward as outward. During this period for the first time in the century the haciendas had become self-sufficient as to the work force, which permitted them to expand territorially (Cramaussel, 1990). For the hacendados, the land in Chihuahua became a northern extension of their southern properties: in 1706 Blas Cano de los Ríos, son and heir of Diego Cano de los Ríos made a request for a grant that would enlarge his original holdings of 123 1/2 *sitios*; one part of the request was for land in the Valley of San Bartolomé, the other portion within the jurisdiction of Chihuahua (AGI Contaduría 873, Composiciones de Tierras 1706). This made Cano one of the largest latifundistas of Nueva Vizcaya at that time.

Immediately after the opening of the mines of Santa Eulalia and San

Francisco de Cuéllar in 1707 and 1709, respectively, the region of Chihuahua, already well populated, attained great fame. This is the period when much more information becomes available than previously, because, thanks to the mining boom, local documentation began to appear. The first census of *vecinos* (family heads and land owners) known for Chihuahua was taken in 1716 by order of the governor Manuel de San Juan de Santa Cruz. At that time 289 *vecinos* were censused for the jurisdiction (AHP 1716, Lista de vecinos de Manuel de San Juan de Santa Cruz). The size of numbers indicate that with the opening of the mines the population of Chihuahua began to increase at a pace never before experienced. As we have noted, in 1722 it is said that around 8,000 persons lived in the region (BPG, 58-4-733 [1722]), whereas in 1723 the priest Juan Bautista de Lara estimated the population of the town of Chihuahua at 9,000. In 1725, 303 *vecinos* were censused (AHP 1725a, f. 276) and in 1730 the total for the region was estimated at 2,000 families. In short, between 1716 and 1724 the region of Chihuahua contained a total population between 8,000 and 9,000 persons, whereas the count of *vecinos* fluctuated between 300 to 400 heads of families or owners of a house lot in the town. In 1736 it was estimated that 25,000 inhabitants lived in the entire jurisdiction of Chihuahua (Bargellini 1984, 26).

Although it is well known that during periods of mining booms numbers of all kinds of things customarily increase, the above references give us an idea of the extent of settlement: in 20 years Chihuahua had been transformed into the richest region of Nueva Vizcaya. But even more important than stating such facts is the attempt to understand the way in which the process developed. The censuses of 1716 and 1724 provide us with firm data on what was occurring during the first big mining boom, but the question of how, when, and where all these people settled in Chihuahua remains to be resolved. To find an answer we have made an investigation of all the *vecinos* and miners listed in the censuses; registers of mines, house lots and lands; civil documents; account books of mercury; lists of haciendas; and government inspections in the region between the years 1700 and 1733. This kind of treatment was previously used in a study of the Parral region (Alvarez 1989).[11]

The initial conclusive result of the experiment, to a certain degree, was a surprise: contrary to what had been previously thought, Cusihuiriachi was not the place from which most of the Chihuahuan colonists came. It was the Valle de San Bartolomé that became the main provider of people for Chihuahua. Of the 340 immigrants who settled there, 148 were from San Bartolomé, 73 from Parral, San Diego and Santa Bárbara, and the remaining 79 from Cuishuiriachi and its surroundings. In other words, 65 percent of the colonists came from the Province of Santa Bárbara and only one in three from Cusihuiriachi. Nonetheless, the result should not actually be a surprise, especially if we consider that Santa Bárbara was the most densely peopled area of Nueva Vizcaya, and that we have verified the arrival of hacendados and their people from San Bartolomé since the last part of the seventeenth century. Thus we are dealing here with a large population movement that began fairly early; in fact, Chihuahua was a point of convergence of movements resulting from territorial and demographic expansion of northern Spanish society, in which Cusihuiriachi represented no more than a stage. The idea that most of the miners of Chihuahua came from Cusihuiriachi arose from the fact that first Almada (1959) and then Hadley (1979) analyzed only the documentation relative to the founding of San Felipe de Chihuahua. The main hacendados who settled in the vicinity of the new town did come from Cusihuiriachi, as did Trasviña y Retes and the priest José García de Valdéz, among others. Even so, the *real*, or mining center, of Cusihuiriachi was settled by people from the Province of Santa Bárbara and to a less extent from Sonora. Of the 79 *vecinos* from our sample who arrived in Chihuahua from Cusihuiriachi, 19 had previously resided in San Bartolomé and another 14 in Parral.

The Economic Organization of a New Mining Region

The Supply Network

At the dawn of the eighteenth century Chihuahua was already being transformed into a well established agricultural region: the opening, or reopening, of mines did nothing but accelerate this process. We know that even before the opening of Santa Eulalia and San Francisco de Cuéllar, mining occurred on a small scale in two small *reales* called Chihuahua el Viejo and San Pedro de Alcántara (Cramaussel and Alvarez, 1992). Although the extant documentation cannot affirm it definitely, it is very possible that these were the old mines exploited in the 1640s (AAC, microfilm rolls 1 and 2, 1710-1720). We have found accounts of dual agricultural-mining haciendas in full operation before 1716, that is, at the time when the *reales* of Santa Eulalia and San Francisco de Cuéllar were being opened. Taking careful heed of this chronology, we can report that the opening of mines on a large scale was made possible only when a true local supply system was in place, for labor as well as for agricultural provisions. It was not by accident that the large contingents of migrants came from the Valley of San Bartolomé. Most were hacienda owners or persons associated with them; between 1716 and 1728, 40 haciendas and ranches of San Bartolomé were sending grain to Chihuahua, but most of that grain was intended for the provisioning of the agricultural and mining operations that the same land owners had in Chihuahua. In fact, we know that almost all of the important miners of Chihuahua were from San Bartolomé, Parral, or Cusihuiriachi and were owners of the haciendas that provisioned them.[12]

The formation of the town or *villa* of San Felipe el Real, in the very heart of the Chihuahua region, followed a similar process (fig. 1). The new *villa* was not built on virgin land. Its site was on the banks of the Chuvíscar River, a place intermediate between the *reales* of San Francisco de Cuéllar and Santa Eulalia. The site had been occupied for some time by various agricultural haciendas, and the reason for the town's founding did not stem from the mining boom: the project was considered much before. On August 10, 1701, a royal edict was issued to the effect that a new Spanish settlement, which would be named "San Felipe," should be constructed on the road to New Mexico to shelter all kinds of vagabonds, delinquents and poor people in general who infested the large cities of New Spain (Porras Muñoz 1989; 1980a, 190; 1980b, 202; Almada 1984, 83-89). In time, San Felipe became much more than a refuge for undesirables; instead, an opulent "city" was formed. However, it should be remembered that in 1702 the mineral wealth of the region was not yet fully exploited; the only factor that might have led to the founding of a "city" in the area at such an early date, was simply that the site was known for its spaciousness, availability of water and wood and, above all, of food and labor at an accessible distance, thanks to nearby haciendas and Indian pueblos.

Indeed, apart from the supply of grain the haciendas presented a sure source of labor. In this regard the displacement of Indians continued to be an essential factor for the stabilization of settlement in Chihuahua with the establishment of new reductions. In 1707 the Indian pueblo of San Gerónimo was formed for Conchos Indians (Griffen 1979, 77). Later, in 1716, a large group of Conchos was concentrated in the old Franciscan *visita* of Nombre de Dios; nearby a new pueblo called Santa Ana de Chinarras was made and placed under the custody of the Jesuits. The natives who were concentrated in this place came from an area some 300 kilometers north of Chihuahua near Paso del Río. Some 77 of them were Conchos from the plains between the hacienda of Encinillas and the area of El Carrizal, where later the hacienda of Manuel de San Juan de Santa Cruz was established; another 35 Conchos belonged to the pueblo of Nombre de Dios. There, in 1717, 144 persons were recorded and a year later the "*chinarras*," as they were called, numbered 200 (AHP 1718a; Griffen 1979, 77-88). This was perhaps the best example of how such reductions were made; more than true missions, these were concentrations of captives of war destined to work in the haciendas and mines of the area. In 1725 padre Güendulain reported that among the 38 Indian families he counted were Conchos, Tobosos and Sumas, besides some of the original "*chinarras*"; all were deserters from the reduction of San Pedro de Conchos (Griffen 1979, 80). Another important reduction was formed in the Franciscan convent at San Felipe el Real, where 40 Coyame families were gathered in 1726 (Aldama,

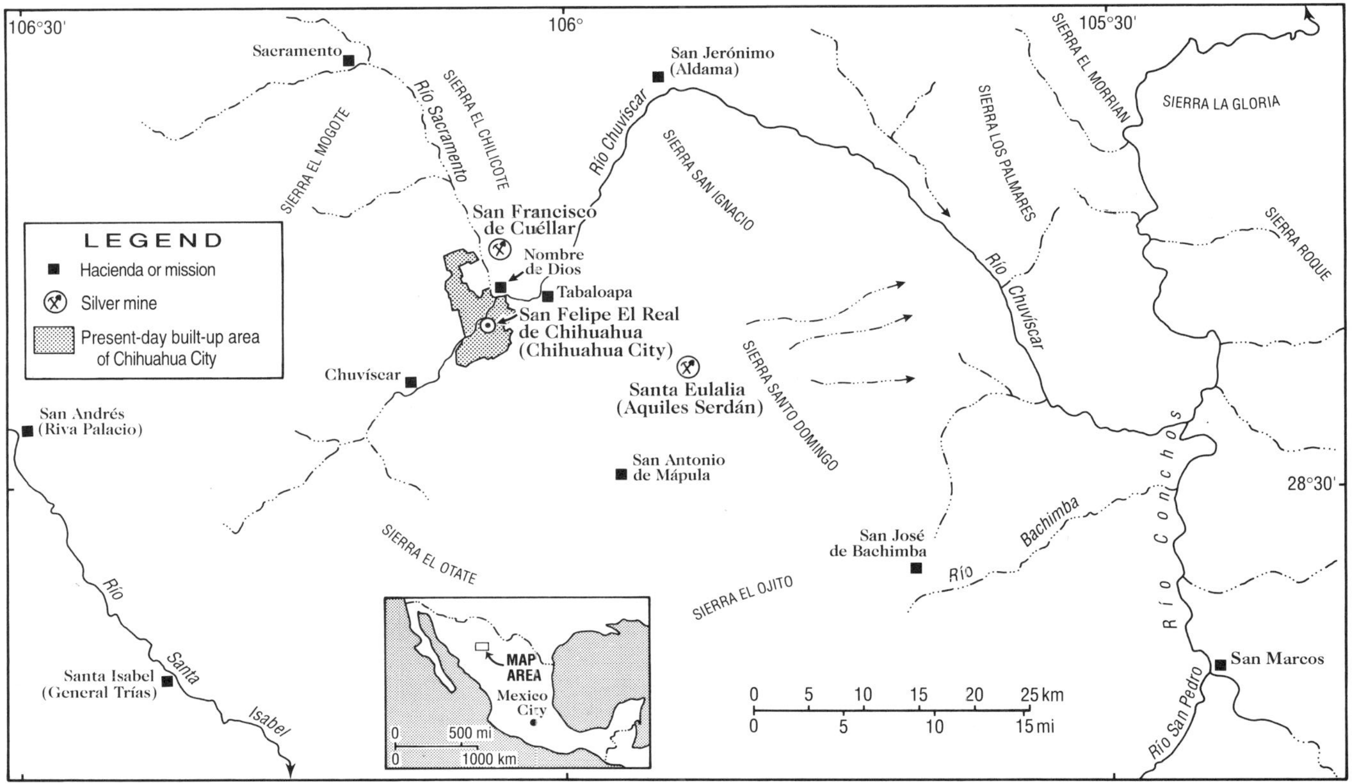

Fig. 1. Central portion of the Chihuahua area in the first third of the eighteenth century. [Map provided by the volume editors, who take full responsibility for its accuracy. Data on names and location of haciendas or missions are based, in part, on the *visita* of Manuel San Juan de Santa Cruz, governor of Nueva Vizcaya, made in 1716 and commented on by Guillermo Porras Muñoz in his introduction to "Las minas de Chihuahua," *Boletín del Archivo General de la Nación* 7(3): 638-740 (1966)].

1959, 39-40). Other reductions took place nearby at the mission of Guadalupe in 1709 (Bergellini, 1984, 15).

Soon the agricultural haciendas occupied most of the land, especially along the Chuvíscar River and near Indian pueblos and reductions where native labor was available. Besides those in Chihuahua proper, in Bachimba the old hacienda of Fernando Colomo was using Cholome Indians around 1730. In Babonoyaba Joseph de Corral owned an hacienda for which he petitioned for the service of 12 Indians through the *repartimiento* system (AHP 1723d). The *villa* of San Felipe el Real was built on land already occupied by farms of the powerful hacendado of Cusihuiriachi, Juan Antonio de Traviña y Retes, and by the college of the Jesuit order, founded in 1715. The *villa,* as mentioned earlier, thus had to compete for space and water with the haciendas. From an urban point of view, the presence of these two establishments caused difficulties that vexed the *villa* throughout its early history (Cramaussel and Alvarez, 1992).

As seen from another angle, these difficulties for San Felipe formed one of the secrets for the success of the mining industry. Missions and haciendas assured the supply of the two essential items for the mines: labor and food. A widespread thesis in the historiography of mining societies affirms that the great mining centers of the North stimulated the development of areas that specialized in the production of grains to supply the mines; frequently these areas were far from the mining districts (Florescano 1969, 47). The case of Chihuahua, however, belies many of the main points of this hypothesis: its supply network preceded the opening of its mines. But this does not mean that its supply system was free of difficulties. Although many of the leading miners of Chihuahua provided their operations with grains and other necessities from their own haciendas, it is true that not even the most expedient among them could ensure completely the needs of the entire area. Even during the period characterized by the maturity of colonial New Spain, Spanish society could not avoid completely the Indian world that surrounded it. The very existence of a prosperous mining society would not be possible without the help of missions and local Indian pueblos in terms of both food and labor. Such had been the case of Parral in the seventeenth century (Cramaussel 1989; 1992b), and in the eighteenth century it occurred again in Chihuahua.

The best available source for responding to the question of how Chihuahua was fed, are the account books of the *alhóndiga*, or public granary, that was established in San Felipe in 1730. The most complete available registers are those of 1733 (AAC, microfilm roll 8). This is an exceptional document which details, month by month, the receipts of grain fed into the *alhóndiga*, indicating the quantity sent and the names of each of the senders. We have used this source to carry out an investigation of 201 suppliers of grain for one year and we have tried to locate the origin of some of the persons through a method similar to that mentioned earlier in the case of immigrants: fortunately, it was possible to identify 139 of the suppliers. During 1733 a total of 14,532 *fanegas* of maize and 16,050 quintals of wheat flour arrived in the storehouse of the *alhóndiga*. The 139 identified suppliers contributed 12,271 (84 percent) *fanegas* of maize and 13, 751 (85 percent) quintals of flour. Through the process of identifying each of the suppliers we can confirm that the grain arrived from 10 distinct areas:

(1) Chihuahua, including the following pueblos and haciendas: Julimes, Chuvíscar, Bachimba, Nombre de Dios, Sacramento, Santa Isabel and Encinillas.

(2) Papigochi, with the pueblos of Papigochi, Yepómera, Basuchil, Namiquipa, Temeichi and Santo Tomás.

(3) Cusihuiriachi and its jurisdiction.

(4) Province of Santa Bárbara, including the jurisdictions of Parral, Valley of San Bartolomé, San Diego and Santa Bárbara.

(5) Casas Grandes, which also included Carretas and San Buenaventura.

(6) Conchos, with San Francisco de Conchos.

(7) El Paso, which included the missions of Paso del Río.

(8) Missions of Sonora, with Baserac and San Ignacio, and including those of Ostímuri and Yécora.

(9) Tarahumar missions with Babonoyaba, Satevó and Nonoava.

(10) Province of Durango.

Again, this investigation produced some surprises. First, and most important, Chihuahua never completely relied on either the Valley of San Bartolomé (granary of the North), or the missions of Papigochi (famed for their efficient agricultural organization), or even the Cusihuiriachi area for supplies: most of the grains consumed in Chihuahua came from within its jurisdiction. The Indian towns of Nombre de Dios, Chuvíscar, Bachimba and Santa Isabel (all almost forgotten today), together with the haciendas along the Chuvíscar and Sacramento rivers, were the main sources of maize and wheat for the region. More than a third of the grains came from the neighborhood immediate to San Felipe el Real. Outside the jurisdiction of Chihuahua the main suppliers were, in part, the Papigochi missions (28 percent of the maize, but only 11.5 percent of the wheat flour) and the Valley of San Bartolomé (21 percent of the wheat flour, but only 8 percent of the maize). For its part, the Cusihuiriachi area was fourth in importance (10 percent of the maize, only one percent of the wheat). As can be seen, just these four regions furnished 81 percent of the maize and 66 percent of the wheat that entered the *alhóndiga* of San Felipe.

We note that outside Chihuahua wheat flour came from areas of old Spanish colonization, such as the Valley of San Bartolomé, whereas most of the maize originated in areas of more recent colonization: the missions of Papigochi and Cusihuiriachi. This indicates that there was no real case of agricultural specialization directly linked to the opening of the Chihuahua mines. As one of the oldest farming regions of the North, the Valley of San Bartolomé owed little to Chihuahua for its prosperity (Cramaussel 1990). The cases of Papigochi and Cusihuiriachi were more complex, for they were areas of more recent colonization, where Indian pueblos and the missions were important suppliers of grain for the Spaniards. Frequently it has been said that the Jesuits forced Indians to cultivate mission lands, the products of which the Jesuits sold. Although this was a customary practice, today we know it was not the only one used by the Jesuits. In the Papigochi area, because of difficulties in congregating the Tarahumar, the priests preferred to obtain maize from the Indians in exchange for various items such as cloth, tools, and so forth. The maize thereby traded was sold immediately to the miners. We conclude that this case also does not represent true economic specialization on the part of the missions, for it concerned a crop produced by the Indians themselves, not by the missions (Alvarez, in press).[13]

The case of the opulent priest José García Valdéz of San Felipe illustrates the position of the big hacendados in the supply network. These personalities, especially the more powerful ones, customarily established themselves both as miners and hacendados in Chihuahua and used the resources of their haciendas located in other areas to supply their mining operations. The best example of this type we find in Pedro González de Almoina, who eventually became the leading supplier of the *alhóndiga*. Almoina owned four haciendas in Chihuahua, one called San José, and a complex of mines, ore refineries and farms; he also had another hacienda called Namiquipa in the Papigochi area and still another two farther north, El Torreón and San Miguel. From these, Almoina obtained enough grain to contribute 20 percent of the maize and 10 percent of the wheat to the *alhóndiga*. The same can be said of Miguel Moreno, who placed second among the individual suppliers of wheat (1,307 quintals, or 10 percent of the total); he possessed haciendas in both San Bartolomé and Chihuahua. Other cases include Salvador de Lemus and Baltasar de Arechavala (second and fourth, respectively, as suppliers of maize); Domingo de Escandón, hacendado of Buenaventura; and Alonso Rubí of El Paso del Río, to mention only a few. Table 1 and figures 2 and 3 summarize the supply network for grains consumed in Chihuahua for the year 1733.

But the Indian pueblos supplied Chihuahua not only with grains; they also furnished labor under the repartimiento system. San Gerónimo, Chuvíscar, Santa Isabel, Satevó, Babonoyaba, Julimes and others were all

Table 1. Amount and percentages of grains received in the public market (*alhóndiga*) of San Felipe el Real, 1733, from identified regions.

Region	Maize		Wheat	
	fanegas	percent	quintals	percent
Chihuahua	4,188	34.12	4,384	31.88
Papigochi	3,482	28.37	1.624	11.81
Cusihuiriachi	1,224	9.97	130	0.94
Prov. of Santa Bárbara	1,064	8.67	2,897	21.06
Missions of Sonora	917	7.47	1,042	7.57
Tarahumar Missions	634	5.16	341	2.47
Casas Grandes	557	4.53	1,280	9.30
Conchos	185	1.50	977	7.10
Durango	20	0.16	203	1.47
El Paso	—	—	873	6.34
Total	12,271	100.00	13,751	100.00

subject to the repartimiento (Cramaussel 1991). In short, there were two parallel supply networks: one dealt with labor, the other with food. These were among the developments that help explain the growth that the region experienced since the end of the seventeenth century. When the Spanish colonists from the Valley of San Bartolomé and Cusihuiriachi first settled in Chihuahua, they were accompanied by their workers — Spanish and Indian servants and black slaves, with whom they established farming and mining haciendas throughout the region. In addition, well after its founding at the end of the 1730s, the new *villa* of San Felipe el Real extended for two leagues along the Chuvíscar River, but more than a real pueblo, it consisted mainly of a series of haciendas in close formation along the river floodplain.

Mining: A Domain of the Big Hacendados

Almost like a review of the past, the region of Chihuahua became, as it were, an example of one more of those great mining discoveries for which Spanish America is famous. Nearly a century had passed since the founding of Parral before northern New Spain relived a mining boom of such proportions. Like Parral during its time, Chihuahua, within a few decades contained the largest agglomeration of people in Nueva Vizcaya: already by the 1730s twice as many people were in Chihuahua as were in Parral in 1635, that is, at the end of great boom of that *real*. But beyond these analo-

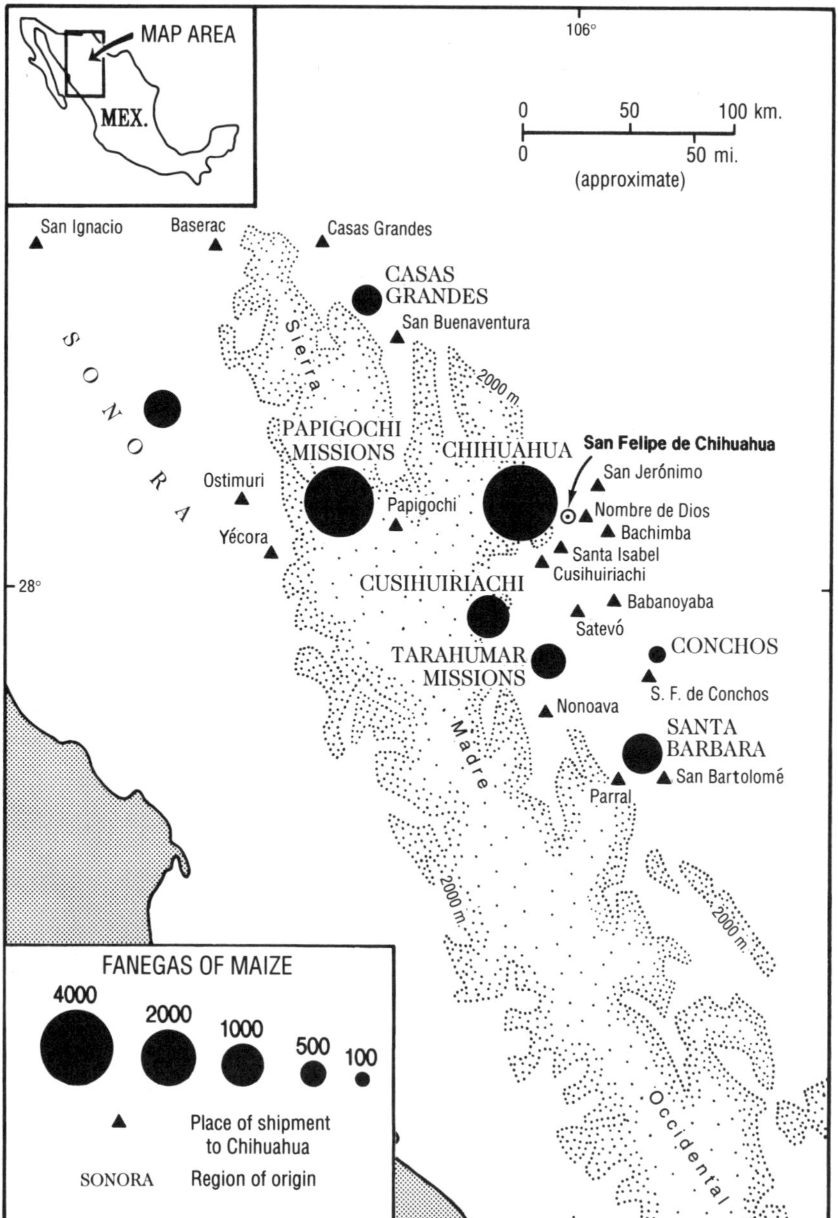

Fig. 2. Supply of maize for the public market (*alhóndiga*) of San Felipe el Real de Chihuahua, 1733. Size of graduated circles is proportional to amount of grain supplied.

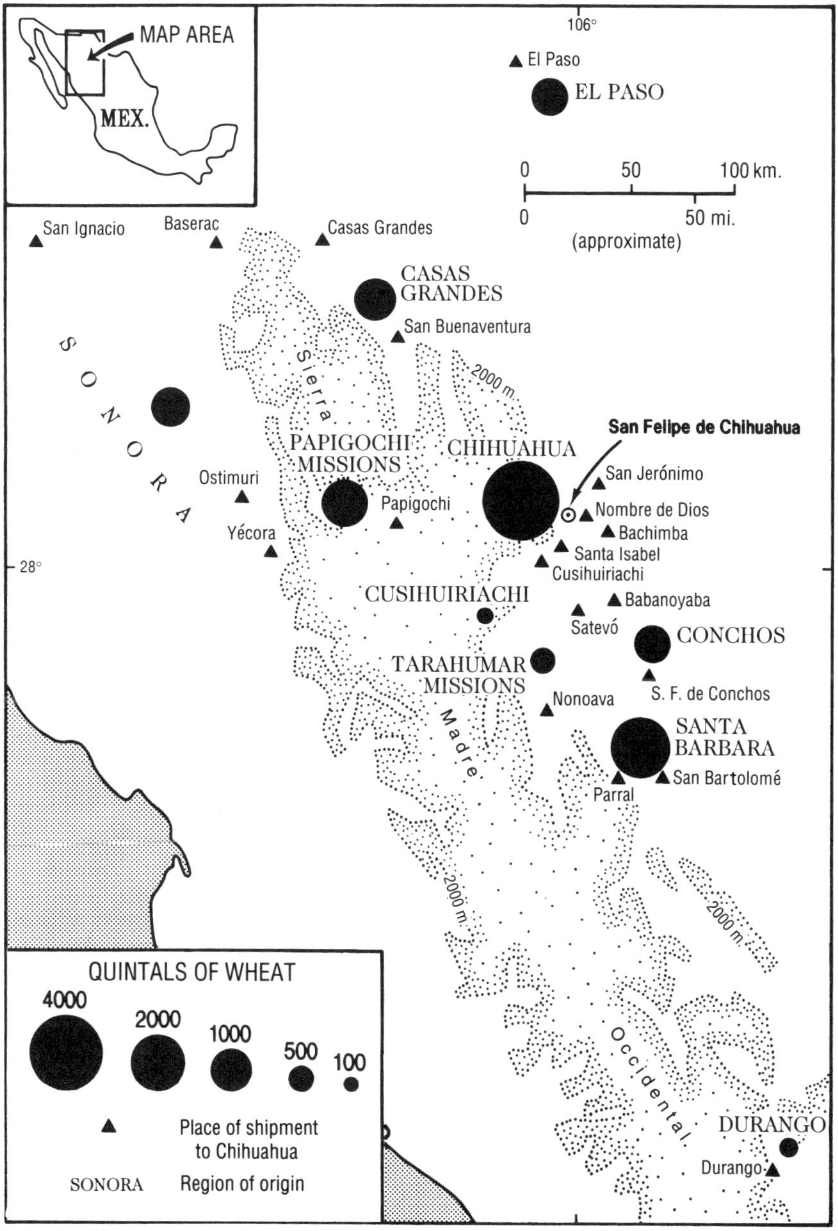

Fig. 3. Supply of wheat for the public market (*alhóndiga*) of San Felipe el Real de Chihuahua, 1733. Size of graduated circles is proportional to amount of grain supplied.

gies, if the beginnings of these two mining districts are compared, we can see how both evolved in completely different ways. From a strictly mining viewpoint, Chihuahua's beginning was much less spectacular than that of its illustrious predecessor. Only four years after the opening of the Parral mines the total silver output of Nueva Vizcaya doubled, and by 1650 it had increased ten times over that registered in 1600 (Alvarez 1989). When the mines of Chihuahua were opened the effect on the total production in Nueva Vizcaya was much more modest.

In 1700 the silver output for all Nueva Vizcaya was 149,170 marks [1 mark = 8 ounces]; in 1714, 7 years after the opening of Santa Eulalia and 5 years after that of San Francisco de Cuéllar, production was 165,479 marks, that is, hardly 10 percent higher than that at the start of the century (AGI Contaduría 925). Almost 30 years later, during the administration of governor Juan Bautista de Belaunzarán, the annual output of the province was around 199,300 marks (BPG, civil 53-6-670). Of course, we cannot rely solely on these numbers as a guide for trends in silver production in Nueva Vizcaya in the eighteenth century, but they do show that the opening of the Chihuahua mines did not provoke a boom in silver output that was in any way comparable to that caused by the Parral discoveries a century earlier. This, however, does not mean that production of the Chihuahua mining district was not relevant.

A strong shift of mining activity into Chihuahua was taking place at the expense of other districts, especially Parral and Cusihuiriachi. Gradually, Chihuahua monopolized much of this activity, which resulted in usurping from Parral its rank as the foremost mining district of Nueva Vizcaya, a position it had occupied for a century. In 1738 a royal counting house (*caja de marca y ensaye*) was established in Chihuahua and that of Parral was closed. The *villa* of San Felipe became the center of the most powerful people in the entire province; hacendados and merchants built sumptuous homes and the construction of an aqueduct was started. The Jesuits established a college, while the Franciscans in turn opened a new convent in the *villa*. As early as 1727, funds were collected to construct an elaborate parochial church of stone (Bargellini 1984, 23). Nonetheless, as mentioned previously, the truly urban part of the villa continued to remain quite small as a space reserved for the influential (Cramaussel and Alvarez, 1992).

In the same way that the hacienda dominated the physical landscape of Chihuahua, it was the hacendados who prevailed within the social scene. Just money and initiative was not enough to become an hacendado, miner, or merchant in a place like Chihuahua at the beginning of the eighteenth century. The essential element in that context was to arrange for the necessary materials to develop any one of these activities; and in its practical aspects commercial activity was not much different from mining and agriculture. Necessary were animals, wood for making carts, leather for harness, drivers, and, of course products of all kinds to transport and sell. Two different commercial circuits were formed: one involved transport of items in long-distance trade, such as clothing, iron tools, exotic foods, and so forth; a second type of trade included things that were produced locally but had to be taken to market such as lumber, firewood, charcoal, hides, tallow, meat, salt, grains, and many others. Mainly hacendados were engaged in these commercial activities; of the 20 merchants registered in the jurisdiction of Chihuahua in 1716, 15 had been hacendados in Cusihuiriachi and San Bartolomé, and in addition all were local mine owners.

The tendency to concentrate economic activity within the scope of trade also occurred in a similar manner in mining. During a boom almost everyone aspired to become a miner. A small excavation followed up by a simple legal proceeding was enough to claim ownership of a mine, but it was quite another problem to obtain a concession to work it and to actually begin to exploit it. In a strict sense, a true miner had to be capable of extracting ore from a vein and refining it (Romano 1970). Between 1705 and 1720, 368 mining concessions were registered in Chihuahua (AAC, microfilm roll #1), but most of these early concessions involved only small *catas*, or exploratory excavations; these were soon abandoned. Of the mines that actually prospered most were small shallow workings from which rich superficial ores were taken. Few mines resulted in true producers of profitable ore. During an official visit of the mines of Santa Eulalia in 1727, only 11 were considered worthy of inspection, these being the only ones that possessed shafts and other installations worthy of the name. These 11, which were the most important ones in Santa Eulalia, were operated by only six owners, all of whom were big hacendados.[14] These owners of both mines and ore refineries were sustained almost always by products from their own agricultural haciendas.

Operated on a small scale, the mines did not need large numbers of workers or many materials. As in all young mining centers, in Chihuahua small workings abounded, most of them small *catas*. These served to extract small amounts of superficial ores with the hope of finding a rich vein. Generally, when the fineness of the ores decreased excessively, the mine was abandoned, until someone returned later to claim it and to renew the cycle. The few mines that were considered profitable can be classed as (1) those that consisted of a large group of small but numerous superficial workings and (2) large mines exploited intensively. To exploit rich ores at depth necessitated an enterprise that was never in reach of the small miner. To work such a mine large numbers of laborers were needed, as were lumber for beams and cribbing, hides and leather thongs, tallow for candles, and iron tools for extracting ore; in addition there was the cost of feeding workers and paying salaried operators, freight charges, and costs of refining ore. Very few owners could deal with the expenses by relying only on the profits from a poor mine. In reality, the tales of miners who enriched themselves overnight after finding a rich vein are usually nothing more than fantasies. As a rule, those who did not have at hand the resources to enrich themselves before undertaking such expenses would inevitably fail, however great their discoveries might have been. A good example of that certainty is that of Juan Rangel de Biesma, discoverer of the Parral mines; he was never able to enrich himself with his discovery despite the fact that Parral became the most important *real* of northern New Spain in the seventeenth century (Cramaussel 1992a). Thus, it was not by chance that the owners of the big agricultural haciendas acquired control of the mining industry in Chihuahua.

Those who arranged ways to undertake major work in their mines exploited them intensely, intending to extract the maximum amount of ore. In Chihuahua this kind of exploitation was based mainly on a large work force without much attention to technical refinements. The exploitation of large mines equipped with a complex of shafts, ventilation, and drainage was a slow and expensive process, adequate for old and well organized mining districts. This was not the case in Chihuahua. In 1727, after more than two decades of operation, the best mines of Santa Eulalia continued to be worked with primitive methods; none contained a major adit or installations for ventilation or drainage. At that time the mine of San Judas Tadeo, property of Joseph de Zubiate, the most important in the *real*, was intensely worked; various *labores*, or large galleries, had been opened, some measuring 15 yards high and 20 wide, and others as large at depths of 7 or 8 fathoms. The workmen descended by means of simple shafts a few yards wide and only two galleries were equipped with wooden reinforcements (Hadley 1979, 228-229). The remainder of the mines inspected were in even poorer condition; several of those owned by Manuel de San Juan de Santa Cruz had collapsed because of overexploitation and lack of improvements.

In regard to refining ores, need for labor, ingredients and installations was much greater than in working the mines. Furnaces for the heating and smelting of ores, storehouses, dwellings, mills and other structures were necessary. More than installations, however, the most important features to ensure the operation of the process were first, the labor force and second, a constant supply of various ingredients used in smelting as well as in amalgamation, such as mercury, litharge, ash for cupels, firewood, charcoal, tallow, lumber, and metal tools. There was in addition the need to feed the workmen maize, wheat flour, vegetables, beverages, and so forth. Few account books of refineries exist for this period, but we have located some that help analyze the internal functions of the facility. The most complete are those of the refinery of Juan Pérez de Bustillos for 1730, in which ore was processed either by amalgamation or smelting. We have evaluated the consumption of the main ingredients listed in the accounts, especially charcoal and firewood, as well as maize and other foodstuffs used in the rations of the Indian workers. In that year the refinery produced 2,194 marks of silver by smelting and 2,679 marks by amalgamation; these figures translate to a total of 39,729 silver pesos (*pesos de ocho reales*) (AAC, microfilm roll #50, Libro de gastos y platas que se sacan en la hacienda de Benito Pérez de Bustillos, 1730).

From these accounts, the high consumption of firewood and charcoal needed in both smelting and amalgamation is evident. In 1730, 3,008 *cargas* (120,320 kilograms) of charcoal and 4,272 *cargas* (170,880 kilograms) of firewood were used.[15] In volume those ingredients were the most important ones used in the refining process; together both represented an expense of 6,198 pesos, 4 reales. Another expense was maize, of which 657.45 *fanegas* (59,707 liters) worth 3,824 pesos were consumed. Some 1,406 pesos was spent on food and other items dispensed to the 186 Indians who were brought in at given periods to work in the mines and refineries. Also, 10 quintals of mercury at 82.75 pesos per quintal were utilized.

A brief resume of these accounts gives a total expenditure of 34,110 pesos, showing that exploitation resulted in little profit, which was only 6,619 pesos: great effort for little return, barely equivalent to the expense for firewood and charcoal. High production costs formed an endemic problem for mining in New Spain; costs were always greater than those experienced by their counterparts in Europe. But it is erroneous to suppose that all items consumed in the mines actually passed through the market (Romano 1970, 143-144). Evidently, most of the costs did not pertain to actual expense for the hacienda. Thus maize, for instance, was listed as an expense two different times in the accounts: once when it arrived at the *alhóndiga* and again when listed in costs of rations for the workers. The same is true for firewood and charcoal; we do not know if these items were purchased from outsiders or obtained from agricultural haciendas belonging to the mine owner.

Consideration of the problem of provisioning is essential in order to understand the functioning and profitability of mining. The profitability of a mining enterprise depended less on the intrinsic value of the product (silver) than on the availability and cost of provisions, and these varied from mine to mine, according to the wealth and importance of the owners. The method of accounting for provisions that came from the proprietor's own agricultural haciendas was radically different from that used when it was necessary to purchase from an outsider. Prices assigned to charcoal, firewood and maize varied. Thus, for example, in the accounts of Domingo Pérez de Bustillo the maize consumed in the mine was accounted for at the average market price (5.9 pesos per *fanega*), which indicates that the maize was obtained from the *alhóndiga*. In contrast, during 1737 in the mine of the powerful grandee, Manuel de San Juan de Santa Cruz, maize was listed at 3.25 pesos the *fanega*, whereas the current price was above 6 pesos the *fanega* at the *alhóndiga*. It is evident that maize was obtained either from his agricultural hacienda or from the Indian pueblos nearby (AAC, microfilm roll #2, 1737-1739, Libro de las cuentas de toda la semilla que entra en la *alhóndiga*).

But perhaps one can better appreciate this kind of difference with products like firewood and charcoal. In the accounts of Pérez de Bustillos the prices of these items in 1730 were one peso per *carga* for charcoal and 6 *reales* per *carga* for firewood. In the same year in the mine of La Regla, property of widow María Ortíz de Campo, both items were estimated at the same price: 5 *reales* the *carga*. The difference in price is explained by the fact that the widow obtained charcoal from her properties in Santa Isabel (AAC, microfilm roll #50). It is interesting to note the small difference in price between charcoal and firewood despite the great contrast in the amount of labor expended in processing each; this may indicate that most of the price of both stemmed from transport costs rather than the labor involved (Romano 1986). During the early years, mining exploitation in Chihuahua, both charcoal and firewood were obtained from the area surrounding the mines, but due to overcutting, the supply lasted for only a short time and eventually it was necessary to transport the items from afar.

The greater the capacity of the hacendado to obtain gratuitously all kinds of agricultural products, the greater was the profitability of his mining enterprise. Until now we have spoken exclusively of the items consumed in the mines, leaving aside a fundamental problem: that of labor. There were some mines that operated on a large scale with numerous workers. In the case of the mining hacienda of Pérez de Bustillos, besides the 186 Indians obtained through the *repartimiento*, there were at least another 35 mestizo and Spanish employees in the hacienda. These supervised transport of necessities especially the tons of firewood and pine charcoal that was consumed in the refining process. The fact that still at this period most of the workers in the hacienda were *repartimiento* Indians underlines the importance of the role that the Indian reductions continued to play in the growth of Chihuahua. In this case the Indians came from the pueblos of Carichí, Santa Isabel, Paso del Norte and Bachimba. Twenty to 24 Indians from the first two pueblos were sent to work for a period, or *tanda*, of one month, whereas those from Paso del Norte and Bachimba came in sporadically. Those from Carichí in particular were obliged to work preferably in the mines perhaps because of specific ordinances. Wherever they worked, all of the Indians arrived under the charge of the native governor of their respective pueblos and they were given rations of maize and meat during their stay in the mines. They were also given some items that they handed over to their governor. The *repartimiento* system was, in short, another gratuitous resource for the mines.

Only in this way can one explain the success of miners such as Joseph Antonio Vergara who in 1734 consumed 19 quintals of mercury to refine 525.5 marks of silver by amalgamation, a return hardly a fourth part of the production considered normal for that much mercury at that time (100 marks per quintal). In other cases the productivity of the ore was much higher than average, as was the case of Pérez de Bustillos who obtained an extraordinary return: 267 marks of silver per quintal of mercury. Nonetheless, his mine would have hardly been profitable had he been obliged too purchase in the "market" all of his necessities. In Chihuahua it was not possible to clearly separate miners from hacendados or from merchants in the commerce of regional products. In time there developed an active trade in bulky items of low unit value, a type of commerce dominated by hacendados who bred mules and horses used in transport. Among these Manuel de San Juan de Santa Cruz, with others, was outstanding. Doubtless, there developed eventually a stratum of independent merchants, although both hacendados and merchants overlapped in their functions for some time.

But a sphere in which the dominance of the big agricultural hacendado was absolute and indisputable was in mining. By acquiring the materials necessary for the extraction and refining or ores, the hacendado controlled the complex of mineral production. Even the small independent miners had to depend upon them. The ledgers of the larger mines clearly indicate that part of the ore processed was brought in by numerous small miners, who paid for the service with a portion of the treated ore, or a toll in kind (*maquilas*). In time, this custom resulted in a growing dependence of all the small miners on the refineries, which rapidly fell into few hands. This assertion derives from an investigation of mercury imports into the Chihuahua region between 1720 and 1740 (AHP 1728 G-2; 1727a Protocolos y escrituras de mineros para reciber azogue, 1727-1740). We have tried to identify the miners who received mercury year by year, as well as their social standing and holdings. Of the 47 persons who received the coveted liquid metal during the 20 year period, we could identify 41, and only 10 of these were not owners of refineries. The remaining 31 were in some capacity owners of both agricultural haciendas and mines, and to them had gone 78 percent of the total mercury consumed between 1720 and 1740. It should be noted that 6 of the unidentified persons purchased mercury only once, and it was not uncommon that some acquired the liquid on credit to some big hacendado. In brief, a very small number of hacendados controlled practically all of the extraction and refining of the silver ores, and all of them owed their original wealth to their agricultural haciendas.

Conclusion: The Last Great Establishment in Northern New Spain

The pervasion of haciendas and their owners in all spheres of economic activity forms the essential background of settlement history in the Chihuahua region. After seven decades of slow agricultural colonization and pacification of surrounding Indians, that region was molded and vitalized by the haciendas which became the main instrument of colonization. Mining was not an independent activity capable of self-development without the concourse of a well-established agricultural system and a work force to sustain it. Its role in the history of colonization was to accelerate growth, the result but not the cause of territorial expansion. The characterization frequently made in the historiography of mining districts as eminently urban societies is an extreme simplification that favors the external causes in the long run, but not logical internal economy. The randomness of a mining discovery is incapable in itself of generating territorial settlement. The ostentation of mining cities, the aristocratic life style of its opulent inhabitants, and the display of big business that occurs during mining booms are events dependent on the extraction of silver; in the case of Chihuahua, when this ceased to flow, these events tended to disappear and in the context of modern industrial life, it was the farmer who took the lead.

Chihuahua and its region was the final product of a long period of Spanish territorial expansion in the North. This period began during the second decade of the seventeenth century, before the founding of Parral and culminated around 1730 with the consolidation of Chihuahua. Since then this region has gone through successive periods of "boom and bust," but never again experienced the same intensity and dimension of territorial expansion as occurred in the late seventeenth and early eighteenth centuries. One might say that Chihuahua was the last entity of its type to be established in northern New Spain, the last stop of the seventeenth century North.

Endnotes

1. Documents in AGI Contaduría 926 refer to Cuentas de la Real Caja de Durango, 1654-1656; Contaduría 927 to Cuentas de la Real Caja de Durango, 1666-1673.

2. An excellent example was that of Miguel de Iriate, who in 1671 denounced his claim to land at a place between the towns of San Felipe and San Joseph, called La Natividad, which has been "a former Tarahumar pueblo ... a wilderness and unoccupied ..." (AHP 1671a; AGI Contaduría 927). Another similar case was that of Nicolás Caro, who settled on land that had once belonged to the Indians of Santa Cruz de Tarahumara; he acquired a place called San Javier with three *caballerías* of land and two *sitios* for livestock.

3. During the 1640s, for example, the Jesuits, with the help of the military from Parral, offered peace to the Conchos in order to consolidate San Felipe, the most important mission in the lower Tarahumar country. The move paid off at the time of the great rebellion, for San Felipe was the only mission that resisted the attacks of the rebels, whereas Papigochi and San Francisco Javier de Satevó were destroyed and San Francisco de Borja had to be abandoned

(Naylor and Polzer 1986, 394). The Indians of the surrounding area were finally concentrated in the missions, which began to function as stable establishments.

4. Arlegui (1851) stated that the new mission of Santa Isabel was founded in 1668; although probably the Franciscans were already established there previously, the fact is that only in 1673 did the mission begin to operate with financial help from the *caja real* in Durango.

5. The oldest Spanish land acquisition registered at Casas Grandes is that of Andrés López de García, a resident of Parral, who, between 1665 and 1666, settled a few leagues from the Indian village (Casas Grandes) with his "servants, armed guards, horses, and family," in a place near the San Buenaventura River (AHP 1671a, ff. 354-357); in 1676 he also obtained a grant of two *sitios* for livestock and four *caballerias* of farm land (AGI Contaduría 927).

6. According to Arlegui (1854,97), Bachíniva was founded in 1660 with the name of Santa María Nativitas de Bachíniva; however, this date is improbable since in 1694 it carried the title of "*nueva conversión*" (Hackett 1926, 358-361).

7. Those in the Papigochi area included Matachi, Tojórare, Cocomórachi, Temósachi, Yepómera, Tutuaca, Temeichi, Tomochi, Carachí, Basúchil, Namiquipa and Santo Tomás, among others (González Rodríguez 1989). These comprised a zone that could serve as an example of the efficacy of the mission system.

8. Other nearby Indian pueblos also recorded increases in population: Chuvíscar had 44 families, Santa Isabel, 70; San Andrés, a Franciscan mission of Tarahumar had 120 families, and Santa Cruz, 70 (Griffen 1979, 74-77). Also new *pueblos de visita* were incorporated into this network: Sainápuchi with 70 families (a *visita* of Santa Isabel, 14 leagues away) (Griffen 1979, 74); Santa Clara with 20 families, 12 leagues east of Namiquipa (Griffen 1979, 83); San Juan de la Concepción, another *visita* of Santa Isabel with 70 families in 1693 (Griffen 1979, 74-83). There were other *visitas*, most without population counts, that also belonged within the reduction system; among these we can cite San Bernardino, Las Cruces, Queparipa, San Diego, and Guadalupe, all among the new reductions; and Bachima and Bachíniva among the old ones.

9. When a request for unsurveyed land was made, generally the assessor determined the fee according to the estimated land value. In a previous study that we made of 1060 land requests in Nueva Vizcaya for the seventeenth and eighteenth centuries, a fee of 200 to 300 pesos corresponded to an area of 60 *sitios* for cattle and horses (*ganado mayor*). Although it is not possible to use this estimate as an index, probably the request of Pérez de Rivera might have amounted to at least 10 *sitios de ganado mayor* (Alvarez 1991).

10. For example, Diego Cano de los Ríos in 1669 married Francisca de Grados, relative of the Urbaneja family, one of the oldest hacendado clans of the Valley; her sister Magdalena wedded Juan de Navarrete, member of another established family in the San Bartolomé area (APSB, Registros parroquiles).

11. The initial results of the investigation in good measure confirms the trustworthiness of the census figures and estimates of the time. Although the data are of a time of a mining boom, during which, at least at its beginning, ingress and egress of people are frequent, the total population is relatively stable, especially among the *vecinos*, merchants and miners. In fact, the stability of the numbers of inhabitants in the mining centers is much greater than previously thought (Alvarez 1989). We have been able to identify 585 persons (*vecinos*, miners, and other permanent inhabitants) all of whom lived in Chihuahua between 1700 and 1733. Here we do not attempt to ascertain the length of time particular persons remained in Chihuahua, but to identify the source of origin of human movements, that is, currents of migration that occurred in the peopling of this region. The search concentrated on the most populous areas of the plateau of Nueva Vizcaya, such as Parral, Valle de Allende (San Bartolomé), and Cusihuiriachi. An investigation of the activities of the Chihuahuan colonists in the surrounding areas in the period between 1660 and 1730 has led us to establish the ori-

gin of 310 of the 583 identified persons, that is, 53 percent of the total. This, we believe, can be considered a representative sample that can be applied to the total number.

12. Some of the more important ones included members of the Cano de los Ríos family (Miguel, Nicolás, Pedro), Diego de Maturana, Cristóbal de Orrantia, all hacendados from the province of Santa Bárbara; Pedro Facundo Carbonel from the Chihuahua region; Juan Domínguez de Mendoza from Satevó; José García de Valdéz from Matachi, Cusihuiriachi and Chihuahua; Domingo del Valle from Chihuahua; and Antonio de Trasviña who had haciendas in San Bartolomé, Chihuahua and Cusihuiriachi (AHP, 1716a, Vista de San Francisco de Cuéllar, 1723d, 1728a).

13. Thanks to such practices, the priests and missions combined with the big hacendados to produce part of the grain supplied to Chihuahua. It was a supply network that included nearby villages of Santa María Baserac and Yécora (Radding 1992, 64). In addition, the priests and missions contributed 21 percent of the maize and 10 percent of the wheat flour to the *alhóndiga* of San Felipe. We should point out that from the viewpoint of supply there did not exist a sharp difference between hacendados and priests, all of whom operated similarly. A clear example is seen in the Papigochi area, where, together with the missionaries, there was one Pedro de Almoina, a major supplier of grain for Chihuahua and owner of haciendas, including that of Namiquipa; in the same area Padre Navarrete also exported sizable quantities of maize to San Felipe. Another example is that of Juan Yarro and Eugenio Ramírez, respectively missionary and hacendado of Santa Isabel, both of whom exported maize and wheat flour to Chihuahua. Perhaps the most interesting example is seen in José García Valdéz, priest of the villa of San Felipe El Real, who was owner of the famous hacienda of Canoa in Chihuahua and two others in Matachi and Yepómera (Papigochi area), from which he supplied 8 percent of the wheat to the *alhóndiga* of San Felipe. An owner of mines and ore refineries as well, he succeeded in becoming one of the most prominent persons in local society.

14. Manuel de San Juan de Santa Cruz, María Ortíz de Campos (widow of Juan Antonio de Trasviña y Retes), Tomás de Duero, Joseph de Aguirre, Domingo Basoco, and Joseph de Zubiate (AHP, 1727, Visita a las minas de Santa Eulalia).

15. As a general rule, one *carga* is equivalent to 138 kilograms; however, in the case of products that are difficult to transport, such as firewood and charcoal, the traditional weight of a *carga* in northern Mexico varied between 30 and 40 kilograms, that is, approximately that which a beast of burden could carry. We have used the equivalent of 40 kilograms per *carga* in order to avoid the excessively high conversion rate of 138 kilograms (Carrera Stampa 1949; *Dirección General de Estadística* 1933). Calculating a *carga* at 138 kilograms the volume would have reached 415,040 kilograms of firewood and 509,536 kilograms of charcoal.

References

Archival Sources

AAC — Archivo del Ayuntamiento de Chihuahua [Chihuahua City]

AGI — Archivo General de Indias [Sevilla, Spain]

AHP — Archivo Histórico del Parral [Parral, Chihuahua]

AIPG — Archivo de Instrumentos Públicos de Guadalajara [Guadalajara, Jalisco]

APC — Archivo Parroquial de Cusihuiriachi [Cusihuiriachi, Chihuahua]

APMN — Archivo Parroquial de San Diego de Minas Nuevas [Parral, Chihuahua]

APSB — Archivo Parroquil de San Bartolomé [Valle de Allende, Chihuahua]

BPG — Biblioteca Pública de Guadalajara, Archivo de la Real Audiencia [Guadalajara, Jalisco]

Published and Other Materials

Almada, F. 1928. *Diccionario de historia, geografía y biografía chihuahuenses*. Chihuahua: Talleres Gráficos del Gobierno del Estado.

——. 1984. *Guía histórica de la ciudad de Chihuahua*. Chihuahua: Ediciones del Gobierno del Estado de Chihuahua.

Alvarez, S. 1989. Minería y poblamiento en el norte de la Nueva España: los casos de Zacatecas y Parral. In *Actas del I Congreso de Historia Regional Comparada*, 105-109. Ciudad Juárez: Universidad Autónoma de Ciudad Juárez.

——. 1991. Tendencias regionales de la propiedad territorial en el norte de la Nueva España: siglos XVII y XVIII. In *Actas del II Congreso de Historia Regional Comparada*. Ciudad Juárez: Universidad Autónoma de Ciudad Juárez.

——. 1992. Chiametla, una provincia olvidada del siglo XVI. *Trace* 22:9-24.

——. In press. Indios misioneros y tentaciones: escenas de la vida cotidiana en una misión. In *Ensayos de historia colonial del norte*. Ciudad Juárez: Universidad Autónoma de Ciudad Juárez.

Arlegui, J. 1851 (1st ed. 1737). *Crónica de la provincia de N.S.P.S. San Francisco de Zacatecas*. Mexico City: Cumplido.

Bakewell, P. 1976. *Minería y sociedad en el México colonial. Zacatecas (1546-1700)*. Mexico City: Fondo de Cultura Económica.

——. 1990. La minería en Hispanoamérica colonial. In *Historia de la América Latina, América Latina colonial: Economía*, vol. 3, ed. Leslie Bethell. Barcelona: Crítica-Cambridge University Press.

Bargellini, C. 1984. *La catedral de Chihuahua*. Mexico City: Universidad Nacional Autónoma de Mexico City.

Carrera Stampa, M. 1949. Evolution of weights and measures in New Spain. *Hispanic American Historical Review* 39:2-24.

Cramaussel, C. 1989. Haciendas y mano de obra en Nueva Vizcaya. El curato de Parral. *Trace* 15:22-30.

——. 1990. Evolución en las formas de dominio del espacio colonial: Las haciendas de la región de Parral. In *Actas del Segundo Congreso de Historia Regional Comparada*, 115-140. Ciudad Juárez: Universidad Autónoma de Ciudad Juárez.

——. 1991. Encomiendas, repartimientos y conquista en Nueva Vizcaya. *Historias* 25:73-91.

——. 1992a. Juan Rangel de Biesma, un descubridor en problemas. In *Chihuahua, las épocas y los hombres*, vol. 8. Ciudad Juárez: Meridiano 107, Universidad Autónoma de Ciudad Juárez, and Gobierno del Estado de Chihuahua.

——. 1992b. Haciendas agrícolas y abasto de granos en el Parral del siglo XVII. In *5 siglos de Historia de México*, 347-354. Mexico City: Instituto Mora/University of California at Irvine.

Cramaussel, C., and Alvarez, S. 1992. El plano de 1722 de San Felipe el Real de Chihuahua. In *Arte y Coerción*, 45-69. Mexico City: Universidad Nacional Autónoma de México, Instituto de Investigaciones Estéticas.

Dirección General de Estadística. 1933. *Censo agrícola ganadero de 1930*. Mexico City: Secretaría de Economía Nacional.

Florescano, E. 1973. Colonización. Ocupación del suelo y "frontera" en el norte de la Nueva España 1521-1750. In *Tierras Nuevas: expansión territorial y ocupación del suelo en América (siglos XVI-XVII)*, ed. Alvaro Jara. Mexico City: El Colegio de México.

Galaviz de Capdeville, M. E. 1967. *Rebeliones indígenas en el norte del reino de la Nueva España (siglos XVI y XVII)*. Mexico City: Campesina.

Gerhard, P. 1982. *The North frontier of New Spain*. Princeton: University Press.

González Rodríguez, L. 1984. *Crónicas de la sierra Tarahumara*. Mexico City: SEP.

Griffen, W. 1979. *Indian assimilation in the Franciscan area of Nueva Vizcaya*. Tucson: University of Arizona Press.

Hackett, C. W., Ed. 1926. *Historical documents relating to New Mexico, Nueva Vizcaya and approaches thereto, to 1773*. Vol. 2. Washington: Carnegie Institution.

Hadley, P. 1979. *Minería y sociedad en el centro minero de Santa Eulalia, Chihuahua (1709-1750)*. Mexico City: Fondo de Cultura Económica.

Naylor T. and Polzer, W., Eds. 1986. *The presidio and the militia on the northern frontier of New Spain: A documentary history*. Vol. 1. Tucson: University of Arizona Press.

Porras Muñoz, G. 1966. Las minas de Chihuahua. *Boletín del Archivo General de la Nación*, 2a serie, 7:635-660.

——. 1989. La fundación de Chihuahua. *Cuadernos del Norte* 7:3-11.

——. 1980 (1st ed. Pamplona 1966). *Iglesia y Estado en Nueva Vizcaya*. Mexico City: Universidad Nacional Autónoma de México.

——. 1980. *La frontera con los indios de Nueva Vizcaya en el siglo XVII*. Mexico City: BANAMEX.

Radding, C. 1992. La economía misional y la subsistencia indígena en Sonora. Siglo XVIII. *Trace* 22:59-71.

Romano, R. 1970. Sens et limites de l'industrie minière en Amérique espagnole du XVIème au XVIIIème siècle. *Journal de la Société des Américanistes* 59:129-143.

——. 1986. Fundamentos del funcionamiento del sistema económico colonial. *VI Simposio de Historia Económica del Consejo latinoamericano de Ciencias Sociales de Lima*, 23-27.

American Mining and European Mining, 1750–1820: a Comparative Perspective

Julio Sánchez Gómez and Guillermo Mira Delli–Zotti,
Universidad de Salamanca, Spain

Abstract

A reevaluation of the innovations in mining which were transmitted by Spain to its colonies offers a revision to the widespread idea of the backwardness of American colonial mining with respect to that of Europe. A comparative perspective on the different contexts of Central Europe and Spanish colonial America allows us to fix our attention on administrative and juridical–institutional aspects that until now have been neglected and which have strongly influenced the development of other factors, such as the technology or financial infrastructure, that are no less important. Even if the conclusions are not definitive, they indicate the interest of the Spanish government in transmitting to America not only the techniques but also the central European models of organization and administration of mining matters.
Key words: mining, technology, mine administration, Spanish America, Central Europe.

Resumen

El artículo plantea una revisión de la difundida idea del retraso de la minería americana colonial respecto de la europea. Para ello se adapta una perspectiva comparativa, que permite poner en relación los diferentes contextos de Europa central y la América española. Esto permite fijar la atención en aspetos hasta ahora descuidados, como el administrativo y el jurídico–institucional, que condicionaron fuertamente el desarrollo de otros factores (como la tecnología o la estructura financiara) no menos importantes. Si bien las conclusiones no son definitivas, se advierte el interés del gobierno español por trasmitir a América no solo las técnicas sino los modelos de organización y gestión centroeuropeos en materia de minería. Las puntualisaciones de este breve trabajo estimulan a reevaluar los resultados de tales iniciativas.
Palabras claves: minería, tecnología, administración, América española, Europa central.

In Quest of Mineral Wealth: Aboriginal and Colonial Mining and Metallurgy in Spanish America, edited by Alan K. Craig and Robert C. West, 1994. Geoscience and Man, vol. 33, pp. 205-212. Dept. of Geography & Anthropology, Louisiana State University, Baton Rouge, LA 70893-6010, U.S.A.

The present paper is a preview, still very limited, of an on–going investigation that attempts to establish a comparative perspective between European mining and that of colonial America. The motive that brought us to choose this work was the widely held idea that American mining was backward and suffered in comparison with that of Europe. The origin of this idea goes back in time to the perception that some qualified witnesses transmitted in the field of mining–metallurgy — the majority being Europeans — such as Humboldt ([1811] 1984]), Nordenflicht, Helms or Daubuisson (1802), who invariably criticized and even were full of irony about the rustic techniques used in Latin America. Their opinions have been taken up again, at times without enough criticism, by some of today's specialists. However, that primitiveness seems to contradict the success attained in the production of precious metals during the epoch to which we direct our attention.

One of the developments considered in this comparative perspective is the fact that during the second half of the eighteenth century contacts in the sphere of mining notably intensified between America and Europe. Fruits of this open policy on the part of the Spanish monarchy were well known scientific expeditions, among which those specifically on mining have already been the object of various investigations, whose development and results need not occupy our attention.

From the whole of reformist policies, together with improvements of essentially technical character that began to be transferred from Europe, a series of administrative and judicial–institutional measures were taken whose inspiration and roots we believe have not yet attracted sufficient attention. An approximation of the legal and administrative picture in which mining in Central Europe evolved can help us to find them, since among events which unfolded for colonial America suggestive coincidences appear.

In Central Europe, the district of Freiberg in Saxony was one of the principal producers of silver. There, as in the territories of the Spanish Crown, the subsurface was royal property, given for exploitation to his subjects in exchange for payment of a right that in principle symbolized recognition of his eminent ownership of it. Up to this point there is parallelism between the systems. However, in Saxony nothing altogether escaped the vigilance of the Prince, while in America the King wanted nothing to do with the process of production and his presence appeared only at the moment of taxing the production, of which he withheld a part. It is certain that in the Indies there existed a motley body of law destined to regulate mining activity, although of uncertain compliance.

In Saxony the administration was more centralized and effective, and the authority of the Prince was manifest in every phase of the production process. The management of the mines was reserved for the Elector of

Saxony, who exercised it by means of a pyramidal group of entities (Private Council, Council of Finances, High Council or Mining Court of Freiberg, and Councils of the District) and a corps of functionaries at the orders of a General Director named by the Prince.

This corps was set up for the sake of direct intervention of the royal administration in the processes of production, but only parts of the operations ended up in their hands. Such is the case of the metallurgical phase that was carried out in the only foundry belonging to the Prince, which according to contemporary testimonies, had reached high levels of planning, resulting, for example, in significant cost savings. Near the end of the eighteenth century all miners were obliged to have their ore treated in the royal smelter, and it was here the minimum for incoming ore quality was established, as well as payment for it and corresponding taxes. The establishment of minimum fineness for the ores obliged the miners to deliver concentrates that permitted the royal assayers to obtain the most appropriate mixtures. The concentration of all the metallurgical activities of the district in a single treatment plant produced a cost savings derived from the economy of scale, to which was added the savings of fuel obtained from the Crown forests.

The capacity for decision that this organizational framework gave to the Prince, together with his financial possibilities, facilitated the adoption of technical innovations that exceeded the political desires of the Spanish monarchs. We cite for example the introduction at Freiberg of the new process of amalgamation in barrels, the result of a simple royal decree. By the same means, the most important drainage tunnels were constructed under the direction and financing of the state.

Decisions of such scope did not obey, however, the caprices of the Prince. Instead, they were taken only when they had the careful assessment of the whole group of expert functionaries graduated from the School of Mines of Freiberg, whose creation goes back to 1766. Even before that, there existed specialized studies destined to create a qualified functionary. Such officials not only advised the Prince in technical matters, but they also played an advisory role to the mining companies. For example, mining surveyors made plans of the mines while the assayer tested the different minerals in order that directors in charge of exploitation could anticipate the richness of them.

On the contrary in America, in spite of the Crown's interest in the production of precious metals, the administration of mining presented very different characteristics. If indeed the state intervention could have had such weight as in Central Europe, its mechanisms were different and, in consequence its results. The Crown was concerned with supplying hand labor and cheap mercury to the American miners, but did not intervene in the internal governance of the mines, nor had it a notable role in the technical decisions that marked the day–to–day exploitation, nor in their financing.

Another notable difference between European and American mining resides in the form of capitalization and direction of the mining company. In the former, financing and exploitation were separate processes. Ownership consisting of companies of stockholders existed on the one hand, and on the other the technical directors to whom they entrusted all questions relative to exploitation. This division contributed as much to maximizing the technical quality of underground works, always supervised by a specialist, as to insuring the continuity of work as a result of a regular flow of capital. This steady flow was insured by two means. First, all the stockholders had the legal obligation to provide more capital in proportion to their holdings in the company in those times when the mine was not profitable, under penalty of losing their stock. Secondly, when the mine was making a profit and produced enough to declare dividends, a part of this was placed in a contingency fund that allowed the stockholders to weather possible losses in bad years. By these mechanisms Saxony mining seems to have resolved one of the obstacles that held back the development of mining activity in America: the lack of investment caused by the attitude of commercial capital that hoped to grab the benefits from mining without assuming the risks.[1]

Throughout the colonial period, as much in Mexico as in Peru, the mining sector experienced a chronic lack of capitalization, because of various factors. To begin with, there was the separation between commercial capital and the mining company, whose interests were frequently divergent; while big business sought to capture precious metals to transfer them to Europe, the miner found himself left out of the capital necessary for reinvestment. This situation was aggravated by the excessive fragmentation of mine ownership and the reduced size of the companies. In search of resources, the miners ended up falling into dependence on usurious loans made available by their suppliers or financiers, typically representatives of the large commercial firms in the capital. In spite of repeated complaints of the miners, this symbiosis was inevitable, given the absence of credit institutions or more developed finance companies. Within this general description, naturally there were exceptions: wealthy businessmen who invested in mining were not unusual in eighteenth century México; also there were fortunate miners who diversified their risks by investing in agriculture or urban property. But this does not invalidate the picture presented above.

Also, from the point of view of the application of technology to production, the contrasts between Central European mining and colonial American mining were very marked. In the first case, the following and removal of veins was in response to a rigorous, previously made plan. The system of shafts and galleries was practiced in quadrangular form; winzes were made at predetermined intervals; and the advance along the veins was done according to the technique of forming steps, as many ascending as descending. This rationality in the removal of ore is seen to be complemented by the rapid removal of material from the galleries a task which was entrusted to ore carts riding on wooden rails.

In Saxony mining had reached a high degree of mechanization, and manual labor was largely displaced by the massive use of hydraulic energy that required heavy initial investments, but was amortized fairly soon thanks to the practically free energy source. In Saxony two factors favored the alternative employment of hydraulic force. On one hand, there was the high cost of human labor. On the other hand, a series of conditions permitted the advantageous use of the water: its abundance in the region, easy access to it, and the proper stream gradients.

The work of mucking out and draining of mines depended in large part on hydraulic energy applied to large waterwheels with buckets. This system not only enormously reduced the costs but also increased productivity in a notable way. In an eight–hour shift, the commonest machine in ore extraction, the hydraulic mine hoist or *malacate*, was able to lift 480 cubic feet of cargo compared to only 200 in a hoist operated by horses, and barely 120 in the case of manual labor. The reduction in costs became equally noteworthy: it could reach 48 percent compared to a horse–operated hoist and 84 percent compared to a manual labor shift. Somewhat similar savings occurred in the case of drainage accomplished by hydraulically–actuated pumps.

The importance of employing hydraulic power in the Saxony mines does not rest so much on its originality — this was utilized in other regions and even in some American mining centers [most notably Potosí -Ed.] — as in the high level of technology obtained that permitted its use in an intensive and rational form. For example, a single water current could successively serve to operate the wheels applied to the drainage pumps, the extraction machinery, the crushers and the concentration tables, serving at the same time an ample grouping of uses. At times, by means of tie–rods and levers that remotely transmitted the movement, it was possible to obtain the versatility to actuate machinery situated in places quite distant from the water wheels. These were constructed with great skill at different levels, often in the interior of the mines with the object of also using the underground seepage water.

Hydraulic energy constituted the base for mechanization of the tasks of preparing the ore outside the mine, and contributed indirectly to betterment of health conditions. The mill consisted of grinders of hydraulic stamps attended by a single operator. The ore transformed into fine particles was swept away by a water current that deposited it in successive decantation basins. From there it was transported to the concentration tables, also activated by waterpower causing vibration. At the end of the process a significant concentration of ore was reached that facilitated the development of the later metallurgical step. Throughout these operations, the workers were much less exposed to the common lung diseases that threatened their health as a consequence of the dry method, a phenomenon that continued to be habitual in American mining even during the nineteenth century.

Finally, it should be pointed out that the accessibility and cheapness of hydraulic energy notably slowed down the introduction of steam in the district of Freiberg until very late in the nineteenth century. The extreme example of the Hartz mountains, where it was never introduced, demonstrates the excellent adaptation of these regions to the natural resources available to be transformed into energy.

In contradistinction to the picture described, perhaps the most notable element of American mining was its poor recourse to mechanization. However, it is necessary to point out that the technical development in some Mexican mines exceeded, during the eighteenth century, that of Peruvian mining. In the viceroyalty of New Spain some drainage tunnels reached spectacular dimensions; the use of the *malacate* became progressively widespread in drainage and ore extraction, even if indeed it was animal powered, which implies a major improvement in vertical haulage. To this was combined the introduction and rapid diffusion of blasting in breaking down of ore and opening of galleries, about the time of the appearance of incipient companies of stockholders. Together with these traces of modernization there remained elements openly archaic, such as the poor prior planning in the delineation of underground works, and underground transportation which was completely manual or with little recourse to any source of energy other than animal, and even this was only applied to power the mine hoists.

In Peru the technical anachronisms were much more pronounced than in Mexico. Subterranean excavations were made with greater disorder and less understanding than in the viceroyalty of New Spain. In the mines visited by the mariner Malaspina, for example, the excavation always followed the direction of the vein, producing sinuous tunnels, irregular and dangerous that made it difficult to work, which increased the costs of extraction and diminished productivity. There was a shortage of capital to begin large scale underground works, caused in many cases by the disposition and quality of the veins, with the result that in Peru the use of vertical shafts served by mine hoists was almost unknown. In mining centers such as Potosí or Hualgayoc, where mineral deposits were found in the extreme upper part of the mountain, veins were attacked by horizontal adits that served for drainage at the same time. But even in the case of different configurations, as in Cerro de Pasco, the shafts in use did not reach excessive depths (no more than 100 yards). The extraction of ore and drainage were accomplished manually, by crude winches turned by workers who went down by narrow passageways and rough leather ladders. The cost of animals in the Andean world, added to the deeply entrenched practice of compulsory labor, might have been the basis of this alarming lack of mechanization in the extractive phase. This state of affairs was attenuated by recourse to hydraulic energy in the grinding of ore in some mining camps, especially at Potosí.

The roots of the situation described, as much in Mexico as in Peru, should be looked for in a multitude of factors such as the quality of the veins, the juridical organization, the location of the mining camps in their natural surroundings, and the kind of production relations that were established between the Spanish and creole entrepreneurs and the indigenous workers.

The greater use of technology in Germany came from factors we have already indicated — better possibilities of recourse to hydraulic energy, and greater costs of hand labor which were an incentive to find a substitute. Added to these were other factors such as: a greater investment capacity afforded by the more widespread existence of stock companies; seams with greater silver content; use of the associated industrial metals found with silver (lead, copper, and others}; greater technical capacity of those in supervisory capacities as a result of the institutionalizing of instruction in mining and metallurgy; and possibly greater benefits in the differential in the price of silver.

When these marked contrasts were recognized by the reformist sectors in the Spanish administration, they sought to rectify them in the last third of the eighteenth century, which would give rise to a group of reforms on two fronts: the technology and the administration. The first translated itself into sending scholarship holders to Central Europe in order to incorporate the most modern techniques in mining and metallurgy. These Spaniards, together with a handful of German specialists, would later be sent to America with the object of spreading there the techniques developed in Central Europe.

The second front became institutional reform, conceived for all of American mining — especially in Mexico and Peru. It was José de Gálvez, from the Ministry of the Indies, who imposed this policy with the purpose of conceding certain autonomy to the mining sector and elevating its social status. The formalization of a mining guild, the creation of courts and provincial local governments, the modernization of the judicial system and the founding of mining schools constituted the nucleus of the reformist project in the institutional area. If it has been accepted until now that the technical innovations were of Central European origin, we believe that the institutional model that they attempted to apply in the Indies also sank its roots in the juridical–administrative framework that regulated Germanic mining, an aspect notably ignored until now and which we offer here only in the form of an hypothesis. From this perspective, the Bourbon methods were instituted to bridge the abyss that separated American mining from that of Central Europe, of which the reformists were very much aware, taking as a point of reference the organization and the institutions that the latter had developed. The evaluation of the results of this policy in America still constitutes an open debate, but that is beyond the limits we have imposed on this research.

Endnote

1. Certainly, in the Mexican mines of the second half of the eighteenth century some companies were formed for exploitation. We do not know, however, more details about the form of their constitution and internal functioning.

References Consulted

Brading, D. 1975. *Mineros y commerciantes en el México borbónico, 1753–1810.* Madrid.

Brading, D. and H.E. Cross 1972. Colonial silver mining: Mexico and Peru. *Hispanic American Historical Review* 52:545–579.

Breithaupt, J.F.A. 1825. *Die Bergstadt Freiberg im königreiche Sachsen.* Freiberg: Craz und Gerlach.

Daubuisson, J. F. 1802. *Des mines de Freiberg en Saxe et de leur exploitation.* Leipzig.

Fisher, J.R. 1977. *Minas y mineros en el Perú colonial, 1776–1824.* Lima: Instituto de Estudios Peruanos.

González Claverán, V. 1988. *La expedición científica de Malaspina en Nueva España, 1789–1794.* Mexico City.

Helmer, M. 1969. Deutsche Bergleute in amtlicher Mission in Peru (1788–1798). *Technik Geschichte* 36: 4.

Humboldt, A. [1811] 1984. *Ensayo político sobre el reino de Nueva España.* Mexico City: Editorial Porrua, S.A.

Kasper, H. H. and E. Wächtler 1986. *Geschichte der Bergstadt Freiberg.* Weimar.

Majer, J. 1975. *Sbornik Národního Technického Muzea Praze* (Fire and watercolumn engines in European mining). Prague: Acta Musei, Nationalis Technici Pragae.

Ovando Sanz, G. 1975. *La Academia de Minas de Potosí, 1757–1790.* La Paz: Banco Central de Bolivia.

Tandeter, E. 1978. Potosí y los ingleses a fines de 1826. *Historia y Cultura* 3.

Labor Force and Mercury Production in Almadén, Spain, 1759–1808

Rafael Dobado González,
Universidad Complutense, Madrid, Spain

Abstract

Mercury mining at Almadén, Spain, increased and led to innovations in both technology and labor relations during the long expansive phase that transpired between the mine fire of 1755–1757 and the beginning of the War of Independence. The efforts of the Bourbon authorities to increase the production of mercury at Almadén succeeded in increasing sales to mining in America, especially to New Spain. As a result of introducing the use of amalgamation, there was a consequent reduction of the exorbitant costs that had endangered the production and coinage of silver. The Bourbon reforms gave the mercury mines at Almadén greater financial resources and fostered technological innovation. However, the latter occurred even before the increase in production resulting from reduction in the unitary cost of mercury. The considerable increase in sales of mercury to New Spain made possible the increase in consumption in the colony.

With regard to work, the Bourbon reforms established an original model of labor relations that did away with the "lack of workers" that had lasted for centuries. Absence of compulsory labor at Almadén was a notable difference from American mining locales.
Key words: Almadén, mercury, labor, eighteenth century, New Spain (Mexico).

Resumen

Este artículo pretende abordar conjuntamente la descripción y explicación de los aspectos productivos y laborales de Almadén durante la fase expansiva de larga duración que transcurre entre el incendio de las minas de 1755–1757 y el inicio de la Guerra de Independencia. Los esfuerzos de las autoridades borbónicas para incrementar la producción de azogue en Almadén pueden ser calificados de exitosos y responden al intento de incrementar la oferta a la minería americana, en particular a Nueva España, del principal input del proceso de amalgamación, con el consiguiente efecto positivo sobre la recaudación de las figuras impositivas que gravaban la producción y amonedación de plata. Por lo que respecta a la producción de azogue en Almadén, el reformismo borbónico dotó a las minas de crecidos recursos financieros y favoreció la innovación técnica. Sin embargo, esta última perseguía antes el au-

---

In Quest of Mineral Wealth: Aboriginal and Colonial Mining and Metallurgy in Spanish America, edited by Alan K. Craig and Robert C. West, 1994. Geoscience and Man, vol. 33, pp. 213-231. Dept. of Geography & Anthropology, Louisiana State University, Baton Rouge, LA 70893-6010, U.S.A.

mento de la producción que la reducción del coste unitario del mercurio. En cualquier caso, el considerable aumento de la oferta de azogue a Nueva España hizo posible el aumento del consumo en la colonia.

En cuanto al trabajo, el reformismo borbónico estableció un original modelo de relaciones laborales que permitió superar el obstáculo al aumento de la producción que durante siglos había constituido la "falta de gente" en Almadén. A diferencia de lo que se observa en la minería americana, el trabajo compulsivo no representó ningún papel en Almadén.

Palabras claves: Almadén; producción de azogue en Almadén; consumo de azogue en Nueva España; fuerza de trabajo en Almadén; reproducción de la fuerza de trabajo en Almadén; relaciones laborales en la minería española, siglo XVIII.

Introduction

In contrast to the interest awakened by questions related to mining labor in Spanish America during the colonial period, that field of study in the case of Spain remains scarcely explored. This paper hopefully will fill such a historiographical gap. We will concern ourselves with a mineral exploitation of special relevance to Spain and its American colonies. After diffusion of the amalgamation process in the second half of the sixteenth century, relations between Almadén and America were particularly close.

Because of their size and contribution to the finances of the Crown, the mines of Almadén constituted the principal element of the Spanish mining sector during the colonial period. With thousands of workers and an annual production that one year rose above 1.3 million pesos, the mines of Almadén were far from being unimportant. Their contributions to the Royal Treasury in the form of taxes converted them, in the words of Morete de Varela, into "the most integral jewel and true base, motive force and support of the Crown."

During the 1700s, the Crown paid increasing attention to the mines. After the turn of the century an overall policy to place Almadén on a more solid base intensified. Recruitment of German technicians, increasing financial resources, and proliferation of incentives to attract workers are only some examples of the new governmental attitude. The objective was not so much to increase income from mercury as it was to give strength to the varied price controls on silver in America, especially in New Spain, principal market for Almadén mercury. To achieve this, it became necessary to increase both the price and availability of mercury, an element indispensable to the amalgamation of silver. Consequently, increasing production of mercury at Almadén constituted a basic step in any attempt to increase price controls on silver.

Production of mercury, after the mine fire of 1755–1757, experienced a new impulse culminating in especially voluminous exportations at the beginning of the nineteenth century. The consumption of mercury in New Spain multiplied fourfold in the course of the 1700s and, after the mid–1770s, Peru became another destination of Spanish mercury exportation.

All of the indicators of silver production in New Spain (coinages, manifests and fiscal income) reflect a century–old tendency to increase.

After the mine fires were extinguished, increasing mercury production became the foremost management objective. But it was not always reached — an example being the decrease in mercury production during the decade of the 1780s — even though the principal techno–economic decisions were subordinated to it. Another priority objective was reduction in cost of production. However, this paled in comparison with the increase in exportation. Given the yield ratio between mercury and silver, the multiplier effect on the royal treasury income for every additional *quintal* [hundred pounds] of mercury was greater than that of whatever reduction in production costs. Technical innovation at Almadén was slow and limited to confronting, with some success, the physical obstacles arising from working the veins at increasing depths.

Almadén Mines and Workers

The years 1759 and 1808 delimit the period of history of Almadén in which there was a chronic "lack of people" to work in the mines. Socio–economic and demographic processes that underlay the expansion of mining are included in the concept of replenishing the work force.[1]

Almadén is located in the extreme southwestern part of the southern Spanish highlands (fig. 1). A sparsely populated, rugged terrain made communications difficult and limited the possibilities of agricultural expansion to livestock raising because there were no markets nearby. However, wood, low scrub and quarries were plentiful. Thus, the mines had easy access to a few essentials for mining, but reduced population, isolation and bad roads were serious handicaps.

The Pozo and Castillo mines exploited three rich and extensive veins in the subsurface of Almadén. Two other mines exploited less productive seams — one of them was quickly worked out— in Almadenejos during the second half of the eighteenth century.[2] Diverse surface installations (smelting furnaces, workshops, offices, warehouses, etc.) were located near the mine entrances where various highly integrated production activities took place. These included everything from excavation of ore to transport of the mercury to Sevilla, as well as fabrication and repair of tools, care of the beasts of burden (oxen and mules), and the accounting. Coordination of all these was complex and often unsatisfactory.

The bulk of the Almadén population, lacking alternatives, worked in the mines. Consequently, to increase mercury production it was necessary to expand the resident population. Almadén also presents marked demographic peculiarities. Before the second decade of the nineteenth century, the population demonstrated little or no capacity for growth. A comparatively high birth rate was frequently exceeded by an exceptionally elevat-

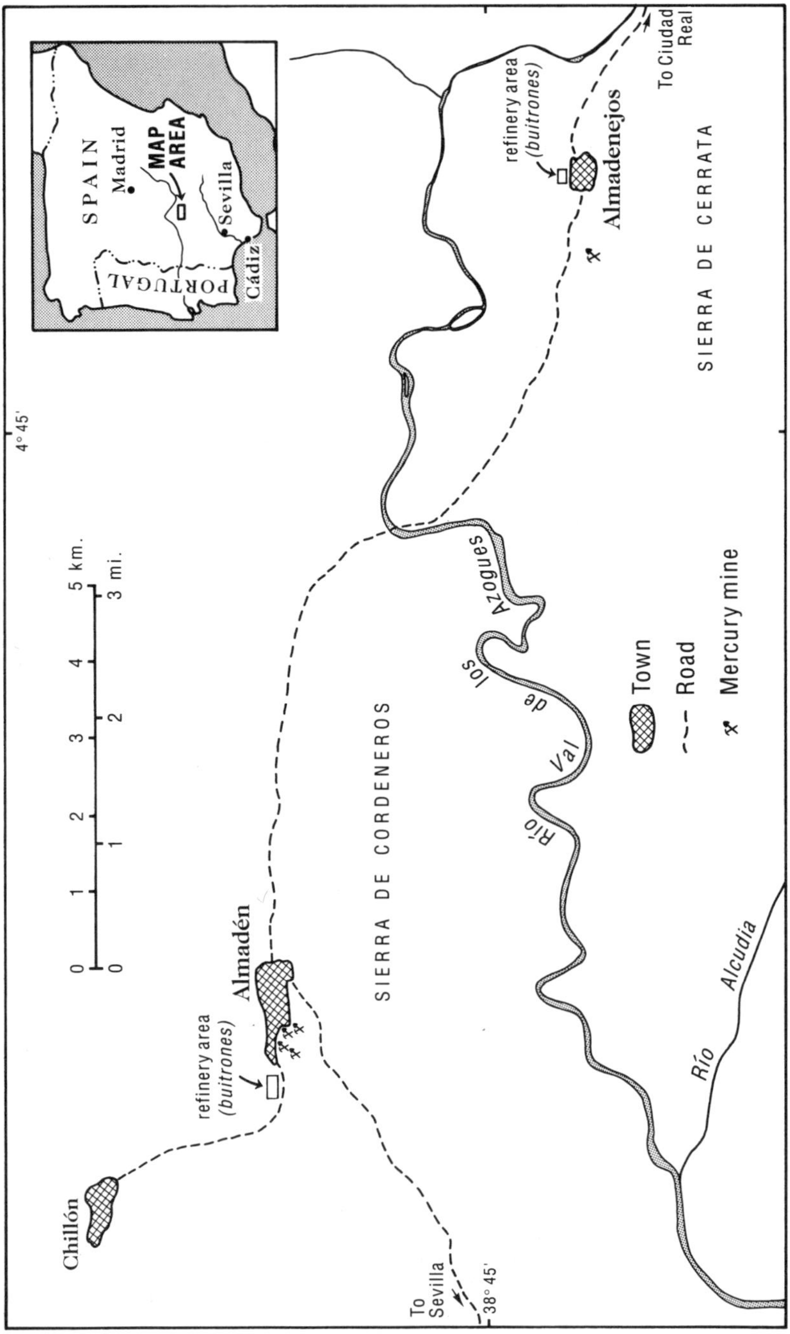

Fig. 1. Almadén mining area. Based on Spain, Mapa Topográfico Nacional, 1:50,000, sheets 807 (Chillón) and 808 (Almadén).

ed death rate. It can be shown that this excess mortality was concentrated in the adult male population involved in mining. The primary cause was mercury poisoning, or *hidrargirismo*, which had such dramatically noxious affects on their health. Other sicknesses of professional origin, such as lung diseases, and accidents, greatly reduced life expectancy of the mine workers. Multiple testimony of contemporary observers agree it was the most notable manifestation of the social costs associated with the mercury production process and one of the relevant factors used to explain the traditional "lack of people" in the mines.

In spite of the high mortality rate, the population of Almadén increased significantly between the middle and end of the eighteenth century. How was such an increase in the number of inhabitants possible? The answer is immigration. During various decades, Almadén as well as Almadenejos, was the destination of a swollen migratory current that amply compensated for losses.

The mercury production process showed a marked seasonality caused principally by a climatic factor. With the arrival of high temperatures, the smelters were shut down and the underground activities slowed to the point of simple maintenance of the mines. Around May or June, the increase in mercury losses in the furnaces and the decline in the chronically poor mine ventilation encouraged the suspension of exports until autumn. These months of inactivity were used to advantage by the resident workers in Almadén to "clean themselves" of the effects of mercury poisoning. Tasks in farming, at their apogee during the summer, were thought to be especially good for this purpose. However, it failed as a mechanism for increasing the work force during the months of greatest activity in the mines.

The temporal complementarity between seasonal unemployment in mining and agriculture is also found in another intersectorial flow that was gaining in importance during the second half of the eighteenth century. Temporary workers gathered preferentially in winter at Almadén in order to work some days or weeks to gain supplementary income during times when there was little activity in the fields.

Mine workers did not constitute a completely homogeneous population; some significant segments are clearly perceptible. In the first place we recognize the existence of neighborhood and seasonal workers in Almadén. The availability of permanent labor — termed "continuing miners" — contrasted with the the second, fluctuating group of temporary help, who had no possibility of access to the most qualified and best paid positions.

In the second place, a relatively small number of workers enjoyed the status of "employees," while the greater part worked as day laborers. The "employees" had the understood assurance of a guaranteed salary, possibility of vertical mobility and seniority — all of which played an outstanding role and provided access to diverse forms of indirect salary and job stability. The day workers lacked these advantages or only participated in them to a limited extent.[3] Under these circumstances, the salary level and other labor conditions such as the offer of subsidized food and staples, took on decisive importance. In summary, the term "mine worker" encompassed a variety of personal situations.

It is not easy to determine the number of workers laboring in the mines. In addition to problems of a documentary nature, the unequal work efforts of the different labor segments makes it difficult to interpret the scarce available data.[4] Seasonality of the work is an added inconvenience. Nevertheless, some figures give an idea of the magnitude of the hand labor that was involved in the production of mercury. Between 1771 and 1776, almost 1,000 workers were involved in mining and metallurgical tasks of medium or high skill in Almadén and Almadenejos.[5] In 1787, almost 1,400 inhabitants of both localities were to appear in a census as workers in the mines. Together with some 400 "employees," they constitute a stable segment of the work force that consisted of somewhat less than 2,000 members. The component that we have called "fluctuating" is identified with the approximately 1,000 temporaries that gathered annually at the sites during these dates. These numbers are compatible with the estimate of Director Hoppensack of at least some 2,000 day laborers by the end of the decade of 1780. No other Spanish mineral exploitation generated such a volume of employment. Everything seems to indicate that between the middle of the eighteenth century and beginnings of the nineteenth century, "mine work" experienced a considerable increase.

Salaries earned by mine workers are a "slowly dynamic" variable; they present a certain inflexibility toward the bottom, little changeability between years and a slight rising tendency that responds to some spot elevations. Their differential with respect to agricultural pay was higher. The distinction among the workers was reflected in their salary income. For the "employees" the assured daily pay rate resulted in a significant and comparatively high annual income. Among the day workers, the discontinuous work and a lower average pay rate implied less annual income. Significant differences between the established salaries for the distinct labor categories of day workers widened the salary stratification.

The variations in salaries scarcely operated as an equilibrium mechanism between supply and demand. Despite what might be expected from neoclassic economic theory, the periods characterized by a major "lack of people" were not accompanied by increases in salaries. Another noteworthy result of the contrast between salaries and prices of bread is that the annual income of the majority of the workers did not guarantee subsistence for a family.[6] It is not easy to understand the absence of salary conflicts. That may have been due to the fact that salary did not play a role in increasing the work force that we might have expected from a superficial observation of a labor model that seems to be dominated by mercantile relations.

The Production of Mercury

In figure 2 we see the amount of mercury production between 1646 and 1808, which allows one to distinguish at least three well differentiated periods (1646–1700, 1701–1755 and 1759–1808).[7] There are two criteria used to make such a distinction: the long–range trends of mercury exports, and their yearly variations. The magnitude of production from the mines, the expectations concerning its future evolution and the intensity of its fluctuations decisively influenced American mining. The absence of exogenous factors that might affect transport, large exports, and fluctuations, favored the production of amalgamated silver. The contrary occurred when the sale of mercury was down or fluctuating, or the prospects uncertain.

The period 1646–1700 is characterized by a decline in production. The first decades of the State arrangement with the administrators were not fortunate. Due to reduced profits in quantitative terms, however, production did not experience excessively sharp interyear fluctuations.

Much more voluminous exports were achieved in the beginning of the eighteenth century. Between 1701 and 1755, production levels exceeded those of the previous period. But intense interyear variations, not always driven by exogenous factors, limited the possibilities of stabilizing

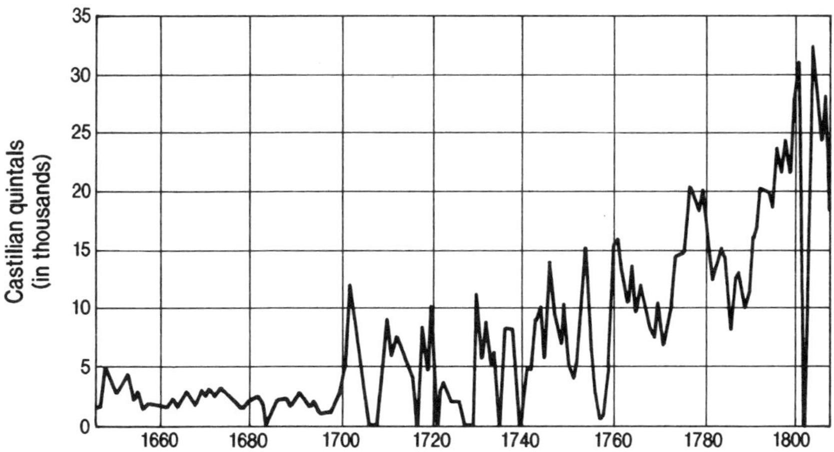

Fig. 2. Production of mercury in Almadén mines, 1646–1809. *Sources*: Bernáldez and Rúa (1862); AHN, Minas de Almadén.

the supply of mercury to the New World. The enormous exports during two five–year periods before the mine fires of 1755–57 — 9,400 Spanish *quintales* in 1745–49 and 8,100 *quintales* in 1750–54 — emphasize the fact that exploitation was seriously compromised by the more obvious inadequacy of the mining techniques and declining ore reserves.

The reopening of production began in 1759. A few years after the fire, the mines achieved exports that constitute historic records. In spite of some down cycles (1762–1773 and 1781–1790), in which, together with other factors, labor played a major role, the middle five–year periods of production in this third phase always were superior to the maximums of the previous phase. The increasing production became, toward 1800, a tripling of exports with respect to the largest at the beginning of the eighteenth century. This made it possible to increase the sale of mercury to New Spain and, after the decline of Huancavelica, to extend it to Peru after 1776. Furthermore, in this third period, together with the increase in production, a decrease was registered in the sudden interannual fluctuations characteristic of the first half of the eighteenth century.[8] The positive effects of a greater and more regular mercury production from Almadén on the American silver mining are unquestionable.

After this superficial picture, we pass on to consider something more detailed that happened in the years 1759–1808. In figure 3 we see the rising trend in production.[9] The average interyear increase is approximately 2 percent. This rate not only surpasses that obtained in 1701–1755 (somewhat less than 1.5 percent), but it was achieved in spite of some of the most

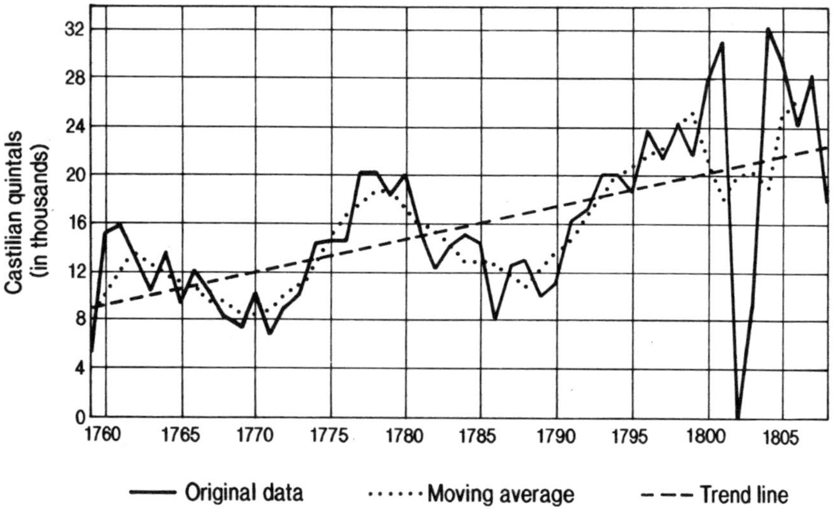

Fig. 3. Production of mercury in Spain, 1759–1808. *Sources*: Bernáldez and Rúa (1862); AHN, Minas de Almadén.

adverse production conditions. In principle, the continuance of work at depth, if it is not accompanied by increased richness of the minerals extracted, implies inevitable decreasing profits. At Almadén, both circumstances coincided: the mines increased in depth and there was no permanent increase in the fineness of the ore being extracted.[10] Work was continued, however, without an unacceptable increase in mercury unit production costs.

The explanation can be found in the technical innovations adopted in the middle of the eighteenth century. After the fire of 1755–1757, German technicians were able to bring about changes in work methods. An improvement in ore extraction by means of room and pillar stoping, together with application of underground surveying to select the best emplacement of shafts, winzes and gallery layouts, allowed for more logical ore extraction. The new physical configuration of the mines also improved drainage as did the introduction of hand pumps and a shaft hoist in the *Mina del Pozo.* The mine supports also were improved, but ventilation, essential to minimize the risks of mercury poisoning in the workers, continued to be very deficient.

A second wave of innovations, adopted half way between the decades of 1780 and 1800, was intended to answer new problems that arose from — or were aggravated by — the deepening of the mines, and others that were never entirely satisfactorily resolved. Among these innovations we point out the following: the importation of a steam engine drainage pump; a connecting tunnel between the two mines of Almadén that allowed the extraction of ore from both by means of the mine hoist built in 1769; the utilization of ore carts for underground transport; the construction of a crane on the outside for loading ore into wagons; the substitution of wooden bents by arches and retaining walls of brick masonry in the principal reinforcement works; the construction of a pair of smelting ovens patterned after those in use at Idria. The adoption of a working system thought up by Larrañaga [the chief administrator— Ed.], whose basic principles remained in force until the beginning of the twentieth century, merits special mention.

About 1808, the mines were far from being a model of technical exploitation, but the adoption of successive innovations at least had permitted continuance that was economically profitable during periods when production was decreasing. Moreover, the results from technical change in the drainage system were already apparent in terms of productivity.

Other factors also contributed to the increase in the production of mercury. One of them was the renewal of the works begun in Almadenejos around 1700. It was not until 1757 that this mineral deposit began to produce significant quantities of mercury on a sustained basis. The *Mina Antigua de la Concepción de Almadenejos* began to make a relatively significant contribution to the exports between 1758 and 1785 (fig. 4). In 1761–1766, 1777 and 1781, they obtained from it between 4,000 and 6,000 Spanish *quintales* of mercury that represented between a fifth and half of the total production. The Almadenejos production fell to nothing between 1786 and 1790. As a result of instituting a policy of excavating "*registros*" (exploration shafts) during the last fourth of the eighteenth century, in 1779 the rich deposit that was named the *Mina Nueva de la Concepción* was discovered. Exploited since 1794, its production surpassed 12,000 *quintales* in 1802 and 1805.

Although generally unfruitful, the "*registros*," excavated more or less systematically in Almadén and Almadenejos, constituted an intent on the part of those responsible in the headquarters to diversify the ore supply, thus enabling them to extract most of the available ore. Also, exploitation of new veins probably offered a temporary relief from the pressure caused by decreasing returns. At the beginning of the nineteenth century, financial difficulties put an end to excavating these exploration shafts.

Another factor contributing to the growth in production of mercury was the increase in the work force. We point out that, even though some demographic and socio–economic mechanisms of Spain of that day contributed to the mines' having the necessary workers, the leadership of the managers of the establishment played an outstanding role. But eventually the mines gave up using the large contingent of workers required by the expansion of exports, when such numbers were no longer necessary thanks to technical changes.

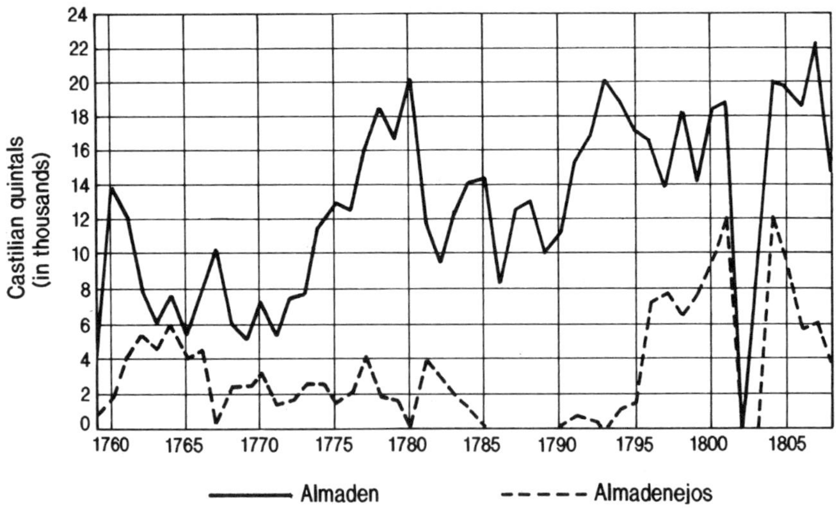

Fig. 4. Production of mercury in mines of Almadén and Almadenejos, 1759–1808. ***Sources*****: Bernáldez and Rúa (1862); AHN, Minas de Almadén.**

Nonetheless, the technical innovations, exploration for new veins and contracting a large number of workers would not have been possible without increasing financial resources. The mines financed their expenses (raw materials, supplies, tools, salaries, etc.) through subsidies received from the Royal Treasury. These amounts were determined by the Government, disregarding any calculation having to do with value of mercury produced at Almadén. Between 1744 and 1779, successive increases elevated the annual expenses of the mines from 1 to more than 6 million *reales de vellón* [6 million would amount to substantially less than 750,000 pesos since the units referred to are in the heavily debased silver–copper alloy known as billon rather than assayed coin silver of relatively high fineness — Ed.] Later on, this rapid increase abated. Only afterward on a few occasions did this expense exceed 7 million. At the beginning of the nineteenth century, the financial crisis of the Crown resulted in subsidies that were judged insufficient.

The reasonably satisfactory financial position of the mines during the second half of the eighteenth century is an indicator of the favorable government attitude. The same can be said for the collaboration received from Madrid to establish the necessary conditions to facilitate the recruitment of workers. Therefore, we can attribute to the Bourbon Reformation a certain efficacy in the matter of increasing mercury production in Spain. The same certainly could be said concerning its commercialization, at least in New Spain.

In figure 3, the recessive phase from the middle and end of the decade of 1760 is associated with a stagnation in the extraction of ore and a decrease in its quality. The change in operations between 1765 and 1773 is seen in the payment of pry–bar men who excavated ore. When their salary was changed from piecework to day work, it became fixed at an average below what was previously obtained. An acute decrease in productivity of the work (volume of ore excavated per shift) followed as a result of this modification. Under the new payment scheme it is probable that the workers reduced their work input at the excavation "sites" with better grade ore which was more compact and "dangerous." This impression is compatible with some indications that suggest the existence of a "disciplinary crisis."

The recessive phase of the decade of 1780 coincides with an initial decline and later recovery of the quantity of ore extracted and with a pronounced fall in its fineness. Also in this case labor problems appear, although of another sort. It has to do with a "shortage of workers" to be understood in relative rather than absolute terms. The little work directed to ore extraction and drainage of the deepest mines limited the availability of hand labor for digging. In some years it forced the neighbors of Almadén and Almadenejos to come together to work a certain number of shifts in the Pozo and Castillo mines, with the inevitable consequence of a strain in labor relations. References to partial floodings and the discontinuous nature of the veins began to darken the outlook. Contracts with Idria allowed the well–founded doubts about the capacity of the mines to sustain the sale of mercury in the Americas to be pushed aside.

Expansion phases show increases in the extraction of ore that usually, but not always, come accompanied by improvements in the quality. But in either case, it is not worse. Nor are there symptoms of unusually grave labor difficulties noted.

The production trends at Almadén described above did not fail to influence the imports of mercury into New Spain, its principal market. As figure 5 indicates, the consumption of mercury in New Spain increased significantly during the latter part of the eighteenth century.[11]

The Work Effort

Every productive process needs to sustain — and to replace — the labor expended. This need became particularly acute in Almadén for three reasons.

In the first place production techniques were always labor intensive. Furthermore, the tendency to increasing production implied an increase in work effort. Thus the decline in unit of work requirements per unit of product that was barely noticeable, and generalized in all of the productive process steps, becomes compatible with the increase in total requirements.[12]

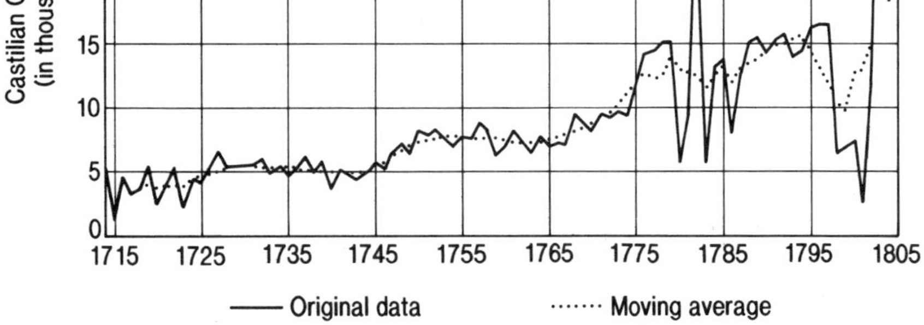

Fig. 5. Consumption of mercury in New Spain, 1714–1805. *Sources*: AGI, Contaduría and México; Heredia 1978.

In the second place, the work in mines and smelters created an especially intense consumption of work effort. Sicknesses (notoriously the *hidrargirismo*) and, to a lesser extent accidents, generated a dramatic and extensive deterioration of the "organic economy" [healthy energy] of the workers. In the different censuses of Almadén various groups of workers are integrated: those definitively crippled or temporarily incapacitated; those in the majority that, although they did not cease to work, were in more or less deficient physical conditions; and those, in the minority, who enjoyed an acceptable state of health. In other words, the effective workforce capacity was less than that expected of a group of this size.[13] Until well into the nineteenth century there was not a clearly perceptible reduction in the incidence of sicknesses and labor accidents.

In the third place, in spite of the existence of bonuses and punishments, there are abundant proofs that, to a greater or lesser extent, the workers frequently eluded completion of their assigned "obligations." With variable success, the managers of the mines tried to become more efficient in their control system, but everything suggests these were transitory and reversible achievements.

Eventually both the company and workers agreed the production process would be improved by: shortening the mine shift (6 hours); reducing the number of monthly or annual labor shifts; rotation of the pry–bar [later hand drill] workers between the "dangerous" and "healthy" jobs; seasonal thinning out of activities; relatively high day wages; and progressive amplification of the right to the "recuperation" shifts above ground for the miners.[14] These peculiarities of the work at Almadén, although they limited somewhat the "damage" received by the workers, did not prevent the high social costs associated with the production of mercury. Only a profound reform in the production techniques, oriented toward reducing the morbidity and victimization of workers, and/or a substantial raise in pay, that may permit subsistence with a lesser number of work shifts in dangerous activities, would have proposed effective conservation solutions. High unit salary costs and low productivity per worker were the inevitable consequences of a wasteful work effort production process.

Toward the end of the decade of 1770, coinciding with historic maximums of mercury production, Superintendent Soler insisted on the necessity of modifying the means of replenishing the supply of workers. In his judgement, only an increase in neighborhood residents would allow high levels of production reached to be sustained, "using extraordinary ways and means because the number of workers is so small." To increase the local population without renouncing use of temporaries was his objective. In reality, the novel element of the plan consisted in application of methods that, in some cases, dated from the sixteenth century. However, it was not until the last years of the eighteenth century and beginning of the nineteenth century that they became a fundamental part of the replenishment model that sustained the material existence of the mine workers.

In the decade of 1770 a great liberalization in the concession of charitable gifts to widows and orphans of the "continuing" workers is noted. The workers charitable fund was organized in 1779. Thus, in 1789 for example, 340 family units of Almadén and Almadenejos, lacking income from a deceased head of family, received payments [extremely small—Ed.] to the mean value of one real of billon (*vellón*).

In 1776 the military exemptions were increased which had been enjoyed since the period of the agreements with the Fugger, a minority group of workers who settled in Almadén. The permanent residents as well as the temporary workers underground and in the foundry were able, with a certain diligence, to take advantage of military exemptions. The peculiarities of matters of taxation at Almadén also date far back, but it was not until 1778 that a series of dispensations were initiated that, in 1792, completed the path to total fiscal exemption for the inhabitants.

Also in 1776 the Miners Hospital of Almadén was founded and provided free attention for the workers. Although some medical assistance already existed, with the establishment of the hospital this was significantly increased, extending equally to Almadenejos. Between 1782 and 1785, an average of 884 sick or wounded workers passed through the hospital. Its capacity for assistance reached a maximum at the beginning of the nineteenth century, when annual expenses exceeded the amount of 250,000 reales [of billon].

In 1781 another outstanding public benefit appeared: the annual distribution of subsistence plots on the *Dehesa de Castilseras* [cattle range] for the residents to cultivate. Superintendent Soler, originator of the idea, saw in it the possibility of increasing the population thanks to an increase in the local production of wheat, a decrease in cost of living, and improvement in health of the workers involved in agricultural tasks. The engagement of the cultivatable surface was supposed to extend subsistence agriculture compatible with work in the mines. This last aspect explains the favorable attitude of the mine managers toward production of wheat by the workers in contrast to their permanent opposition to development of a commercial agriculture that might detract labor from the mining sector. The annual distribution of garden plots on a small scale only constituted a supplementary income and not an alternative to work in the mines. On the other hand, access by the residents to the cattle range restricted the end to which it could be used by livestock of outside owners with transhumance flocks, which generated funds they contributed to the Widows and Orphans Charity and the Hospital. The utilization of forestry resources by the administration also limited the agrarian use of the *Dehesa* by the residents.

The supplying of wheat under non–commercial conditions became an effort to limit the adverse effects of high prices for miners whose salaries did not suffice for food. Thus, the superintendency frequently undertook the purchase, transport and sale of significant quantities of bread. The price fixed by the superintendency seldom exceeded the cost of preparing it and it was also customary that rationing was established according to the number of members in a family unit. In terms of quantity and price, the sale of bread at Almadén and Almadenejos was seen as a palliative to consequences of the lack of a local wheat market. Indeed, the first manifestations of this subsidised supply occur at the middle of the eighteenth century. After 1801, and especially during the extremely grave agricultural crisis of 1804–1805, most local bread consumption occurred as a result of the creation of the Mining Provisions Granary established by the superintendent of the mines. On the occasion of the crisis of 1779–1780 and 1804–1805, the superintendency had to subsidize bread. This mechanism came to be a very effective lure for the temporary workers. At a time when there is no lack of proof of the massive presence of outsiders working in the mines, the high levels of production of these years thus seems to confirm it.

Conclusions

In the second half of the eighteenth century, a better overall perception of the relation between mercury–silver–income to the Crown and a capacity to act in consequence resulted in an increase in the scale of operations to insure the long–term viability of the mines of Almadén. The improvement in the offering price of mercury, especially in New Spain, with the consequent effect on the production of silver and its specific fiscal tax collection, is inseparable from what occurred at Almadén.

The management of the mines was dedicated to a sustained increase in the production of mercury. The direct (and particularly the indirect) high profitability of the mercury explains management's indifference to reducing the costs of production.

After the reinitiation of production that followed the mine fires of 1755–1757, we see an expansion of mercury exportation, reaching a maximum at the beginning of the nineteenth century. In addition to the intensification of increased production, a considerable diminution in interannual fluctuation is recorded. However, two recessive phases can be observed of some ten years of duration that begin at the middle and beginning of the decades of 1760 and 1780 respectively. The second had worse consequences for American mining. It occurred when Peru had converted into an habitual importer of Spanish mercury and in conjunction with rising amounts of consumption in New Spain.

The increase in financial resources available to the mines permitted technical innovation, the exploitation of new veins in Almadenejos and the progressive resolution of the scarcity of hand labor. These three are the principal factors of the tendency to increasing production. Downturns seem related to labor problems and decreases or stagnation in the quality and quantity of ore.

The "lack of workers" was a persistent limiting factor on exportations. Internal and external obstacles to the production process slowed down the process of replenishing the work force. The intense and rapid deterioration of the welfare of the permanent workers stands out above others as first in importance. Second in importance was poor articulation with the subsistence market and the scarcities caused by specialization of the labor force in mining. The demographic structure of Almadén, and in particular the slight (or null) population increase in the absence of immigration, plus the difficulty in recruiting outside workers, reflect the gravity of the labor recruitment problems.

Special labor practices incorporated into the production process (such as short labor shifts, seasonal employment, furloughs, etc.) were able to soften — not to say turn around — the worsening of the workers welfare even though deepening of the shafts took place without significant improvement of ventilation in the mines.

Transfers of salary income to the local domestic economies, free hospital care, military and tax privileges, annual distribution of garden plots to residents, and the supply of wheat at cost, made up an effective subsidy. The increase in available family income, less recourse to the market for subsistence goods and control of shortages were the principal results of the new policies.

At the beginning of the nineteenth century, there was a coincidence — not accidental, in our judgment — between the disappearance of references to "lack of people," the greater exportations of mercury, and the open replenishment model. In these years increased productivity resulted in high levels of production.

Endnotes

1. We use the concept in the original Marxist meaning; that is, we understand work force to mean the potential capacity, collective or individual, to accomplish a job. The work effort is exchanged for a salary when, as in Almadén, a market exists for such kinds of transactions. The work force of a population is determined by its size and its physical condition; job related illnesses and food scarcity have a negative impact on the work force. Intermediate states exist between ample availability and absolute lack of the work force. Consumption of goods (food, clothing, housing, etc.) restores, with varying grades of efficacy, the capacity to work, thus increasing the work effort.

2. As its own name indicates, Almadenejos was a small locality near Almadén, and was nonexistant before the beginning of permanent mining activities in the decade of 1750.

3. The monthly payrolls of the mines allow a clear appreciation of the pay rate differences between "employees" and day workers. The former received a set daily pay which was multiplied by the actual number of days in each month regardless of sick days, authorized leave, etc. Day laborers only were paid for days actually worked.

4. The "employees," as much in the interior of the mines (mine supervisors, their assistants, bent–makers, etc.) as in the exterior (accounting or foundry officials, foremen of the warehouses, *sentadores*, shop masters, etc.), needed to report for work every working day. The day laborers, although they were "continuing miners" only did so intermittently. Thus, for example, the pry–bar men were only in the habit of working a maximum of some 10–15 days, in shifts of six hours, during each one of the 9 months that came to comprise the active annual production cycle. The danger of prolonged contact with mercury is the commonly deduced reason for explaining the absence of more continuous work. Labor records for "continuing miners" demonstrate that their active life, generally initiated at the end of infancy and concluded four decades later, was interrupted by prolonged periods of inactivity. In the case of the temporary workers, the discontinuities were even much more marked.

5. This figure does not include various tens of "employees" and artisans (carpenters, blacksmiths, etc.), nor various hundreds of day workers doing various unskilled jobs (drainage, ore extraction, smelting, warehousing, etc.).

6. The calculation was made as follows: we evaluated the annual cost of bread consumption by worker's families from demographic and price data, assigning a pound a day to each family member. The number obtained is divided by the most representative daily salary. The result is the number of annual work–days necessary to obtain income equivalent to the cost of the family's consumption of bread. Since this number exceeds that of income from the amount of labor known to be habitual among the workers, this result throws serious doubt on the viability of medium and long term family subsistence supported from salary alone at Almadén.

7. Because of their unusual nature [i.e., mine fires— Ed.], we leave out the years 1755–1757 from the periods proposed.

8. The notable exception established in 1802 needs to be pointed out, when the accumulation of mercury in Spain, as a consequence of the marine blockade, made the suspension of activities at Almadén advisable. Although in smaller measure, the vicissitudes of Atlantic traffic also made their effects felt on the export of 1803. The exceptionally short duration of the smelting efforts for the mining year 1807–8 that placed the effective production for 1808 below the potential, also obeys the politico–military happenings of the moment.

9. The least squares method has been used.

10. The available data, although a bit unsatisfactory, point out a decreasing long–term tendency in the average fineness of the ore being extracted.

11. Mercury consumption in New Spain was determined as follows. Given the known yearly imports (A), and existing reserves at the end of each year (B), and the reserves of the previous year (C), we derive the formula for the calculation: A+C–B. I acknowledge the unpublished work on the mercury trade by K. A. Brown, copies of which he kindly furnished me, and the published work of Heredia (1978).

12. One must remember the effect of decreasing returns and the limited reach of technical innovation. In this respect, it is convenient to point out that no continuity of significant and durable production increases exists in underground work (shoring up or excavating) or in the metallurgical activities and others developed in the above ground installations (quarries, workshops, warehouses, offices, etc.).

13. Note that the distinction between the capacity to work, complete and partial, does not match theoretical planning a priori but in the reality observed in Almadén. The incomplete character of the capacity to work is manifested, for example, through the low productivity, the alternating phases of activity and inactivity and/or the achievement of a low number of work days per relevant unit of time (weeks, months, or years).

14. There are many proofs that the preservation of basic health was a preferred objective of the workers. From whatever point of view, these minimum health regulations were clearly insufficient for guaranteeing a satisfactory existance. Varied attitudes can be detected (reduction of labor effort, lack of interest in material incentives, opposition to work shift prolongation, reluctance to occupy especially unhealthy jobs, etc.) that can only be interpreted as mechanisms to defend their precarious health.

References Consulted

Archival Sources

AHN — Archivo Histórico Nacional [Madrid]
Minas de Almadén, legajos 1190 and 1274

AGI — Archivo General de Indias [Sevilla]
Contaduría, legajos 1044–1048
México, legajos 2174–2184, 2196

Published and Other Materials Consulted

Assodourian, C. S. 1979. La producción de la mercancía dinero en la formación del mercado interno colonial. In *Ensayo sobre el desarrollo económico de México y América Latina, 1500–1975*, ed. E. Florescano, 223–292. Mexico City: Fondo de Cultura Económica.

——. 1983. *El sistema de la economía colonial. El mercado interior: regiones y espacio económico.* Mexico City.

——. 1989. Base técnica y relaciones de producción en la minería de Potosí. In *Ciencia, vida y espacio en Ibero–américa*, ed. J. L. Peset, vol. II, 185–205. Madrid.

Bakewell, P. 1989. *Mineros de la montaña roja.* Madrid.

Bernáldez, F. and Rúa Figueroa, R. 1861. *Memoria sobre las minas de Almadén y Almadenejos.* Madrid: Imprenta Nacional.

Bradby, B. (in press). The "Black Legend" of Huancavelica: The mita debates and opposition to wage–labor in the colonial mercury mine. In *Hombres, técnicas, plata. Minería y sociedad en Europa y América (siglos XVI–XIX)*, ed. J. Sánchez Gómez and G. Mira. Madrid.

Brown, K. W. n.d. (a) The Spanish Imperial Mercury Trade and the American Mining Expansion under the Bourbon Monarchy. (mimeo)

——. n.d. (b) La distribución de mercurio a fines del período colonial y los trastornos de la independencia hispano–americana. (mimeo)

Carmagnani, M. 1963. *El salariado minero en Chile colonial.* Santiago de Chile: Universidad de Chile, Centro de Historia Colonial.

Chavez Orozco, L. 1978. ed. *La situación del minero asalarido a fines del siglo XVIII.* Mexico City: Centro de Estudios Históricos del Movimiento Obrero Mexicano.

Dobado, R. 1982. Salarios y condiciones de trabajo en las minas de Almadén. In *La economía española al final del Antiguo Régimen*, vol. II, ed. P. Tedde, 337–440.

——. 1984. Actitudes intelectuales frente a la condiciones de trabajo en las minas de Almadén. *Revista de Historia Económica* 2(2): 59–89.

——. 1989. El trabajo en las Minas de Almadén, 1750–1955. Tesis doctoral presentada en la Universidad Complutense de Madrid.

——. 1990. Las Minas de Almadén, el monopolio del azogue y la producción de plata en Nueva España en el siglo XVIII. Memoria de investigación presentada al Banco de España.

——. 1991. La minería estatal española, 1748–1873. In *Una historia de la empresa pública en España*, ed. F. Comin and P. Martín Aceña, 89–138. Madrid.

——. 1992. El azogue en Nueva España, 1709–1805, *Actas del X Congreso Internacional de Historia de América* 2:561–575. Sevilla: AHILA.

——. (in press). Las técnicas productivas en Almadén, 1759–1808. In *Hombres, técnicas, plata. Minería y sociedad en Europa y América*, eds. J. Sánches Gómez and G. Mira. Madrid.

Fisher, J. 1977. *Minas y mineros en el Perú colonial, 1776–1824*. Lima: Instituto de Estudios Peruanos.

Heredia Herrera, A. 1978. *La renta del azogue en Nueva España, 1709–1751*. Sevilla: Escuela de Estudios Hispanoamericanos.

Klein, H. 1982. The state and labor market in rural Bolivia in the Colonial and early Republican periods. In *Essays in the political, economic and social history of colonial Latin America*, ed. K. Spalding, 94–106. Newark, N.J.

Ladd, D. M. 1992. *Génesis y desarrollo de una huelga. La lucha de los mineros mexicanos de la plata en Real del Monte, 1766–1755*. Mexico City.

Morete de Varela, J. 1804. Descripción general de Almadén, Almadenejos, minas y demas de su territorio. *Revista Minera* 8(1857): 338–384.

O'Phelan, S. (in press) Vivir y morir en el mineral de Hualgayoc a fines de la colonia. In *Hombres, técnicas, plata. Minería y sociedad en Europa y América (siglos XVI–XIX)*, eds. J. Sánchez Gómez and G. Mira. Madrid.

Tandeter, E. 1992. *Coacción y mercado. La minería de la plata en el Potosí colonial, 1692–1826*. Cuzco.

Velasco, C. 1989. Los trabajadores mineros en la Nueva España, 1750–1810. In *Historia Económica de México*, ed. E. Cárdenas, 563–587. Mexico City.

The Reception of Modern Scientific Mining in an Enlightened Mexico

José Luis Peset, Centro de Estudios Históricos
Consejo Superior de Investigaciones Científicas, Madrid

Abstract

One of the great debates on the history of American technology concerns the role played by Spanish efforts in the eighteenth century to diffuse new European technology into the colonies. It is evident that the Spanish Crown, by sending technicians and scientific expeditions, and through viceregal politics, made certain that in America there was an understanding of the status of European mining and metallurgy among those who were still using old techniques, yet discovering rich mineral deposits. This produced a dialogue that continued uninterrupted for centuries, one in which practical experience encountered a new science; one with angry creoles set against distrustful Spaniards; and a flowering economy against one in crisis. New World technology encountered an important revitalization, the creoles found independence, and the industrial revolution obtained the silver it needed.
Key words: mining, technology, Mexico, mining college, school of mines.

Resumen

Uno de los grandes debates de la historia de la técnica americana es el papel que jugaron los esfuerzos españoles en el siglo XVIII por difundir allá nuevas técnicas europeas. Es evidente que la corona española, a través del envío de técnicos, de la política virreinal y de las expediciones científicas, consiguió que se conociera en América el estado de la minería y la metalurgia europeas y que los que allí fueran descubrieran las viejas técnicas y los ricos minerales americanos. Se producía la continuación de un diálogo interrumpido por siglos, en que experimentada práctica se encontraba con una nueva ciencia, unos enojados criollos con unos desconfiados españoles y una economía floreciente con otra en crisis. La técnica encontró una importante revitalización, los criollos la independencia y la revolución industrial la plata que necesitaba.
Palabras claves: minería, tecnología, México, Colegio de Minería.

In Quest of Mineral Wealth: Aboriginal and Colonial Mining and Metallurgy in Spanish America, edited by Alan K. Craig and Robert C. West, 1994. Geoscience and Man, vol. 33, pp. 233-251. Dept. of Geography & Anthropology, Louisiana State University, Baton Rouge, LA 70893-6010, U.S.A.

Spanish efforts to introduce European mining technology to the American viceroyalties in the eighteenth century met with considerable resistance; one of the places where scientific, political and economic confrontations were most lively was in the viceroyalty of Nueva España (Alperín 1985). It took place in the important scientific experiment creating the Mexican School of Mines, which opened its doors on January 1, 1792, under the direction of the Basque scientist Fausto de Elhuyar. This event was part of a broad plan of the Spanish Crown to improve mining in Spanish America. It had been initiated with some administrative reforms which, as a result of the efforts of Minister José Gálvez, were strengthened by sending scientists to Peru, New Granada [Colombia] and New Spain [Mexico]. Here they attempted to relieve the problems of lack of capital by creating the Bank of Loans (Banco de Avíos) and initiating fiscal reforms. The lack of adequate mining legislation was resolved by the creation of a Board of Mining and by the issuance of mining ordinances. Technology was introduced to improve some mines whose exploitation was increasingly massive and complex. From here on, once José Gálvez had assumed power, you may be sure that, from the Ministry of the Indies to the viceregal authorities, indispensable mandates were given to reform the existing system which had been in place since the beginning of the colonial era. The orders of Gálvez and the initiatives of the Mexican miners themselves, led by Joaquín Velázquez and Juan Lucas Lassaga, promoted the important improvements that allowed a new age of silver production in Mexican mining during the last decades of the colonial period. However, I suggest that these happenings were widespread throughout the colonies, and we must not forget the earlier dispatch of Juan José Elhuyar together with German technicians who were sent to the principal mining regions of Spanish South America.

The School of Mines

At the end of the summer of 1788, the frigate *Venus* arrived at the port of Veracruz bringing the Basque chemist Fausto de Elhuyar who, trained in Paris and in Freiberg and with ample experience in European teaching positions and laboratories, came to replace Joaquín Velázquez de León. On assuming his post at the head of the Board and of the Tribunal of Mines [vacated by the death of the creole miner Velázquez], he received from the hands of the most senior delegate "a baton as a sign of his authority and jurisdiction."[1] Thus a Spaniard assumed charge of various aspects of the Mining industry, and it is only logical that the operators did not view with pleasure the newcomer who represented such rash Spanish politics as handing over high level jobs to the *peninsulares* [Spanish–born], thus beginning a moderately tense period in the history of the Tribunal. If all these happenings had had their origin in the interaction between Gálvez and the Mexican miners, especially Lassaga and Velázquez who had enjoyed the use of power until then, now with the arrival of the wise Basque, the balance was tilted toward the Spanish side. Aware of his power and of his role, Elhuyar wished to become President of the Tribunal in spite of the miners. They argued, and the chemist left the room abruptly to appeal to the viceroy, and even the King, who had conferred the required presidency and should therefore punish those who opposed him.

These beginnings did not seem to predict tranquil times, but rather that the opposition would continue. But Elhuyar, known to be a skillful politician, completely accepting the inheritance of Velázquez, on formulating in January of 1790 his "Wish of the Director General of Mining on Arrangement of his Tribunal," used masterful language to draw upon the plans of his predecessors. He unsuccessfully defended the Bank of Loans, which was created to provide capitalization for the anemic Mexican mining administrative efforts, but which was closed because of some scandalously ruinous loans; he accepted with some small changes the organization of the Tribunal, without giving up his own private jurisdiction over mining, as if he wished to limit it to cash settlements, leaving the first hearings to the territorial courts. Without doubt the pressure of the High Court (*Audiencia*) was able to influence him as well as the fear of running into the same luck as Velázquez, who proclaimed that he left his scientific chores to deal with political and legal tasks. It seems that the viceroy did not look favorably on this "Wish," and only after much paper shuffling and a year's delay, was it sent to the King who agreed to it in a royal decree of February 5, 1793.[2] Much as Roberto Moreno has written: "From this moment until the Declaration of Independence, Fausto de Elhuyar was the leader of the Tribunal, although with much opposition from the creoles."[3]

The political skill of the Spanish chemist was important in order to function in a world which without doubt was not fond of him.[4] Elhuyar was a well trained technician, sent by the Spanish Crown in search of the best possible returns for the homeland exchequer. However, he demonstrated his sagacity by comprehending that it was precisely the betterment of both mining and the miners that would achieve the ends for which he was responsible. He did this by combining many of his ideas with the old projects of the miners' representatives. Since all of them were searching for a means to obtain the largest quantity of the best silver possible, in many aspects their ideas coincided. The differences would have been in their opinion about the benefits to the Crown, or concerning the control of the Tribunal and of the college. In the first instance, for example, Elhuyar always defended the Crown's monopoly on mercury, and in the second case he always encountered opposition when he tried to appoint a Spanish professor, while leaving unresolved the delicate matter of jurisdiction, seeking to maintain himself alone as the lower court of appeal.

Although this first encounter with the personality of Elhuyar was not convincing for the creoles, the loyalty of this wise man to the Crown was always beyond question and it was obvious from his attempts to improve teaching and his advantageous use of the institutions he controlled, that he did so for the betterment of mining, and this must have been viewed with approval by the miners as a whole. In this sense his new "Plan of the College of Mining," presented to the Tribunal in January 1790, differed very little from that of his predecessor. The teaching would be divided into four years: the first was pure mathematics, the second, descriptive mining geometry, the third was chemistry, and the fourth was underground topographic surveying, or theory of mountain formation. Each course had its special professor and the students were required to study drafting and French as well. There must have been a display of models of machinery and furnaces, another of mineral specimens and another of chemical products. There would be public lectures, individual mid–course examinations and field trips to the nearby mining centers accompanied by the professors. At the end they spent two years of practical training in more distant mine fields which was a great importance as we shall see; then, after final examinations, they were given their degrees. Without doubt Elhuyar's experience in Vergara and Freiberg would be useful in getting the new school running, but the Mexican heritage was no less important in his decisions.[5]

The creation of the new Ores Seminar is one of the most interesting scientific enterprises of the Spanish presence in America. The seminar became an encounter between European science and Mexican practical experience, between theory and practice, permitting analysis in a privileged context. In the seminar, knowledge of new European technical practices of ore reduction, such as the Born method of amalgamation, was introduced. The most common thesis about the possible efficacy and consequences of such introductions is that proposed by the mining historian Brading (1983, 215–217), who claimed that they had little or no influence in improving the scientific or practical exploitation of silver. This hypothesis is well supported in the book by Federico Sonneschmidt (one of the German technicians who accompanied the new Director) which, with the title *Tratado de la amalgamación de Nueva España* (1983, ix–x), attempted to prove the advantages of the traditional system of obtaining silver. It seems the method of Medina and Barba was considered more useful in Mexico in spite of its slowness and the inferior quality of the silver obtained, because its use was more economical. This was a theme repeated by Humboldt in view of the existence of enormous veins whose quality was no better than those of Europe. And we may add, in agreement with the noble traveler, that given the system of ownership of the mines in hands of private parties, the desires of the state would only have been achieved with difficulty.[6] The extensive use of the old method by the richest mine owners allowed them to obtain great quantities of precious metal thus compensating for the slowness of the process and low grade of the ore.

However, it is evident that the most important dialogue in Mexican mining history of the time began in the School of Mines, established in Mexico City in 1792. The dialogue was between between science and technology, and the students, according to Alzate, began to prepare themselves to carry the enlightenment to the mining centers (Peset 1987). According to testimony of the often–cited Humboldt, modern science was well understood in the new technical school. "The principles of the new chemistry, which in the Spanish colonies are designated with the somewhat mistaken name of *New Philosophy*, are more widespread in Mexico than in many parts of Spain. A European traveler will be surprised to find, deep in the interior of the country toward the borders of California, young Mexicans pondering on the decomposition of water during the patio amalgamation process." And after pointing out aspects of the new institution (and also mentioning the Mexican edition of a very clever translation of Lavoisier, and of the notable works of del Río), von Humboldt insisted the scientific instruction was of quality. "The teaching of mathematics is less adequate at the university of Mexico than in the School of Mines; the students of this last establishment are more advanced in mathematical analysis and they are taught differential and integral calculus. When peace is established and there is free communication with Europe, astronomical instruments (chronometers, sextants, Borda's circles) may be more common and will be found in even the most remote parts of the kingdom as will be young men capable of making observations and calculations by the most modern methods. The government would be able to select a large group with such skills in order to produce a map of the country" (Humboldt [1811] 1984, 80 ff.).

There are equally abundant testimonies, as we shall see, of Elhuyar's interest in technical teaching and in betterment of working conditions in the mines. Since both he and his students were involved in commercial mining, they wanted their installations to be as prosperous as possible.[7] This allowed him to write from first–hand experience important books on the traditional method of amalgamation, such as the one cited, in order that the students of the new college might have a detailed understanding of the process. Sonneschmidt makes this clear when he writes in his preface: "Various amalgamators have indicated to me their desire to learn exactly how the mercury and other ingredients function in the patio process, being convinced that only by this means will they be able to identify the best way to improve this method. They suffer at the same time from a total lack of books that deal with this form of beneficiation, and declare that those that do are very deficient in this subject. More than this, the students of the Royal Mining Seminary need fixed points from which to begin in their study of this amalgamation."[8]

In all these areas the creators of the mining college proved to be rapid renovators. Even before the arrival of the new Director, professor Andrés Joseph Rodríguez was already appointed by a royal order of April 18, 1788. He was a leading expert in underground surveying and mineralogy with experience in Almadén and studies in Madrid. The order indicated that he might be installed after an examination by Elhuyar, who assigned him the chair of mathematics, being confirmed in it by the viceroy and the Tribunal, in spite of the fact that this continual arrival of Spaniards would not be viewed favorably.[9] In this regard, in July of 1791 the Tribunal made known to the viceroy the political disadvantages of continually bringing over professors from Spain. The protest was based upon the *Ordenanzas*, but the Director did not change, and in September at the same time he asked for Spanish teachers for three more classes, choosing them from among young pensioners sent to France and Hungary during the last years of Charles III's reign (Ramírez 1890, 77–78). And following this line of action, he also asked a short time later that the viceroy name as professor of physics Francisco Antonio Bataller, a miner and former professor in San Isidro.[10] The viceroy as well as the Minister of the Indies demonstrated that they were in agreement with the changes, first naming Rodríguez and Bataller to these two classes and Francisco Codón to chemistry, although he did not arrive to accept it.[11] Later on Andrés Manuel del Río arrived to take over the teaching of mineralogy, arriving at Veracruz on board the warship *San Pedro de Alcántara* on October 20, 1794; it docked after costly expenses and long delays, loaded with his books and instruments. On the 18th of December he entered the capital of Mexico.[12]

After all these reforms, when in 1796 the nine years of Elhuyar's appointment were completed, the college openly asked for his reappointment. It is a moment clearly understood when dean, assistant dean, professors and students asked the Tribunal in February for this new mandate, given that "...there is no other person of the merit and circumstances so accommodating to the institution of this Royal Seminar as that of don Fausto de Elhuyar. Undoubtedly Europe may have someone with the education and talent this position requires; but whoever that may be, he will not have an understanding of the character and genius of the natives of this country, will not know the difficulties that it has been necessary to overcome precisely in order to build this School of Mines, nor is it easy to foresee those that are remaining to overcome...."[13]

In January of the following year the college repeated this petition; the Tribunal and managers of the mining companies were united in this general appeal to the viceroy. They wished Elhuyar to receive a raise and for the establishment of a pension for his widow in case of his death. On the seventh of November 1796 a royal order named him member of the Junta of Commerce, Coinage, Mines, and Foreign Affairs, and on March 16, 1797, the Valencian architect Manuel Tolsá presented the plans for the new college, which the lower court accepted on June 27, no doubt marvelling at its splendor (Ramírez 1890, 135, 137–139). In a certain sense, Elhuyar became one of them, having integrated himself into the business world of mining.

Some more changes were produced in the teaching curriculum, broadening and dividing the studies, or perhaps adding new subjects in order to change the direction of study. Wishing to improve the profession of mining, they resorted in the first place to a thorough and up–to–date study of the sciences. There developed an open attitude toward those disciplines in the humanities that, although ancient, were able to contribute to the distribution of a greater distinction and quality to the illustrious personalities that were being formed there. This direction can be clearly seen in the introduction of French, a most useful language in which to triumph in society as well as to understand modern science. But one can understand even better this intention in the introduction of logic with which it was hoped to foster culture and the classics in the students. These novelties were part of the well defined road to upward social mobility that these miners began on, supported by the new institutions that the Crown continued to create. And the same can be said for the teaching of Latin or of the Spanish language, which in time were introduced.[14] It is a path which recurs in all the technical schools where, after a first stage of pure scientism, it enters into a second in which the humanities serve to broaden the training, desiring as much to better understand the world of science as to allow a stronger and more enlightened social consciousness for the future technicians.[15] A note of Elhuyar's, written on his return to Spain, shows us that he understood this route perfectly well, appearing to consider the use of good Spanish to be a distinct attribute. Allowing for the difference in the language that exists between the two countries, we are surprised by his *Memoria sobre el influjo de la Minería en la Agricultura, Industria, Población y Civilización de la Nueva–España... en sus diferentes épocas*, published in Madrid in 1825 where it says: "in the rest of the publication the language as it is used in Mexico has been preserved; this being noted in order that the difference may not be surprising."[16]

Of greater importance for the study of the sciences, and as basic pillars that have supported them since the foundation of the great technical schools, were the changes introduced in the teaching of mathematics and physics. The first ones particularly depended upon constant instruction, increasing the content and continually changing the textbook. In June of

1797, the Director proposed to divide the mathematics classes into two parts, which was immediately approved by the Tribunal and somewhat later by the Crown. Without doubt it was to improve the teaching, such that in November, in spite of the opposition of the chair of physics, the Director recalled that he was able to eliminate practical geometry, which was incorporated into the second course of mathematics together with integral calculus, mechanics, electricity, optics and astronomy.[17] Mathematics and physics were reenforced by these while chemistry remained as always, unprotected, perhaps because of the competition from del Río in his mineralogy course. The excellent chemical knowledge of the wise Spaniard was counterbalanced by the direct recognition of minerals by their characteristic taste and smell.

The teaching of all these subjects, as was logical, was very different from that in universities steeped in Latin and preoccupied with the preservation of classical learning.[18] In the School of Mines the instruction was in Spanish — French was taught, the professors understood German — and the latest up–to–date manuals were used. These were imported from Spain and even translated or rewritten for the students as happened in the case with del Río. The teaching was always practical and the School was always looking for an opportunity to visit an installation, collection, or suitable museum. Some books were published in Mexico but the majority were ordered from Spain. The School purchased instruments in Europe and even brought French instrument makers over to Mexico to demonstrate their use. At times they resorted to purchases from individuals or depended upon the generosity of Baron Alexander von Humboldt, who obtained some instruments for the school and maintained correspondence with the faculty for many years.[19] The same noble had admired the scientific instruments that were available there, writing concerning his visit: "The School of Mines has a chemical laboratory, a geological collection classified according to the system of Werner, and an equipment display in which there were not only valuable instruments such as Ramsden, Adama, le Noir and Luis Berthoud, but also scale models made in Mexico with the greatest realism and from the finest woods in the country....I cite these happenings separately because they give an idea of the ardor with which the study of exact sciences has been embraced in the capital of Nueva España and to which they dedicate themselves with much greater determination than languages and ancient literature" (Humboldt [1811] 1984, 81).

The preoccupation with practical teaching was very early, such that already in May of 1789 the Tribunal placed small furnaces in the patio for the assaying of ores, before it was done in the larger laboratories. And in February and March of the following year there were important expenditures on books and instruments and the ordering of mineral collections and slide rules.[20] The arrival of Andrés del Río resulted in an important increase in books and materials, as did the death of the Public Prosecutor of the Mining Tribunal, Juan Santelices Pablo, the good friend of Alzate, and avid collector of books, minerals and antiquities. His library was purchased in 1793 and his mineral collection in 1794, both being the beginnings of the college collections.[21] In the same manner the collection that Federico Sonneschmidt sold on leaving for Spain, to Martín Sessé, was purchased in 1798 for 500 pesos. And the year before that they began to use the instruments made by Lachaussé, first of the European artisans to work for the School of Mines. Humboldt had the highest regard for his apparatus used in the study of mechanics and hydrodynamics as well as for his collaboration in various mining enterprises.[22]

The introduction of books was also accomplished in an outstanding manner, as is demonstrated by the first translation in Spanish in Mexico of Lavoisier, and also by the pressure exerted by the viceroy that the School should teach the doctrines of Werner.[23] In 1790, physics and mathematics were taught according to Siguad de la Fond and the 1779 edition of Benito Bails, who in 1793 was substituted for Juan Justo García, and in 1815 by José Mariano Vallejo.[24] Together with the great mineralogist Andrés Manuel, del Río collaborated with his own contributions to the science of mineralogy. Perhaps the insistence of the viceroy on teaching the doctrines of Werner, and the enormous delay with which books arrived in America were the reasons why the learned Spaniard decided to write his *Orictognosia*, an extremely important contribution to American mineralogy. In August of 1795 the Tribunal sent the first part to the viceroy, whereupon the attorney general decided immediately that it should be published. Already in February of the following year the Director sent to the Tribunal a bill for 816 pesos and 4 reales for the cost of printing an edition of 1,000 copies. In 1804 the Tribunal asked the viceroy for permission to publish the second part, which was granted immediately.[25] Other responsibilities seem to have fallen to del Río, who was always one to take charge, in contrast to Elhuyar who was more dedicated to administration. It was del Río who obtained the books and instruments, prepared the practical exercises, searched for, bought, and classified the mineral collections. For this reason in 1802 Elhuyar laid before the Tribunal the fact that although the professors of mathematics Bataller and Rodríguez, responsible for teaching subterranean geometry [underground surveying], were preparing notes on this subject, they did not seem adequate and as a result he wanted del Río to translate some appropriate German text "whose language he possesses with the same perfection as the material" (Ramírez 1890, 184–185). And not only did this illustrious chemist contribute equipment and knowledge; he also utilized the presence of von Humboldt. On one hand von Humboldt collaborates in the teaching, examinations and laboratory exercises of the students, but also on the other he continued for many years to send information and instrumentation. During his visit, in November of 1803, the Director proposed that some instruments be purchased that cost the School of Mines the sum of 1,551 pesos. In January of the following year, Elhuyar wished to buy more instruments in Europe and notified the Tribunal that von Humboldt had offered to take care of the matter at no charge in order that the money be given to the Five Guilds of Madrid.[26] The travels of von Humboldt into Mexico had been very fruitful, as much for mining, whose scientific level he helped raise, as for himself for the rich information he obtained and carried to the United States and Europe.

The first intent of the founders of the School of Mines, as much for the locals as for the Spaniards, was to create an adequate technical school in which to train the sons of mine owners who came either on grants or by paying tuition according to ability.[27] Elhuyar was to continue cautiously on this path, always alert to the economic condition of the parents and the legitimacy of the sons. In his attempt to strengthen the mining lineages he proposed to the Tribunal that the first eight students be admitted in December of 1791, adding that "having thought about the abuses that may occur in the repositories of the baptismal registries regarding the legitimacy of the children by those who administer this Sacred Sacrament... it requires the declaration of at least three witnesses..."[28], a condition which also will be asked in order to demonstrate the quality of Spaniard or Indian of noble lineage. Without doubt there is a certain similarity with the proofs of pure blood or of nobility in his wish to strengthen the old and new mining lineages.

When the School of Mines was inaugurated on January 1, 1792, and the courses begun on the 7th, the Director wished to maintain discipline, and asked for authority over both students and employees. To the former he was able to apply light punishment and of the latter he required obedience. There is a certain parallelism with military discipline as a consequence of this authority he took on, which we see in the words he used to bolster his order. "In every entity, composed of members of diverse graduating classes, it is indispensable for its good governance that juniors be subordinate to, and dependent upon, their seniors; without this arrangement, everything becomes confusion, arbitrariness, and disorder" (Ramírez 1890, 100). In exchange for this imposition, great privileges were secured for the School of Mines, which was generally considered the best technical college in the Americas. By 1798, the Ministry of the Indies, being certain of the excellent training given for working gold and silver deposits in the mining settlements, affirmed "...to this end we shall omit no diligence or expense, that once these young men pass their examinations and, having been nominated by their teachers, endorsed by the Director and this [Tribunal], and approved by your Excellency..., they will have no difficulty in becoming Directors and Mining Engineers in the viceroyalties of Peru and Buenos Aires and the Provinces of Guatemala, Quito and Chile, where they will establish the best, most certain and economic method of the extraction and beneficiation by controlling salaries and regulating travel expenses" (Ramírez 1890, 143). There is no doubt that by means of this elitist societal group it was desired that the college be converted into the technical and administrative center of the Spanish empire in matters of mining. Because of the resentment with which José Celestino Mutis contemplated from Nueva Granada the increase in the institutional and technical power of Nueva España, as well perhaps by the presence of Juan José Elhuyar in the future Colombia, this viceroyalty seems to have been omitted as a future destination of the Mexican engineers.

With these perspectives, Fausto de Elhuyar wished to assure himself even more about the training of the students, and they were tested before leaving for the mine fields in which they were to perform practical exercises. In May of 1798 the first practitioners left for the three largest mining centers, Guanajuato, Zacatecas and Catorce, where later on they were sent books, instruments and tools. The new teaching method was of great usefulness in that it trained the students, introduced technology into the mines and supplied information for the School of Mines. As we shall see, the youngsters were also offered an adequate study of mines, the minerals and the terrain, thus beginning the making of a map of Mexican mining. They were sent, as an exercise, to gather information about the deposits and their beneficiation, including location plans of mines, changing mining centers or going to study the beneficiation method when they had finished their first tasks.[29] It is obvious the intent was to realize a complete training program for the students as well as to obtain complete information in order to regulate and improve Mexican mining and that of the New World. As a result, training improved in the college, which still suffered from a lack of an adequate collection of mineral specimens typical of the zone in spite of the purchases made. Von Humboldt ([1811] 1984, 337) wrote: "The traveler must not expect that he is going to encounter in the School of Mines of Mexico a complete collection of those minerals. As all the workings are in private hands, the Mexican government has but the weakest influence in the administration of the mines; it has not been expedient for the professors to bring together everything related to the structure of the veins, the blanket deposits and other concentrations of minerals. Thus in Mexico as in Madrid, one sees in public collections the rarest minerals of Siberia and Scotland while at the same time they look in vain for that which might illustrate the mineralogical geography of the country. It is to be hoped that the cabinet of the School of Mines will be enriched once the disciples of this excellent establishment are sent out to the most distant provinces from the capital, and these must make known to the mine owners how interested they are in facilitating the means of instruction. Without personal knowledge of the sites and without a study of the minerals that compose the bulk of the veins or the content of the concentrations and blanket deposits, so many changes will be proposed to perfect the amalgamation that they will remain unrealistic projects."[30]

This text of Humboldt's, as long as it is interesting, gives us an insight into the interest already manifest by Alzate in seeing to it that students connect their college learning with the business of mining. The collections then lose their beautiful and noble character, based solely on rarity, and become interesting for their utility and proximity, based as much on economic advantage as in the rising nationalism. It consists of teaching the useful, the appropriate and the agreeable. Such were the wishes of Lassaga and Velázquez, that the students must be sent to deposits developed by the Bank, "in order that they enter the mines, measure them and finally, that they practice as a pastime, those things that are to be their destiny" (VV. AA. 1984, 168–170). In order to carefully arrange this kind of practice, the way in which the students would render their reports was established and smaller and nearer deposits were considered, although they were never to abandon the three big mining areas already mentioned. On returning, they were to be examined by all the professors sitting as a tribunal, thus meriting the necessary title for this exercise. Students as well as professors then knew the tricks of their trade, being intimately involved in mining business. The two Spanish chemists had their own mines in the Real del Monte where they introduced new techniques and sent their own students. We find frequent mention of the establishment in their own mines and those of others, of "water" pumps and of "animal motors" counting on the help of the artisans that worked for the School of Mines, especially Lachaussé. It is evident that, as Santiago Ramírez writes, the mining claims of del Río had served for "the establishment of his machinery, the application of his principles, the large scale execution of his experience and methods and [to] directly apply himself to his profession."[31]

The faculty also undertook other missions more concerned with the stringencies of the Crown and the necessities of the miners. As a result of the war with England, there was a great scarcity of iron. After a report by Elhuyar about the necessary minerals, the Tribunal sent del Río to Coalcomán, Michoacán state, to establish an iron smelter. With this they were not just looking after the interests of students and professors, or of the mining community, as with the study and practice of technology we have seen occur. Now it has nothing to do with designing and building a pump for some rich miner, but rather it is to insure a supply of iron, essential as much for the mining industry as for defense. Del Río left the capital in November of 1805 and introduced the French furnaces that were better than the Spanish kind. The foundry began in April of 1807, and the mineralogist left for reasons of health early in 1809. It was kept in active production until it was destroyed in October of 1811 by two Spanish divisions. In effect, although it was constructed to supply iron for the war, the insurrectionists captured it to obtain cannons, munitions and other combat supplies. With the fear that Colima would fall, royal troops captured the smelter and destroyed it (Ramírez 1891, 29–31; 1890, 200, 210 ff.).

The final days of the reign of Charles IV were still brilliant moments for the School of Mines, years before the beginnings of the wars for independence. In July of 1808, when the war of the Frenchman [Napoleon] had already begun, Fausto del Elhuyar boasts of the good name and success obtained by the School of Mines. The number of students increased, and through the favorable public opinion it deserved "it has attracted the attention of the public and motivated many subjects of the first rank in this capital and all the kingdom, to make solicitations to place their sons in it in preference to other colleges in order that they may be given a good education and instruction in the sciences which are taught there." Without doubt it had at last managed to avoid the bad reputation from which the profession suffered. For this reason it is not strange that, a short while later in October of 1807, the students beseeched the Tribunal (and through it the viceroy) for permission to carry swords, an important symbol not only of nobility but also of war.[32] A war soon began, one in which the students of the School of Mines — who would always have posts and important skirmishes in all situations — intervened with valor. In any event, the School of Mines did not succumb and the union established between science and technology, capital and politics, creoles and Spaniards, survived as a center for the formation of a good many Mexican engineers. The work of Lassaga and Velázquez, of Elhuyar and del Río, remained as a tangible example of an important dialog between the two continents.

The Students during Independence

With the war of Napoleon, the afflictions of Fausto de Elhuyar began. If I am not mistaken, in this same month of July 1808, the Tribunal refused for the first time one of his proposals, and the rancourous disputes began. The Director wanted to omit the positions of Principal and Assistant Principal, proposing to substitute for them a chaplain and two laymen who knew something of science and could help students in their studies. They would occupy the positions so frequently vacated. Although this was an improvement, since the old tasks often fell to clerics ignorant of the sciences, the Tribunal denied the change in January of 1809 and in February, ignoring the triad of names presented by the Director, appointed a new Principal. Elhuyar repeated his previous request, angrily leaving the salon and the meeting, refusing to allow the new appointee to take possession of the job. In view of this rebuff, the Tribunal appointed the most senior deputy to the post, but Elhuyar took recourse before the viceroy with whom he was in favor. However, still in July of 1810 the Council of the Regency of Cádiz was to intervene to dismiss the Principal and order that a new election take place among the triad to be presented by the Director.[33]

They were difficult times, in which the insurrection of Hidalgo and Morelos was not long in coming. Elhuyar himself has left us masterful descriptions of the alterations the war caused in the mining industry, wounds that were unable to heal during the colonial period. His vision, it is clear, was always loyal to the Spanish Crown, viewing with horror the latest happenings of the *Independentistas*, to whom he referred as "the furious storm that since the year 1810 has caused so much ruination, changing in a painful way the brilliant aspect of these lands." The rebel troops turned toward the rich mineral districts and held Guanajuato in their hands. "The sources of gold and silver cannot but attract with particular inclination the attention of the promoters of this unexpected movement, and of the rabble that gather with the bait of rapine." They began with the mines of Guanajuato "in order to satiate with its sacking and the blood of innumerable victims, their greed, their hate and to run wild," later ruining Zacatecas and Catorce. With these three mining districts, they wounded half the Mexican silver production.[34]

The Director of the Mexican Mining Tribunal describes the steps in the destruction of this industry in a gradation of effects that is quite intelligent. He begins with the ruination of Mexican mining capital by the driving out or extermination of the residents and most accommodating neighbors of greatest influence; then comes the seizing of the ores and installations, causing grave damage; later on work blockage in the mining settlements that were resistant; and, finally, an increase in strikes and with it the tumult that people without work cause. "Lastly...what with the confusion and the violence the common man had acquired, among which are found the same operators of mines and refining mills, these and their workers are seen without an occupation nor any means to find steady honest work for their support and that of their families, and they do not hesitate to come together and follow the torrent of those that have been dragged down to such pitiful state, increasing the number of unruly mobs, to begin again in new surroundings the same scenes. All this necessarily produced a general paralyzation in the mining districts that were invaded, and a dispersion of capable mine workers"[35]

But that which Elhuyar tells us about is the attitude of his students when the uprising begins to arrive at the mines. There being weak viceroyal authority that far from Spain (which was at war with the Frenchman), the calamities of the year 1810 obligated the priest of Dolores [Hidalgo] to rise up against the throne. Peasants, some officials and some disaffected creoles responded to the revolt, heading toward the nearby Guanajuato.

The Intendent Juan Antonio de Riano took refuge in the Granaditas warehouse with the militia and some Spaniards, supported by the Town Council. The defenders of the silver intended, unsuccessfully, to guard the treasury funds which had attracted the insurrectionists who, on entering the new municipal granary, began to slaughter those trapped there. "The *Alhóndiga de Granaditas*, project of an enlightened government to insure the inexpensive supply of wheat to the public of Guanajuato, in reality was converted into a mausoleum of the Spanish colonial empire, and in the cradle of independent Mexico. The remembrances of these killings prevented a sensible reconciliation between creoles and *gachupines*, who, from then on were no longer cousins who quarreled, but enemies lying in ambush and full of suspicions" (Brading 1983, 454).

The students of the School of Mines, sons of small mining families, participated very actively in the insurrection, Casimiro Chovell organizing the defense of the plaza with an active group of miners. When Brigadier Félix Calleja, victor over Hidalgo, broke through, he ordered Chovell to prison and execution, together with other students who had collaborated, such as Ramón Fabié and Rafael Dávalos. Hidalgo had distinguished these three and José Mariano Jiménez with the rank of colonel — and to them came other students and miners. The last named was the only one to escape with the rebel army toward the north. He was captured and shot the 26 of June, 1811, his head being carried to Guanajuato to be displayed next to those of Hidalgo, Allende and Aldama.[36] His companions in insurrection had not encountered any better luck.

On the 26th of November of 1810 by order of [general] Calleja, Rafael Dávalos was killed in the patio of the Granaditas Warehouse (Ramírez 1890, 202, 218). In the afternoon of the 28th, Casimiro Chovell, now manager of Valenciana mine, was hung in front of the principal gate of this same building.[37] The same luck fell to Ramón Fabié, a Filipino native of Manila, son of a lawyer of the Royal Court of that same capital.[38] A short time later, they shot engineer Isidro Vicente Valencia, who worked in the mine already mentioned and to whom del Río was to dedicate the mine *Valencia* or *Valencita* (a double manganese–copper deposit with zinc chloride). His teacher was to speak of him many years later in the Mexican edition of 1848 of his *Suplemento de adiciones y correcciones de mi Mineralogía impressa en Filadelfia en 1832*. He was to refer to his extremely useful discovery realized 35 years before that decreased wastage "at a time in which Chemistry was not able to provide the help that it can today, and thus it was twice as meritorious." The words of del Río seem to be the keys to understanding the loyalty of these students to their college and to their teachers — thus to their class — when he writes: "Valencia discovered the way to save wastage...in a way so simple that I wrote that *I fear the workers will catch on*; but by the next mail he was already shot for indications of being an insurgent" (Ramírez 1890, 86, 118, 137–138). They were loyal, as such, to a group of miners in social ascent, that was to conclude, years later, by supporting without hesitation the insurrection of Iturbide. But in any case when these first victims knew of Hidalgo's "*grito de Dolores*," they gave their blood in the company of miners, peasants and petty officials in their struggle for Independence. Santiago Ramírez, the first and passionate biographer of the institution, wrote of them in his first pages: "From the School of Mines came the first professors, the first engineers, the first wise men which will be honored forever by our Nation" (Ramírez 1890, 9–10).

Endnotes

1. The classic work by Santiago Ramírez (1890, 56 ff.) has been fundamental for this article as have those of F. Sonneschmidt (1826) and Alexander von Humboldt (4th Spanish ed., [1811] 1984). Among more recent literature, we must point out the book by D.A. Brading (1983) and that by VV.AA. (1984). For a study of Juan José Elhuyar, brother of don Fausto, one must always consult B. J. Caycedo (1971). This present essay is an advanced version of a major study I am undertaking on the scientific politics of the Bourbons in America. See also J. L. Peset (1987). [Ed. note: Another fundamental work on this topic is by Walter Howe, 1949, *The Mining Guild of New Spain and Its Tribunal General, 1770–1821,* Cambridge, Mass.: Harvard University Press.]

2. Without doubt Fausto de Elhuyar had a strong personality, which can be clearly seen in contrast to his brother's in the work of Caycedo (1971, 68 ff., 137 ff., 182). His quarrels with the Vascongada Society, the departure of his brother, his expenses, his enterprising spirit and all aspects of his character allowed him to dream of a trip to America and thus he planned a possible trip to Mexico after 1784.

3. VV.AA. 1984, 152–154. Elhuyar got rid of the fiscal and defender of the High Court in order to avoid friction and problems. See Ramírez 1890, 6o.

4. It does not seem that Fausto cared much for the system but as a good politician he took good advantage of it as is demonstrated in some letters to his brother, Juan José. See Caycedo 1971, 220 ff.

5. Ramírez 1890, 61–74 (especially 62–64). Concerning the proposals of Velázquez, see VV.AA. 1984, 167 ff. and Izquierdo 1958.

6. For Humboldt ([1811] 1984, 319 ff.) the veins seemed very large but of low grade and the system of private initiative in which the state could not intervene did not seem adequate to him.

7. It is clear that Elhuyar was involved in mining business and it is said that in April 1799 the Director went to Oaxaca in order "to visit his mining affairs" (Ramírez 1890, 151).

8. See introduction by Roberto Moreno in Sonneschmidt 1983, xii–xiii. It cannot be left unmentioned that José Garcés y Eguía (1802) has written about amalgamation, which work received eulogies from the Tribunal and the College of Mining.

9. Ramírez 1890, 55–60, 75. Rodríguez assumed the chair of mathematics in April 1789, being confirmed by the viceroy and the Tribunal in March 1790. He died in July of 1803.

10. Bataller died in April of 1800. Salvador Sein was nominated in June, and confirmed by examination in December 1801. (Ramírez, 1890, 79, 158, 172 ff., 181 ff.). Elhuyar was reluctant to nominate Antonio León y Gama, who had offers from Velázquez, and limited himself to asking for more information in December, 1791. In this same year Elhuyar changed the study plan for drafting which had been entrusted by the Tribunal to the Academy of Fine Arts, in order to make it useful for the mining students.

11. When Codón did not arrive, the chair of chemistry was to be occupied by Luis Lindner, one of the German miners, who was nominated as a temporary employee in 1796, was permanently appointed in 1799 and died in 1806. In 1809 they still insisted in bringing Spanish pensioners from Hungary or Saxony (Ramírez 1890, 134, 137, 144, 151, 157, 200, 211).

12. Ramírez 1890, 108 ff.; by orders of May 1793 to del Río, 112 ff.

13. Ramírez 1890, 129–133 (quotation, 132). The Tribunal agreed upon a pension for the widow of Elhuyar, should he die first, as well as pensions for Velázquez family. See VV. AA. 1984, 153.

14. We shall not go into detail about the evolution of the teaching of Latin, logic, French, and Spanish which were always of secondary importance. I only want to point out the role these "humanities" played in uplifting and bettering technical teaching, which the reader may not note in the abundant information that Santiago Ramírez provides on this theme.

15. Concerning the role of these humanities in the maturation of the technical professions, such as at Massachusetts Institute of Technology, see Noble 1979.

16. Concerning the difference in writing, we may point out the criticisms directed at Alzate (1831), his style without doubt being too hasty and filled with Galicianisms. Also excused for his poor writing is Sonneschmidt 1983, x. Also the text of Elhuyar 1825 (Prologue, ix).

17. Ramírez 1890, 139 ff.,136, 184–185, 191–195. In November of 1803 the viceroy approved the dividing of the two classes of mathematics with Ruiz de Tejada and Andrés Ibarra as interim professors. After examinations, in October of 1804, Ruiz de Tejada and Juan José de Oteiza were named in charge of the first and second courses.

18. Concerning the University of Mexico, see Carrancá 1969 and Carreño 1961.

19. Ramírez 1889, 30, writes "The good fortune that was the basis for all the studies of such an eminent man, permitted him to select the School of Mines for the coordination of his data, the classification of his collections, the carrying out of his analyses, the construction of his plans, the editing of his Memorials, and in a word, all his theoretical studies" (30, but see 28–31). Not only did Humboldt perform these tasks there, but he also continued to send letters, books and instruments; with Ruiz de Tejada he maintained great friendship, sending him valuable manuscripts which he then presented to Maximilian [the emperor] who in turn ceded them to the National Museum.

20. However, in February of 1796, there were still complaints about the lack of laboratory experiments and training in the mines (Ramírez 1890, 132).

21. Ramírez 1890, 116, 120–122. Alzate (1831, II, 485–459) stimulated in the Public Prosecutor this interest in archaeology.

22. See Ramírez 1890, 148, 182 ff. on purchase of collections; 153, 193 on foreign machinists. See also Humboldt [1811] 1984, 363.

23. An order of the viceroy arrived in April of 1793 demanding that doctrine of Werner be explained to the students (Ramírez 1890, 112, 114, 125). This trend was to result in some adoptions or translations and above all in the editing of the *Orictognosia* by Andrés del Río. In general, the disciples of Werner were reticent to adopt Lavoisier and abandon his flogistics [a concept that every combustible substance has a hypothetical ingredient that is expressed as flame], as well as the preferred chemical analysis rather than external examination of minerals, in seeking an adequate synthesis.

24. Concerning the different textbooks, see Ramírez 1890, 60, 74–78, 99–103, 106–108, 112–114, 193, 238–239.

25. Concerning the edition of the *Elementos de Orictognosia*, see Ramírez 1890, 129–134, 193–194. A useful facsimile edition was made in Madrid in 1985 by the Universidad Complutense, with an interesting introduction by José Luis Amorós. The first edition in Mexico was published in 1795 and 1805 in two parts.

26. Concerning the collaborations of von Humboldt, see Ramírez 1890, 192–196, 227.

27. There are small modifications or adaptations [to the admission standards]; thus Filipinos such as the Fabié were admitted, and the entry age was raised from 12 to 14 years, and mestizos were accepted... See Ramírez 1890, 83, 90, 181.

28. Ramírez 1890, 86; even in 1806 the Director was against admitting of a postulant (*op. cit.*, 204).

29. The four possible dissertations for which the students needed to study were the following: No.1 "Instructions concerning what must be arranged D.N. in order to form the Geognostic Description of the mining center of N. with a geographic plan of its situation and of the surrounding mountains." No. 2 "Instructions concerning what must be arranged in order to form the description of a mine and its plan." No. 3 "Instruction that will be necessary for D. N. in order to compose the description of the beneficiation of silver and gold by use of mercury as it is practiced in the northern mining centers." No.4 "Instructions as to what must be done by D.N. in order to make a description of the beneficiation by fire [smelting] of silver ores in the northern mining centers" (Ramírez 1890, 158–171).

30. Interest in useful museums, with application to the sciences and the arts, starts at the end of the eighteenth century. According to J. Peuchet, when he affirms that even though there are museums of natural history," nowhere can be found a display of industrial history that would make known the works of industry such as those of natural history allow us to know the productions of nature." (*Dictionaire universel de la géographie commerciale,* 6 vol., Paris 1799–1800, I:10).

31. In January of 1799 del Río filed claims on the mines of Morán and annexes in Real del Monte before the Tribunal. (Ramírez 1890, 22, 40). In August of 1805, Elhuyar proposed to the tribunal that the student Rafael Dávalos go to practice in Morán's mine and that he install the water column machine constructed under the guidance of del Río, and that the students Pedro le Chaussé and Juan José Rodríguez "go to the mine of Jesús, property of the Director, where he is going to install an animal powered pumping system of his own invention." (Ramírez 1890, 198–199). Elhuyar goes to the mine of Morán in the Real del Monte to continue the construction of his pumps in February and April of 1808 (ibid, 206). Even in January of 1822 the Tribunal asks for a report on the pumps invented by Elhuyar (*op.cit.* 151). Humboldt ([1811] 1984, 362–363) admired these pumps and longed for the miners to become interested in them.

32. Ramírez 1890, 205, 207. In 1810 del Río is given the task of revisiting some mercury deposits (*ibid.* 216 ff.).

33. There had already been some uproar in November of 1800 on promotion of the Assistant Principal to Principal without following regulations, to which the Director agreed at the time in order to avoid problems (Ramírez 1890, 175–177, 209–217).

34. Elhuyar 1825, 57. Without doubt the mines chosen to which students were sent were also the richest, for which it is not unusual, given their proximity and problematic population, that they sided with the insurgents. "The three districts that we have had motive to cite various times, to wit: those of Guanajuato, Zacatecas and Catorce, render more than half of this sum [50 percent]. A single vein, that of Guanajuato, produces almost a fourth part of all Mexican silver and the sixth part of all that is produced in America." Thus Humboldt ([1811] 1984, 331) expressed himself after calculating Mexican silver production as two and one–half million marks [1.25 million pounds] annually.

35. Elhuyar 1825, 57; he also continues about the problem of interrupted communications (ibid., 60).

36. Earlier there had been the case of the college student, José Antonio Rojas, who was accused of heresy and materialism in 1804 by the Inquisition in Guanajuato, where he was giving a course in chemistry after having learned from Linder. Found guilty, he was able to flee to the United States from whence he exhorted his compatriots that they must follow the example of the nation that had received him [the independence of the United States]. He was

excommunicated and declared a traitor by the Inquisition and civil authorities (VV. AA. 1984, 175).

37. Concerning Casimiro Chovell, who was greatly appreciated by his professors, see Humboldt [1811] 1984, 391; also del Río, edition by Amorós, 1985, 47.

38. The Fabié brothers were sons of a Spanish lawyer of the Court of Manila. They arrived in Mexico at the end of 1801, sent by the consul in the Philippines, and were presented by his agent to the School of Mines (Ramírez 1890, 181 ff., 218 ff.).

References

Alperín Donghi, T. 1985. *Historia de la América Latina.* 3 vols. Madrid.

Alzate y Ramírez, J. A. 1831. *Gacetas de literatura de México.* 2 vols. Puebla: Hospital de San Pedro.

Brading, D. A. 1983. *Mineros y comerciantes en el México borbónico (1763–1810).* Mexico City.

Carrancá y Rivas, R. 1969. *La universidad mexicana.* Mexico City: Fondo de Cultura Económico.

Carreño, A. M. 1961. *La real y pontificia Universidad de México, 1536–1865.* Mexico City: Universidad Nacional Autónoma de México.

Caycedo, B. J. 1971. *D'Elhuyar y el siglo XVIII neogranadino.* Bogotá: Editorial Kelly.

Elhuyar, F. de. 1825. *Memoria sobre el influjo de la minería en la agricultura, industria, población y civilización de la Nueva España...* Madrid: Imprenta de Amarita.

Garcés y Eguía, J. 1802. *Nueva teoría y práctica del beneficio de los metales de oro y plata por fundición y amalgamación.* Mexico City: M. de Zúñiga y Ontiveros.

Humboldt, A. von. [1811] 1984. *Ensayo político sobre el reino de la Nueva España.* 4th ed. Mexico City: Editorial Porrúa, S.A.

Izquierdo, J. J. 1958. *La primera casa de las ciencias en México: el Real Seminario de Minería (1792–1811).* Mexico City: Ediciones Ciencias.

Noble, D. F. 1979. *America by Design.* Oxford: University of Oxford Press.

Peset, J. L. 1987. *Ciencia y Libertad. El papel del científico ante la Independencia de América.* Madrid.

Ramírez, S. 1889. *Biografía del Señor Don Manuel Ruiz de Tejada.* Mexico City : Imprenta del Gobierno Federal.

——. 1890. *Datos para la historia del Colegio de Mineria recogidos y compilados bajo la forma de efemérides por su antiguo alumno el ingeniero de minas Santiago Ramírez.* Mexico City: Imprenta del Gobierno Federal.

——. 1891. *Biografía del Sr. Don Andrés Manuel del Río, primer catedrático de mineralogía del Colegio de Minería.* Mexico City: Imprenta del Gobierno Federal.

Río, A. del. [1795–1805] 1985. *Elementos de Orictognosia,* 2 vols. [facsimile edition edited by José Luis Amorós] Madrid: Universidad Complutense.

Sonneschmidt, F. T. 1826. *Tratado de la amalgamación de Nueva España y sacado a luz por don J.M.F., París.* Mexico City: Librería de Bossange Antoran y Cía.

——. 1983. *Tratado de la amalgamación* ...[facsimile edition] Mexico City.

VV. AA. 1984. *Minería Mexicana.* Mexico City.

Processes in the Advance and Retreat of the Mining Frontier in Southwestern Colombia During the Colonial Period

María Clemencia Ramírez de Jara,
Instituto Colombiano de Antropologia, Santa Fé de Bogotá

Abstract

From the sixteenth until the mid–eighteenth centuries, southwestern Colombia formed a mining frontier that presents the phenomenon of advance and retreat as a response to the main interest of the Spaniards: the extraction of gold. In accomplishing this end the Spaniards were confronted by various attitudes on the part of the indigenous population that inhabited this zone — attitudes that ranged from submission by some to outright rebellion by others.

Massive forced migrations of indigenous population used for exploiting gold mines took place, such as the movement of Indians from Mocoa to the lode deposits in the district of Almaguer, south of Popayán; or tribes from the Valley of Sibundoy to the placer deposits of Mocoa. At the end of the sixteenth century Indian migrations occurred from Madrigal (Chapanchica) on the banks of the Río Patía into the territory of Abad, where placers were worked along the Pascula River until the beginning of the seventeenth century. At that time the process was directed from two power areas: one in the Andean highlands (district of Pasto) and one in the upper Amazonian rain forest (Province of Mocoa).

Beginning in 1630, the year in which native groups of the Pacific lowlands were subjugated, the mining frontier expanded toward the west: in the Telembí river basin, where Santa María del Puerto (present–day Barbacoas) was founded; and in the lower Río Patía, plus the level coastal plain behind the modern town Tumaco, which became areas of mining development. By 1684 numerous Indian survivors of the Sindagua tribe and African blacks were working in this area. The end of the mining frontier, however, did not result in the incorporation of the mines in either the eastern or the western Andean slopes into the socio–political organization of the central Andes.

Key words: ethnohistory, colonial period, mining, Colombia, mining frontier, *mita* (forced labor).

In Quest of Mineral Wealth: Aboriginal and Colonial Mining and Metallurgy in Spanish America, edited by Alan K. Craig and Robert C. West, 1994. Geoscience and Man, vol. 33, pp. 253-270. Dept. of Geography & Anthropology, Louisiana State University, Baton Rouge, LA 70893-6010, U.S.A.

Resumen

Durante el siglo XVI hasta mediados del siglo XVIII, el suroccidente de Colombia se constituye en frontera minera que presenta procesos de avance y de retroceso como respuesta al interés principal de los españoles, la extracción del oro, para lograr el cual se ven enfrentados a diversas actitudes por parte de la población indígena que habita esta zona como son la sumisión y la rebelión entre otras.

Se destacan los traslados masivos de población indígena para la explotación de minas tanto desde Mocoa hasta el Distrito de Almaguer (minería de veta) al sur de Popayán, como del Valle de Sibundoy (minería de veta) a Mocoa (minería de aluvión). Desde Madrigal (Chapanchica) situado en las riberas del río Patía, a finales del XVI se hacen traslados de población indígena a territorio Abad, donde se explotaron minas de aluvión en el río Pacual hasta entrado el siglo XVII. Se encuentran entonces dos fuerzas que jalonan el proceso: la andina (Distrito de Pasto) y la selvática (Provincia de Mocoa).

Para 1630, año en que se sojuzga a los grupos nativos de las tierras bajas del Pacífico, se amplía la frontera minera hacia el occidente: la cuenca del río Telembí, donde se funda Sta. María del Puerto (actual Barbacoas) y el bajo Patía y la llanura plana detrás del actual Tumaco, son zonas de desarrollo minero. Para 1684, numerosos indios (sobrevivientes Sindagua) y negros están trabajando en esta zona. El cierre de la frontera minera como tal, no se logra, en cuanto a que se incorporen las zonas mineras de la vertiente oriental y occidental a la organización socio–política central andina.

Palabras claves: ethnohistoria, colonia, minería, Colombia, frontera minera, mita minera.

Introduction

In this essay I wish to discuss the processes involved in the advance and retreat that occurred within what I consider to be the colonial mining frontier in southwestern Colombia. This frontier existed from the sixteenth until the mid–eighteenth centuries, and corresponded to the Spanish occupation of the territory between the area south of Popayán and the eastern and western slopes of the Andes. In this zone the dominant economic activity of the Spaniards was the mining of gold.

The Mining Frontier as a Mobile Phenomenon

One concept of a frontier has been summed up by Guhl (1991, 31): "A frontier results from the action of state power in contrast to a natural limit that forms a zone of contact of one natural region with another." For an analysis of the case at hand it is important to perceive the above definition, which is strongly geo–political, from an historical–cultural perspective: we are dealing here with cultural limits, the frontiers of which are dynamic and are determined by people within a specific time period. In the words of Mellafe (1968, 11), "frontiers are formed by the areas of recent occupation, areas in the process of being colonized." In this way Spanish power took place on a territorial basis, where, in some cases, it did not achieve complete domination; these included certain transitional zones or outlying areas that expanded and contracted according to the interest of the central political power. In this way mobile frontiers emerged.

A case in point is seen in the eastern and western piedmonts of the Andes in southern Colombia. These marked the limit between the Andean highlands and the forested lowlands, two quite different natural regions. Caillavet (1989, 3) points out that in the region of Almaguer north of Popayán and as far south as that of Latacunga near Quito, the belt that separates the Pacific coast from the Andean range is covered by tropical rain forest. This phenomenon produces a "double Amazonia" (eastern and western) and indicates graphically how "the aboriginal inhabitants of the highlands that rise between two symmetrical rain forests had to contend with 'savage' neighbors on either side."

The Spanish government established in the Andean highlands, driven by its basic objective of searching for gold deposits, directed incursions into the eastern and western rain forests. However, the colonial period saw repeated failures to subdue the aboriginal forest tribes and even those of the Andean piedmont, such as the rebellious and warlike Mocoas in the eastern piedmont (Bonilla 1968; Llanos and Pineda 1982) as well as the Sindaguas on the western slope (Calero 1988, Romoli 1963, Caillavet 1989). In addition, the rugged nature of the land and the predominance of dispersed population made it difficult to concentrate Indians of those areas in pueblos where they could be indoctrinated, controlled, and utilized as a permanent labor force. All of these factors affected the advances and retreats of the mining frontier. As Victor Alvarez notes (1986, 51), "When the colonial documents speak of the frontier, they are designating provisional limits of a process of permanent expansion, changing as new areas are occupied and thus incorporated into the territoriality of each population group."

In this particular case the Spaniards tried to take over the territory of the piedmont that extended toward the zone of tropical rain forest. In these areas they tried first to incorporate the Andean highland zone economically as well as politically and ideologically, an area in which they had succeeded in consolidating their domain and from which they could exercise their political administration. It would seem therefore that the geographical limits (the piedmont areas) might coincide with the political frontiers for the southwestern region during the colonial period. However, it should be noted that we are dealing here with social and cultural frontiers, rather than natural ones. In the words of Latimore (1968, 375):

> The changing significance, for changing societies of an unchanging physical configuration which may at one time be a frontier, at another time a frontier of different significance (as when an old external frontier becomes an internal demarcation within an enlarged community), and at another time no frontier at all (as in the case of the western frontier of expansion of European man across the North American continent), leads to the axiomatic statement that frontiers are of social, not geographic origin.

The dynamism and mobility of the mining frontier is therefore manifest in activity involving the search for and extraction of gold. In southern

Colombia placer gold mines were as numerous as vein mines, which were surely exploited by the aborigines but less aggressively than by the Spaniards, whose main objective since the beginning of the Conquest was to acquire the precious metal.

Labor System: *Rescate,* the *Encomienda,* and the *Mita*

It is within the perspective discussed above that "*rescate,*" or trade (Tovar 1989), and the *encomienda* gain meaning as forms of exploitation and subjugation that defined the economy and society of the Conquest during the first half of the sixteenth century, and of the Colony in the second half of that century.

The term "*rescate*" meant not only sacking and theft, but also consisted of the establishment of a system of reciprocity and interchange with pre–Hispanic communities utilizing indigenous economic models based on sound principles: for example, Spaniards exchanged hatchets and other merchandize for gold, pearls, and so forth.

> ... *(R)escate* (trade) became a way of entry through the system of exchange, which culminated in war with, and enslavement of the natives. These activities became destructive factors in native resistance, opening ways for the articulation of two worlds, two civilizations, two mentalities... . When *rescate* (trade) ceased to be the leading element in Spanish expansion, the *encomienda,* with its system of tribute and compulsory labor, became an alternate force in colonial relations (Tovar 1989, 2).

In exchange for labor and tribute that he received, the *encomendero* was obliged to instruct his charges in the Catholic faith. Later, religious instruction was made a priority in the *encomienda* system; in 1570 the visitor Valverde decreed that for the southwest the priests should receive a portion of the first tributes collected (Calero 1988, 96). The priests were required to live in the native pueblos and were prohibited from using the villas, or larger towns such as Pasto and Popayán as their centers of operation. Thus, especially in isolated areas, it was the priests who exercised authority and represented the Spanish government by supervising the collection of tribute, punishing improper conduct of the Indians, and acting as police in disputes (Calero 1988, 97). This was accomplished especially in areas difficult of access and with dispersed settlement, where the *encomenderos* and civil authorities did not live, as in the case of the piedmont region. The priests were charged with the expansion of the frontier toward the rain forest, not only with evangelic intentions but also with interest in expanding the economic and mining frontier.

In 1549 under the reformed New Laws, the tribute payments to *encomenderos* was changed to money and products, rather than in personal services. But the need for labor to maintain Spanish enterprises generated an opposition that could be overcome only by the use of "public compulsion" which resulted in the appearance of the *mita,* or allotment of forced but remunerated personal services, decreed in 1550, and which persisted until the first half of the eighteenth century.

The *mita* could be used in farming, mining, herding, domestic service, or in workshops such as textile mills. The system included the establishment of work quotas that the indigenous communities had to discharge according to an assignment that the *corregidor*, or Spanish magistrate, might make. Workers could be used as much for the service of the *encomendero* as for the land owner, or *hacendado*. Indians from a particular place were drafted periodically for work during a determined time period for Spaniards and were paid an adequate wage that was controlled by the authorities. *Encomenderos* deducted from the daily wages the amount each Indian should have paid in tribute. The duration of *mita* mine work period was fixed at 10 months out of each year and the number of workers for such duty could not exceed one third of the permanent population subject to tribute (Capdequi 1941).

In 1558, Tomás López, the *visitador* for the southwest of Colombia, was in charge of replacing the system of personal services with the *mita*, but it was García de Valverde who, during his *visita* of 1570 actually installed the *mita* for mining, the type of labor system utilized for the exploitation of gold.

Valverde established two types of *mita*. One was applied to the areas that were most productive in gold: the Los Abades area on the western Andean slope and the Valley of Sibundoy on the eastern slope. The other type of *mita* was to be applied to the rest of what is today southwestern Colombia.

The *mita* for the mines was decreed for those areas in such a way that the *encomendero* was guaranteed free of charge a fifth of the available labor force for eight months of the year, from November 1 to June 30; there were two work shifts, each of four months in length with a daily work period of eleven hours (Romoli 1977–78, 26). Service in the mines was performed at a wage paid in blankets or shirts.

This [a shirt] was paid to each Indian who began work in November and finished in February, whereas a blanket was for those who began work in March and finished in June ... in addition, each *mitayo* (*mita* worker) was to receive daily a pint of maize, a half pound of meat and some salt. On fast days [*días de pescado*] each would receive only a few potatoes and beans (Padilla 1977, 84–85).

Mita labor was indispensable for the successful exploitation of mineral wealth and for the generation of income for the sustenance of the local Spanish society as well as the curates. Without this economic aid the region would have been abandoned and the evangelization of the Indians would have ceased even earlier. It can be affirmed that the mining frontier coincided with the mission frontier, for while exploitation of mineral wealth was in progress the priests remained in the same area carrying out judicial and administrative functions; they abandoned the area only when mining became unimportant, as happened in the Valley of Sibundoy.

Location of Vein and Placer Mines in Southwestern Colombia

In the north vein mines in the area of Almaguer, south of Popayán, were discovered in 1551 (West 1972, 23) (fig. 1). Fray Gerónimo de Escobar ([1582] 1983, 291) refers to Almaguer thusly:

> continuing the discussion on the pueblo of Almaguer, I understand that each year more than 30,000 pesos of gold is mined, for the area is rich. Three leagues from the pueblo there is a mining center called La Concepción where there are usually two thousand Indians and negros working in the mines, and labor gangs come in from other pueblos.

Besides the mines of Almaguer, located "on a very large and high hill," there were the famous placers in the Valle de Guachicono in the same area.

Toward the eastern slope of the Andean highlands the lode mines of Sibundoy, Yscanzé, and Agreda (Mocoa) were exploited, and toward the western slope those of Madrigal (Chapanchica) were being worked by the end of the sixteenth century. On Sibundoy, Valverde (1570, ff. 207 and 374) stated "this witness saw that in the same valley there are gold mines next to the Indians' houses, and for that reason they are able to furnish tribute in gold." He ordered that such mines be worked by means of open cuts. Located along the upper Río Cáqueta drainage, was the city of San Miguel de Sucumbíos, founded in 1558, and whose inhabitants "had rich placers of this metal during the mid–sixteenth century along the Uchuarina, Bermejo, Anzeula, Amoguajes and other rivers... .The Indian tribes pan gold from the sands of the Caquetá and from its tributaries, the Curiaco, Cascabel and Villalobos; from the Putumayo and its tributaries, the San Pedro, Guineo, San Juan Oritapunto and Guamués; from the Aguarico those that flow into it, the Cofanes, the Condué and the Dué" (Restrepo [1888] 1979, 80).

Also, various references mention the mines and the placers around the city of Mocoa. Vázquez de Espinosa ([1620] 1948, 334) refers to Mocoa: "The city of Mocoa in this province is on the same parallel as Pasto, but to the east, with few Spanish inhabitants; it has a hot climate, much cotton is harvested in its district and there are rich mines and placers of gold." A fuller statement is found in a document of 1800:

> In the old days gold was worked in the ravines of Santo Domingo, San Pedro, Sardinas and Platanar. Along the Mocoa River near the crossing of the Caquetá one sees remains of old workings, as upriver from Yunguillo and in Escancé and Cascabel; all of these placers are very rich as along the banks of the Caquetá River as far as the mouth of the Río Fragua (AHN, Fondo Bernardo J. Caycedo, Caja 23, f. 91., in Figueroa Rivera 1986, 26).

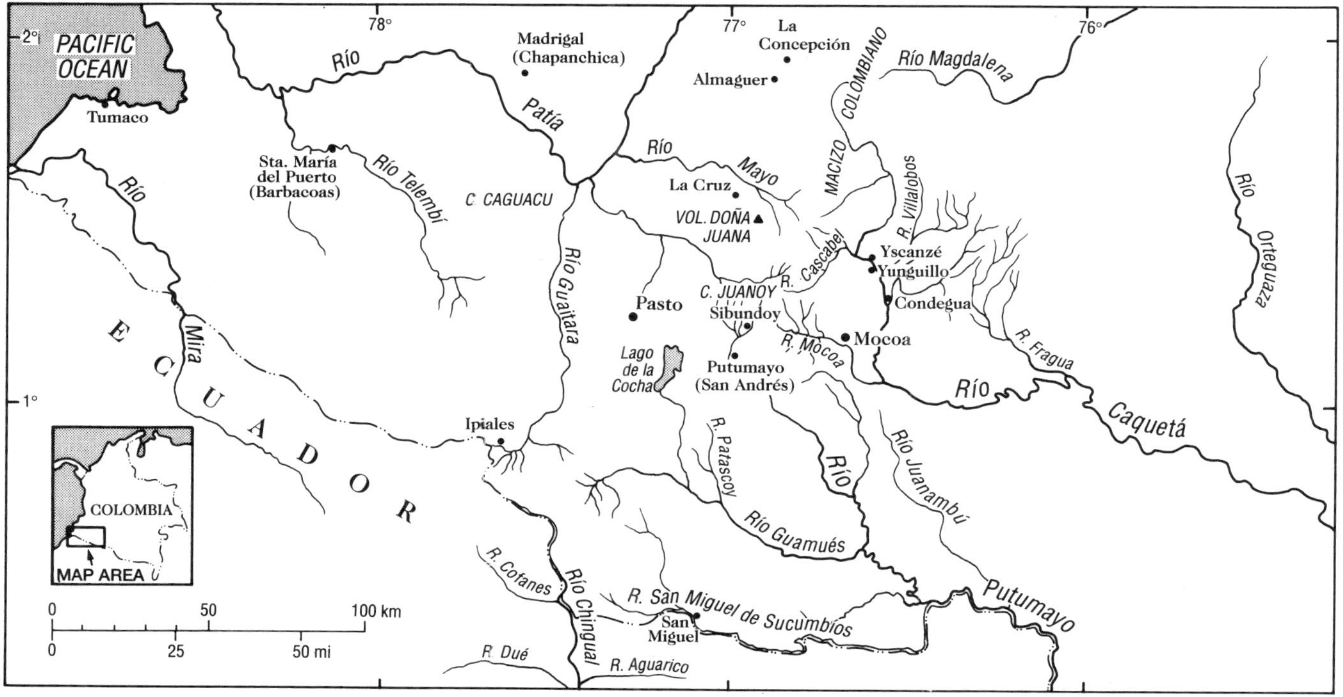

Fig. 1. Colonial mining region in southwestern Colombia.

In Madrigal (Chapanchicha), located on the Río Patía at a place known today as Hoz de Minama, the gold mines were exploited until 1582, when they were abandoned; mining activity was moved to the hill of Nuestra Señora de Caguacara in Abad territory (along the left bank of the lower Río Guaitara to its mouth) toward the east; the Indians of Madrigal were also moved to the new site (Calero 1988, 288).

Fray Gerónimo de Escobar ([1582] 1983, 293) comments on the villa of Chapanchica: "This small town is in very hot country and is thus plagued by sickness. About a thousand Indians are in the town and are commended to twelve town members. It is a town of some twenty men. These thousand Indians live within a radius of ten leagues of the town. It is a land of much gold. The aborigines are few and are worth little." In the territory of Abad, gold placers were exploited along the Río Pacual until the beginning of the seventeenth century, as was the case in the Valley of Sibundoy; as indicated in the *visita* of Armenteros y Henao in 1607, the work period of the *mita* miners was increased from eight to twelve months.

As to the southern Pacific coast:

> The Pacific plains are formed of alluvial deposits left by the streams that descend from the western slope of the Cordillera Occidental (Andes). The northern half of the plain, which extends southward from Buenaventura as far as Guapi, rarely exceeds a width of 48 kilometers and is crossed by short and rapid–flowing streams, such as the Micay, Naya and Timbiquí. In the southern half of the coastal zone the important Patía River and its tributaries have built a wide plain known as the region of Barbacoas (West 1972, 24).

The gold placers of the coastal plain were not exploited by the Spaniards until the seventeenth century.

As can be deduced from the foregoing citations, from Almaguer southward through the drainage basin of the Telembí River, the lower Río Patía (Barbacoas area) and the low–lying plain behind present–day Tumaco, and eastward to the Aguarico basin in the lower Putumayo area, one can ascertain that a fluctuating mining frontier existed during the sixteenth, seventeenth and eighteenth centuries — a frontier linked with the occupation and/or abandonment of areas suitable for the exploitation of gold.

Inhabitants of the Gold Frontier

Having delimited the mining centers south of Popayán and eastward and westward from Pasto, we now proceed to identify the aborigines who lived in these areas (fig. 2).

Those who inhabited this large area were mainly the Quillacingas. The Spaniards classified them into two groups, those of the Andean highlands and those of what was called La Montaña. The first groups (those along the road to Quito, to Popayán, to Almaguer, and to the valley of Pas-

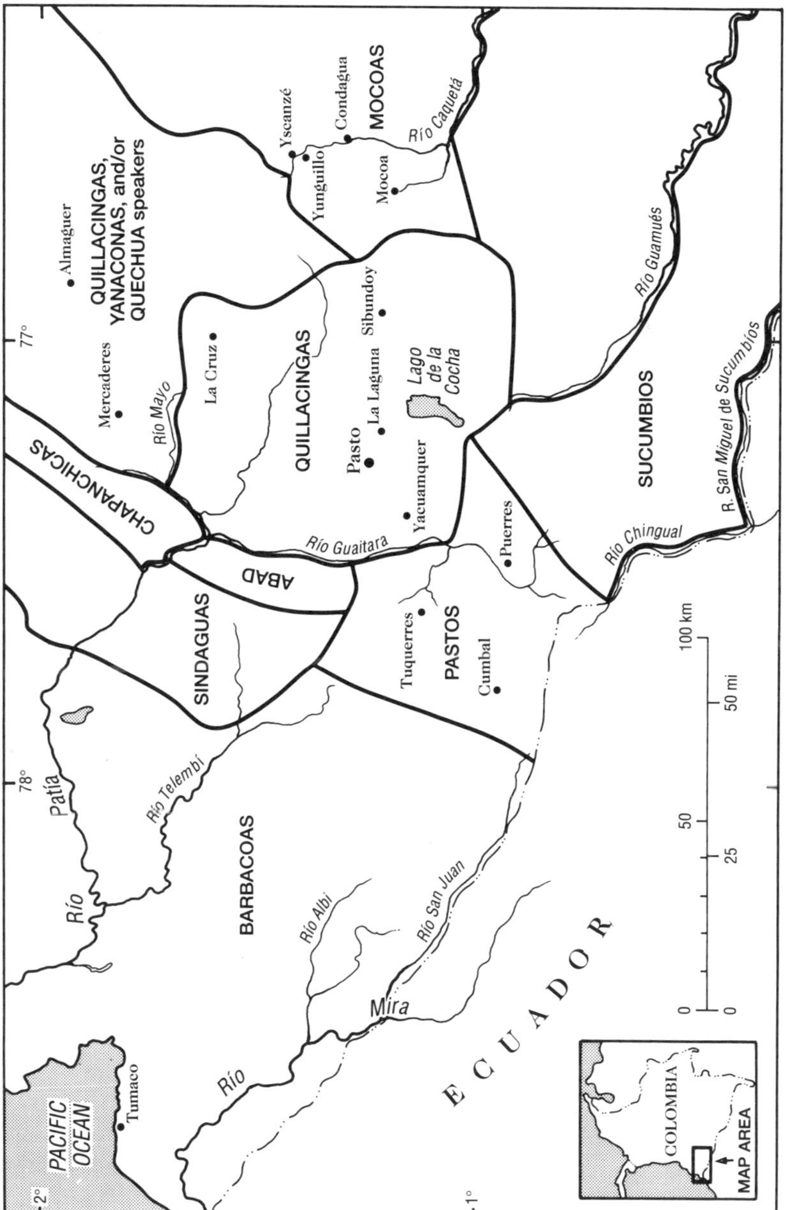

Fig. 2. Distribution of Indian groups, 1535–1635.

to) occupied the lands north of Pasto, along the right margin of the Guaitara River, in the valley of Atris, in the valley of the Juanambú River, and in the high and medium elevations of the Mayo River on the northern limit of the territory. The Quillacinga of the Montaña inhabited "the pueblos of La Laguna (La Cocha or Lago Guamués) and, farther in, those of the Sibundoy Valley and Patascoy ... extended from the summit of the Cordillera Central to the Cordillera Portachuelo east of Sibundoy, and from the drainage divide between the upper Caquetá and the headwaters of the Putumayo as far as the Guamués River" (Romoli 1977–78, 13). We should note that some have suggested the possibility that in the Valley of Sibundoy the pueblos of Patscoy, La Laguna, Putumayo (San Andrés), Santiago, and Genoi may have been inhabited by Quechua–speaking Inka since early times, whereas the pueblo of Sibundoy has been considered a Quillacinga and/or Kansa settlement (Ramírez de Jara and Urrea 1989).

The Sucumbío lived south of the Sibundoy, between the crest of the Cordillera Central and the Putumayo, and between the Guamués River and San Miguel de Sucumbíos. A third group, the Mocoa, inhabited the river valley of the same name and an adjoining stretch along the right margin of the Caquetá River (Romoli 1977–78, 13).

Also within the eastern versant of the Andes farther down the Caquetá was Yscancé (present–day Descancé), but it is not mentioned in the documents dated prior to their pacification (1550–1566). It is not known if the name Yscanzé existed in this area before the entry of the Spaniards, who for long called the upper Caquetá in the same way. Because of their conquest, Yscancé counted 5,000 *indios de encomienda,* but soon were reduced to 2,000 souls, rated as very good people. Later, those who survived were taken to other areas nearby (Romoli 1962, 246). Fray Gerónimo de Escobar ([1582] 1983, 294) stated in 1582 that "Yscanzé was settled ten years ago [1572]" and continues: "... the Indians are very good people; there may be 2,000 of them commended to seventeen town members, most of whom are mestizos and mulattoes." Friede (1967, 240), referring to Franciscans, states "... already in 1635 Franciscans from Popayán and Neiva entered the Fragua and Orteguaza rivers and established the pueblos of Descansé (Yscanzé), Yunguillo, Nuestra Señora de Ecija, and others," but he makes clear that the truth of that statement has not been proven.

The headwaters of the Iscuandé and Patía rivers southward to the Pacific coast formed the locale of the Sindagua, who as warriors, ranged over a vast territory that included the northern part of the present department of Nariño (in the Pacific lowlands), the Patía river valley, the basin of the Río Pacual, and the southern part of the Pacific lowlands of Cauca. The Sindagua attacked the Abad Indians, destroying settlements, an action that seriously affected the mining of gold. Only by 1630 were the Spaniards able to subdue them, which resulted in the possibility of expanding the mining frontier westward toward the Pacific coast. In reference to the *encomiendas* in the Province of Barbacoas it can be deduced that they were composed of various groups of Indians, as Romoli (1963, 272) indicates: "Sindaguas and Nulpes," or "Sindaguas and Cuasmingas." Possibly toward the end of the seventeenth century Indians of the San Juan de Micay river area had been transferred to the Río Telembí.

The Indian Population and the Exploitation of Mines

The Quillacinga and Pasto Indians did not resist the Spaniards, who took over their territory "almost without cost" (Romoli, 1977–78, 17). For that reason, on being forced into labor in the mines, those Indians received the full devastating effects of the Conquest. Referring to labor assessments imposed by Tomás López during his *visita* undertaken in 1558, Moreno Ruíz (1971, 428) notes:

> It appears that the assessment that the judge (*oidor*) Tomás López made previously was not carried out, for neither the Indians nor the Spaniards would agree. The Indians would not, because, except for the Pastos, they had little cotton or knowledge of weaving, and preferred to continue in mine work in the provinces of the Abades, Quillancingas and Sibundoyes, and so doing, satisfy the labor assessment that had been imposed upon them; for that reason 1,405 Indians went to work in the mines. The Spaniards accepted this change in the perception of the assessment.

The displacement of population in order to work the mines was frequent; often not only individual workers but whole families were thus displaced, as attested by Fray Gerónimo de Escobar ([1582] 1983, 294) on referring to the pueblo of Yscanzé: "The Indians neither here nor in their homeland have discovered mines; they have been taken to the mines of Almaguer, where they are today, and for that reason many have died." During the second half of the sixteenth century the decrease of Indian population was accentuated mainly as a result of compulsory labor in the mines. This involved the removal of the inhabitants into the interior of the established frontier. Such displacements were carried out with no regard to ethnicity or to geography. The inhabitants of the Andean highlands, of the piedmont, and of the lowlands were constantly being intermingled, but perhaps maintaining local social standards that were in force before the Conquest.

These population displacements continued until the mid–eighteenth century, as long as the mining *mita* persisted, as is inferred by the following note by Friede (1967, 61): "In the lists of Indians baptized in Mocoa in 1775… one finds surnames such as Chicunque, Mavisoy, Tandioy, Omne, Jacanamejoy, Peña, Quinche, which indicate the introduction of foreign aborigines originally from the region south of the Andean range of central Colombia and from the Valley of Sibundoy."

Regarding the marked decline of population, Calero (1988, 274), referring to the district of Pasto, maintains: "The northern section of the Juanambú River and the Río Guaitara, together with Sibundoy, suffered the largest Indian losses." Comparison of the demographic data supplied by the magistrate Ortega at the end of the sixteenth century (1590) with that of governor Rodrigo de Manosca at the end of the seventeenth (1688–1691) shows that the population of the district of Pasto continued to decline throughout the seventeenth century, a fate from which the area of Sibundoy did not escape.

At the end of the seventeenth century the native population of the lowlands in general continued to decline, while that of the highlands had reached its lowest point and had begun to recuperate.

Regarding the Indian groups of Ecija de los Sucumbíos and Mocoa in the upper Caquetá, Llanos and Pineda (1982) also place the rapid decline of indigenous population within the second half of the sixteenth century, which corresponds with the period of relative economic prosperity brought about by gold mining; this continued until the second half of the seventeenth, when the Spanish centers were destroyed by Indian uprisings. By 1651 the city of Mocoa had been desolated.

The depopulation coincided with the abandonment of intensive exploitation of the gold mines on the eastern piedmont; thus began a process of retreat of the mining frontier, which prevented the eastern frontier as such from closing down and disappearing in the sense that exploitation of this mining zone could be incorporated into that of the central Andes; on the contrary, it maintained its marginal territorial condition. Thus, the third meaning of the frontier as distinguished by Mellafe (1968, 11) is fulfilled:

> A geographic space, in which are given the processes of production of institutional and social structures, has not yet integrated itself into a continuum, but is in the process of a very drastic formation or transformation... .This process also presupposes an impact or fusion and cultural connection of two or more distinct cultural horizons.

After 1610, mining activity expanded its frontier toward the Pacific lowlands. This lowland region extended from the Mira River basin northward to the Río Naya, and from the crest of the Andes on the east to the Pacific coast on the west.

In 1630 the Sindagua warriors were subdued. The survivors were taken to Santa María del Puerto to be used in agriculture and mining. As Calero (1988, 229) points out, "With the subjugation of the Sindagua and establishment of two Spanish settlements — Santa María del Puerto (present Barbacoas) on the banks of the Telembí and Santa Bárbara de la Isla del Gallo — the area was ready for a period of mining bonanzas."

As a result of the decline of population in the lowlands, black labor was imported beginning around the end of the sixteenth century (1592) and by 1684 numerous Indians and Negroes were working in this area.

The eastern slope or piedmont of the Andes (Valley of Sibundoy, Iscanzé, Yunguillo, and so forth) found itself between two forces of attraction: the Andean highlands (district of Pasto) and the tropical rain forest (province of Mocoa). The attraction was seen in the status of production and in the political and religious domain. Despite its geographical position in the piedmont, Mocoa dispatched its activities as though it were part of the jungle, owing to the close relationship with rain–forest tribes, such as the Sucumbíos,Yaguaronjos, and others. Thus, missionaries as well as Spanish conquerors regarded them as jungle people of the rain forest. It was in Mocoa where the Franciscan missionaries exerted their colonizing skills.

Mining within the western versant and the Pacific coast was supervised by Spaniards living in Pasto and Popayán. Around 1699 the mining boom of Barbacoas had diminished, as seen in the decrease of Spanish residents from 40 to 16, an index of decline from the earlier prosperity of the region (Calero 1988, 236); this translates into the beginning of abandonment of the area, thus resulting in another example of mining frontier retreat, with the consequent exclusion and marginalization of the lowlands from the Andean region that belonged to the Gobernación of Popayán.

For the eastern versant there are other notices referring to aboriginal rebellions that broke out in the Putumayo region: "In 1721 a general uprising of the Putumayo and Caquetá tribes within a few weeks resulted in the overthrow of many decades of missionary work" (Friede 1967, 241). Victor Daniel Bonilla (1968, 18) adds: "And the attacks reached as far as Sibundoy Grande, for this town was overrun various times by the Mocoas and Yaguaronjos who were desirous of obtaining the military help of their racial brethren against the Iberian invaders. This desire stemmed not only from the fact that the aborigines had been decimated during the first decades of warfare but also from their entrance into a stage of transculturation that was increasing day by day." This presents a concrete example of the pressure exerted by the Mocoas on the Sibundoyes, who behaved in an entirely different way regarding Spanish domination: whereas the Mocoas rebelled and as a consequence disappeared as an ethnic group, the Sibundoyes have survived to this day.

The intent to subdue the Mocoa Indians and those of the lower Putumayo and Caquetá continued well into the eighteenth century, as the following excerpts indicate. In a document dated 1739 housed in the Archivo del Cauca one reads:

> The Indians of the Province of Mocoa, who in the year 1640 were indoctrinated in the Catholic faith and who recognized the priests [sic] who indoctrinated them until the year [1684], when they rebelled, resulting in the total extermination and death of the

> Christian members of the city of Mocoa, which belongs to the jurisdiction of Pasto and consequently to the Gobernacion of your Grace. Since the year 1724 to the present 1739, they are again being compelled to submit to Christianity, as did their forefathers, through the fervent solicitation of priests belonging to my hierarchical order, who have converted them thus far in the main pueblo of San Antonio de Padua in the great Caquetá region and in other neighboring areas (ACC 1739, Col. El–4g, Sig. 4387).

Apparently this intent to re–evangelize the Province of Mocoa was not a total success, for in this petition of 1740 one reads:

> Señor Governor and Captain General: Don Pedro de Valencia, member of this city, owner of Negro slaves working in the mines, appears lawfully before Your Grace to testify that since the slave gangs of my sons Pedro Agustín and Joseph Valencia, together with mine, in the province of Raposo, are out of work, I have penetrated at much expense the selva of Mocoa; and its inhabitants having lost their priests and living in the rain forest without God, without law, without Christ, for which reason, risking my people to suffer the privations that the said Indians want to cause them, I am determined to serve by taking along a priest ... in order to return to the fold so many lost sheep ... which, if attained will ensure thereafter, to the glory of Our Father, large increases for the royal estate from the fifths [severance taxes] that will come from the rich mine workings, and for the transfer of said labor gangs; and since each of my sons and sons–in–law will transfer and care for their gangs, I would arrange in their names and mine [be known] as discoverers of the said circumstances, to examine from the first ford of the Iscanzé River ... [ed. note: the petitioner, Pedro de Valencia, continues his letter with a detailed plan to survey the rivers, streams, ravines and their banks of the Mocoa area for gold placers] ... and Your Grace helping me with this by giving me title in due form that I and my sons publicly declare to colonize and pay his Majesty, God save him, his royal fifths... (ACC, Col. C1–21mn, Sig. 3972).

Despite the apparent abandonment of [Spanish] mining exploitation during the mid–eighteenth century, from the following notation of Fray Juan de Santa Gertrudis ([1756–67] 1970, 232–234), one may infer that the Indians continue to exploit gold deposits as part of their traditional economy: "These Indians of Mocoa are clothed and they do well, because at the foot of the hill of Mocoa flows a large river that rushes down from the mountains, and one must cross it by boat. It makes a great noise and thus they called it Río Cascabel [Rattle River]. Along its banks the Indians pan much gold that the river carries down from deposits above, and to exploit the gold the Sibundoys bring in iron tools, clothing, meat and flour from San Juan de Pasto."

Around 1784 the Franciscans abandoned the pueblos founded in the Putumayo area. "On balance, for more than a hundred years Franciscan missionary activity on the Río Putumayo was quite negative; colonization during the mission regime had failed" (Llanos and Pineda 1982, 81).

It appears that the abandonment of the missions coincided with the stoppage of gold extraction and the consequent closing of the mines. Mining ceased to dominate and the geographical area that we have called the "mining frontier" of the Southwest was shattered, leaving the eastern and western piedmonts, as well as the Pacific coast, relatively isolated from the central Andean highlands. Around 1806 it is quite clear that the eastern

Andean versant was being depopulated, as referred to in the following document:

> In San Agustín de Mocoa there are 19 men subject to tribute; four old men, two from Mición; 24 women; 30 children; and all these individuals who make up the population are natives of the pueblo of Sebondoy.
>
> The people living in the pueblo of San Agustín de Nieto were eight adult men, seven women and eleven children all of whom have left the pueblo and are now scattered in a place called Pacayaco, among whom I have been able to speak in a friendly way and with the leaders who have promised me to reestablish their pueblo and build a church and house for a priest.
>
> Of those who are colonists in Concepción de Descanzé, there are nine native Indians and tributaries of the pueblos of Sebondoy, seven negro slaves, five free mulattoes, three mulatto slaves, five slave children, five free mulattoes and twelve free children.
>
> The Indians living in the pueblo of San Rafael de Yunguillo are eight in number with their families, all natives and tributaries of the Sebondoy pueblos, who abandoned this pueblo, and they were once members of the first Sebondoy pueblos (AHN, Fondo Bernardo J. Caicedo, caja 23, f. 1 recto).

As one can infer, these reports continue to insist upon the establishment of mission outposts, an effort that apparently was unsuccessful; at the same time Negro slaves were inhabiting the region, indicating that Don Pedro de Valencia had penetrated the area to exploit the gold placers, but not with Indian labor, which further reveals the decimation of native inhabitants. Later, in 1800, the reestablishment of the Putumayo missions was decreed (AHN Fondo Milicia y Marina, vol. 137, f. 888 recto).

Around the end of the nineteenth century the eastern piedmont became a transit area of workers who were penetrating the lowland to exploit quinine and rubber, initiating a process of spontaneous colonization in both the piedmont and the lowlands. Later this colonization was led by Capuchin missionaries who established themselves in the Valley of Sibundoy in 1883, thus beginning the advance of a mission frontier that had the characteristics of a farming and stock–raising borderland.

Conclusion

The exploitation of gold and missionary activity in the Colombian Southwest were intimately related; the mining frontier coincided with that of the missionaries throughout the sixteenth, seventeenth, and first half of the eighteenth centuries. The absence of the *encomenderos* and Spanish lay authorities as residents of the piedmont and rain–forest zones required that royal governance be carried out by the missionary priests. They were the ones who founded pueblos and mission settlements and established their evangelical centers in the piedmont, permitting the introduction of the labor systems used in mining. In the eastern piedmont, abandonment of the mines coincided with the retreat of missionaries from that zone.

The three–stage model of human settlement that Alzate (1989) refers to for the sixteenth and seventeenth centuries is exemplified thusly: missionaries entered indigenous areas to evangelize native inhabitants and prepare the way for lay Spaniards to colonize. In the case of southwestern Colombia the inverse also occurred: Spaniards reasoned that their presence in exploiting gold deposits was necessary for the successful conversion of the "savages" by undertaking to supply "priests and to establish churches."

The uprising of the Indians of Mocoa and Sucumbío, as well as the decimation of their population, hindered the closing of the frontier which had to retreat. To advance again at a productive level the mining frontier was transformed by the introduction of farming and stock raising encouraged by the missionaries.

As for the western versant and the Pacific coast, after the end of the boom around the end of the seventeenth century, mining took on a marginal character, an activity performed by Negro inhabitants of the area, even today uncontrolled by central Andean authority.

Acknowledgments

This paper amplifies that published in the Memorias del V Congreso Colombiano de Antropología, which occurred in October 1989 in Villa de Leyva and was also presented in the 47th Congress of Americanists that took place in New Orleans (7–11 July 1991) at the symposium Colonial Mining in Latin America, coordinated by Dr. Robert C. West and Dr. Alan K. Craig. The present work forms part of the thesis in progress for the masters degree in history, Universidad Nacional de Colombia, Bogotá.

References

Archival Sources

ACC — Archivo Central del Cauca [Popayán, Colombia]
1739 (Col. E–1 – 4g) Sig. 4387
1740 (Col. C–1 – 21mn) Sig. 3972
AHN — Archivo Historico Nacional [Bogotá]
1800 Fondo Milicias y Marina. Tomo 137
1806 Fondo Bernardo J. Caycedo. Caja no. 23 Doc. 2 Misiones

Published and Other Materials

Alvarez, V. 1986. Región y regionalidad en la historia colombiana. (Paper presented to the Banco de la República, mimeo.)

Alzate Angel, B. 1989. Humanización conceptual de la frontera: el caso amazónico. (mimeo.)

Bonilla, V. D. 1968. *Siervos a Dios y amos de indios*. Bogotá: Editorial Tercer Mundo.

Caillavet, C. 1989. Entre sierra y selva: Las relaciones fronterizas y sus representaciones para las etnias de los Andes septentrionales. *Anuario de Estudios Americanos*, 46:71–91. Sevilla, Spain.

Calero, L. F. 1988. Pasto, 1535–1700: The social and economic decline of Indian communities in the Southern Andes. Ph.D. diss., University of California at Berkeley.

Capdequi, O. 1941. *El Estado Español en las Indias*. Mexico City: Fondo de Cultura Económica.

Castellvi, M. 1942. Reseña crítica sobre el descubrimiento de la región de Mocoa y fundaciones de la ciudad del mismo nombre. *Boletín de Historia y Antigüedades* 29(330–331): 367–390. Pasto, Colombia.

Escobar, Fray Gerónimo de [1582] 1983. Relación para los muy poderosos señores del Real Consejo de Indias de su majestad, ansi de la descripción de la tierra que llaman Gobernación de Popayán… . *Cespedesia* 45–46 (supplement 4): 285–307. Cali, Colombia.

Figueroa Rivera, M. 1986. Misioneros indígenas y caucheros, región del Caquetá–Putumayo (siglo XVI–XIX). Tesis de Antropología, Universidad Nacional, Bototá.

Friede, J. 1967. *Los Andaki, 1538–1947. Historia de la aculturación de una tribu selvática*. Mexico City: Fondo de Cultura Económica.

Guhl, E. 1991. *Las fronteras políticas y los límites naturales*. Bogotá: Fondo FEN.

Latimore, D. O. 1968. The frontier in history. In *Theory in anthropology*, ed. R. A. Manners, 374–386. Chicago: Aldine Pub. Co.

Llanos Vargas, H. and Pineda Camacho, R. 1982. *Etnohistoria del Gran Caquetá (siglos XVI–XIX)*. Bogotá: Banco de la República.

López, Tomás [1558] 1989. *Visita y tasación del licenciado Tomás López en la Gobernación y Provincia de Popayán*. Sevilla: Archivo General de Indias.

Mellafe, R. 1968. Frontera agraria: El caso del Virreinato Peruano en el siglo XVI. In *Tierras nuevas: Expansión territorial y ocupación del suelo en América (siglos XVI–XIX)*, 11–42. IV International Economic Congress, Bloomington, Indiana, 1968. Mexico City: Colegio de México.

Moreno Ruiz, E. 1971. Noticias sobre los primeros asentamientos españoles, en el sur de Colombia. *Revista Española de Antropología Americana* 6:426–430. Universidad de Madrid.

Padilla, S. 1977. Tasaciones de encomiendas de Popayán en el siglo XVI. In *La Encomienda en Popayán*, 3–112. Escuela de Estudios Hispanoamericanos de Sevilla. Sevilla.

Ramírez de Jara, M. C. 1989. La frontera minera del suroccidente de Colombia: Quillacingas, sibundoyes y mocoas, productores y comerciantes de oro durante la colonia. In *Historia Comparada de las Selvas Colombianas*, 205–216. Memorias del V Congreso Nacional de Antropología. Bogotá: Instituto Colombiano de Antropología, ICFES.

Ramírez de Jara, M. C. and Urrea, F. 1989. Dinámica etnohistórica, socio–demográfica y presencia contemporánea del curanderismo Ingano Kamsá en las ciudades colombianas. In *Curanderismo la parte*, 181–234. Memorias del V Congreso Nacional de Antropología. Bogota: Instituto Colombiano de Antropología, ICFES.

Restrepo, V. [1888] 1979. *Estudio sobre las minas de oro y plata en Colombia*. Medellín: FAES.

Romoli de Avery, K. 1962. El suroeste del Cauca y sus indios al tiempo de la conquista española, según documentos contemporáneos del Distrito de Almaguer. *Revista Colombiana de Antropología* 11:239–297.

——. 1963. Apuntes sobre los pueblos autóctonos del Litoral Colombiano del Pacífico en la época de la conquista española. *Revista Colombiana de Antropología* 12:261–290.

——. 1977–78. Las tribus de la antigua jurisdicción de Pasto en el siglo XVI. *Revista Colombiana de Antropología* 21:11–55.

Santa Gertrudis, Fray Juan de [1756–67] 1970. *Maravillas de la naturaleza*. Bogotá: Ediciones Banco Popular.

Tovar, H. 1989. El Imperio y sus colonias: Las cajas reales durante el siglo XVI. (mimeo).

Valverde, García de. 1570. Obispado de Sant Francisco de Quito, ynfformación y autos sobre la tassa de Pasto. Folios 206–217 de la visita. Archivo General de Indias, Sevilla.

Vázquez de Espinosa, A. [ca. 1629] 1948. *Compendio y descripción de las Indias Occidentales.* (Transcrito del manuscrito original por Charles Upson Clark.) Misc. Col. 108. Washington, D.C.: Smithsonian Institution.

West, R. C. 1972. *La minería de aluvión en Colombia durante el período colonial.* Bogotá: Imprenta Nacional.

Zavala, S. 1987. Orígenes coloniales el peonaje en México. In *Peones, conciertos y arrendamientos en América Latina,* 7–46. Bogotá: Universidad Nacional de Colombia.

Spanish Colonial Silver Beneficiation at Potosí

Alan K. Craig, Florida Atlantic University,
Boca Raton, Florida, U.S.A.

Abstract

Prehistoric *huayras* (clay smelting furnaces) continued in use at Potosí several decades after discovery (1545) of native silver in veins at high altitude on Cerro Rico. When scarce local fuel was exhausted these primitive smelters were replaced by simple Spanish furnaces using expensive charcoal transported from lowland forests. Small crushing mills of Chilean design were used to process rich hand–sorted silver oxide ores. But miners soon encountered less tractable silver minerals that required grinding and additional beneficiation. Prior to 1572, a few hydraulic grist mills were built nearby at Tarapaya although the location was not convenient. Viceroy Toledo revolutionized silver mining when he introduced the Mexican patio amalgamation process, stimulated construction of numerous hydraulic stamp mills of medieval European design, and had reservoirs built to insure an adequate water supply. Forced labor for mines and the new mills caused silver production to peak in 1590, but a sharp decline in profitable ores and flash floods that destroyed many mills eventually caused widespread decline. Miners, ignorant of the complicated geology, reacted with schemes to improve beneficiation and government–sponsored foreign consultants were hired, but little benefit resulted. Much diminished production has continued into the present century.
Key Words: Silver beneficiation, Potosí, patio amalgamation, *huayras*, stamp mill.

Introduction

Mining has long been neglected by geographers as a field for research and investigation. Social scientists in general tend to view it as a dark and arcane endeavor, largely conducted underground by mysterious and dangerous means. Historians, however, treat the evolution of mining as a rich field for study and have, since the classic contribution of President Herbert Hoover (1950), enthusiastically searched for new insights into the significance of mining, particularly within the New World Spanish colonial empire. Most mining communities tend to be small and ephemeral at best; there are very few places where we can study their evolution and the long-

In Quest of Mineral Wealth: Aboriginal and Colonial Mining and Metallurgy in Spanish America, edited by Alan K. Craig and Robert C. West, 1994. Geoscience and Man, vol. 33, pp. 271-285. Dept. of Geography & Anthropology, Louisiana State University, Baton Rouge, LA 70893-6010, U.S.A.

term effects of driving forces that operate there. It is even rarer to encounter mining camps with good archival documentation. Potosí (fig. 1) is a marvelous exception to the above and it is only the overwhelming mass of information that intimidates scholarly research.

Pioneer Metallurgists

Eighty one years after the publication *De Re Metallica*, the alchemist–influenced work of Agricola ([1556] 1950) was superseded by another book written by an obscure priest who spent much of his adult life prospecting, mining, and speculating as a highly successful refiner of silver ore in some of the most desolate reaches of the Andes mountains.

Padre Alvaro Alonso Barba emerges from the historical record as the single most important figure in the evolution of colonial silver beneficiation in the viceroyalty of Alto Perú. His practical contributions to refining technology as a result of long residence in what is now Bolivia are well known and have been carefully documented by the modern mining engineers Douglas and Mathewson (Barba [1640] 1923). Paoli (1922) has provided equally interesting insights regarding his early life and education in Europe. In 1637 Barba wrote his classic *El Arte de los Metales*[1] while he was the priest assigned to the Parroquia of San Bernardo in Potosí. But his first–hand knowledge of mining and silver beneficiation came from his many years of experience in remote corners of the province of Lipez where he encountered a wide variety of metallic ores with which to experiment. As his expertise increased, Barba even bought the tailings of less efficient refiners and, using his own recipes, claimed to have turned a tidy profit by reprocessing them.[2]

Previous Work

Prominent among modern day investigators of colonial mining technology, particularly of the patio amalgamation process, are Bargalló (1956, 1967, 1969), Greve (1943), Tandeter (1993), Buechler (1981), Tudela de la Orden (1970), Sánchez Gómez (1982), Bakewell (1977) and contributors to the Cátedra de San Isidro (1974) volumes. The outstanding English language study on Potosí of a general nature is Bakewell's (1984) *Miners of the Red Mountain*. It is closely followed by his fascinating sociological account of Potosí's leading miner and refiner of the seventeenth century, Antonio López de Quiroga (Bakewell 1988). These two volumes, richly documented by archival research, present the most accurate and detailed modern account of life in the great mining center and its effect on the economy of surrounding areas within the viceroyalty of Peru.

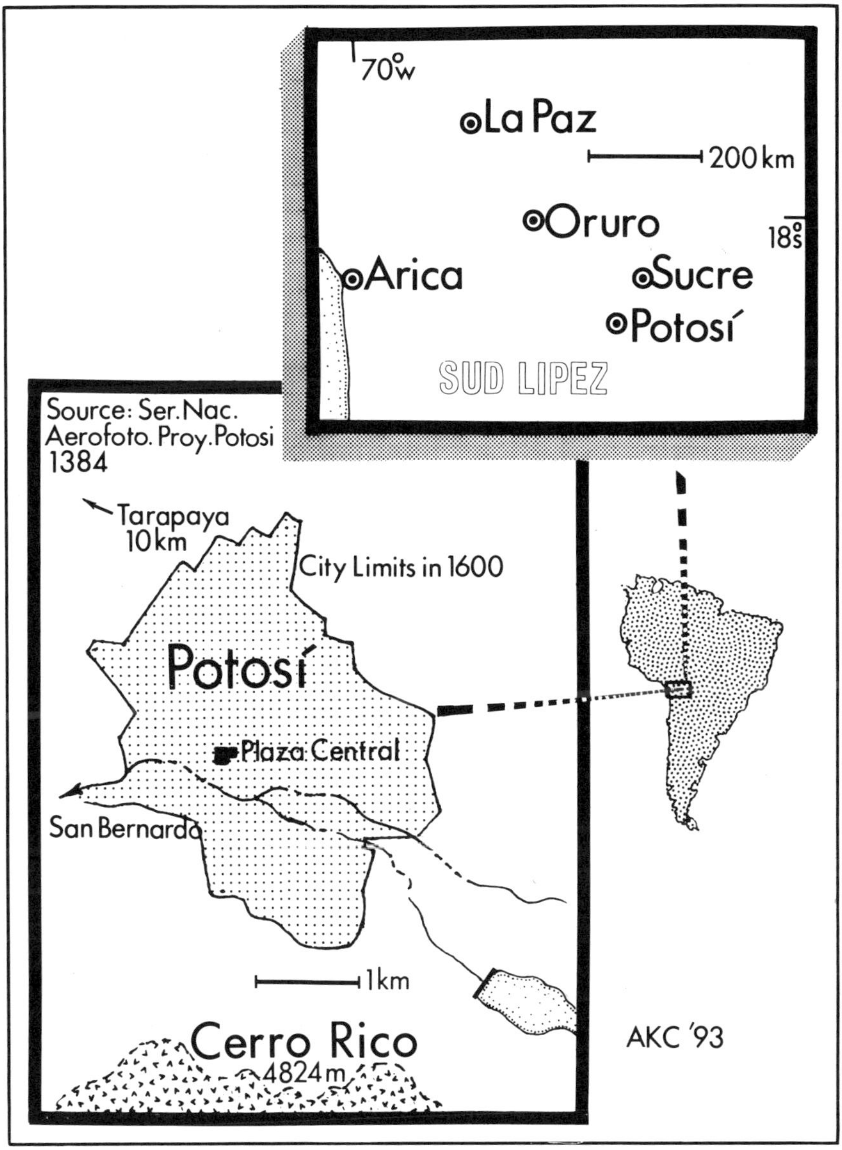

Figure 1 : Study area location map.

Early Beneficiation Techniques

Much has been written about the prehistoric *huayras* method of smelting high grade silver ores in a clay furnace of prehistoric design (Bargalló 1968 and Villalobos 1979, 267, for the large, industrial–size Inka version). These aboriginal smelters consisted of a charge of carefully selected silver oxide ore (*tacana*) mixed with fuel and covered by damp clay into which numerous ventilation holes were punched.[3] Placed on mountain slopes where katabatic air flow would occur, the *huayras* were fired at night and broken open the next day to recover the puddled silver at the base. Some very large (2x2m) *huayras* batteries of a permanent nature can be seen at the Viña del Cerro site where they were installed during the Inka occupation of the Copiapó Valley in Chile (first mentioned in Niemeyer 1981, 112). Analysis of glassy slag collected from the ceramic rubble around the permanent bases confirms they were used to smelt silver ore (Izumi Shimada, pers. comm.). Most of these prehistoric smelters were much smaller and it is clear that designs varied considerably. It is not known where this technique originated, but it is clearly Andean and requires special wind effects (katabatic flow) to function efficiently. If these winds caused by subsiding cold air were not strong enough, the forced draft into the *huayras* openings would be insufficient to raise the temperature to the desired levels and poorly smelted ore would result.

There is nothing in the archival records to indicate the Spanish themselves were directly involved in the use of these smelters, even though it is obvious they encouraged their Indian workers to keep them in continuous operation during the eight months when weather conditions at Potosí allowed mining activities to take place. These nighttime fires "turned night into day" (Cañete y Domínguez [1787] 1952), and Capoche (1959) claims exactly 6,497 were once in operation on the slopes of Cerro Rico.

Incredibly rich specimens of pure wire silver were found on the surface in April 1545 in discovery veins from which they were simply cut out with chisels and smelted without any difficulty. Within three months these fabulous ores exposed in the outcrops were quickly exhausted. As miners followed the veins deeper into the mountainside, they found silver oxides and silver–lead ores that could easily be processed by the same means.[4]

Slowly, but inevitably, the best ores were extracted and the yields from *huayras* smelting could barely justify the cost of imported fuel since Potosí and its surroundings had been virtually devegetated.[5] Within five years most mine owners were forced to adopt other beneficiation steps that were more fuel efficient, especially that of crushing and fine grinding the ore before it was roasted in small ovens of European design. These steps necessarily delayed the refining and slowed silver production from chloride and sulfide ores that were increasingly "rebellious." Even so, an abundance of water power was needed to grind and concentrate the ore in the water mills. But water was in short supply in the semi–arid, high–altitude environment at Potosí where small streams are frozen solid four months of the year.[6]

Arduz Eguía (1985, 104) believes the rich oxide ores smelted during the first 20–25 years were replaced around 1570 by *pacos* (silver chloride), *mulatos* (silver sulfide), and *negrillos* (complex silver sulfarsenide plus silver sulf–antimonides). These ores together created a crisis in beneficiation since they all required some additional treatment beyond simple roasting. The problem was met by using *magistral* — copper sulfate and iron filings, elsewhere (Barba [1640] 1923, 56) called iron sulfate.

By the 1560s the situation was serious. Small mine owners were abandoning their unprofitable works, Indian labor was both expensive and scarce, and attempts to experiment with mercury amalgamation had failed. Worse yet, the King of Spain was not receiving enough revenue from his *quinto* (20 percent tax) on refined silver to maintain his adventures in Europe and he decided to do something about it. Felipe II charged Francisco de Toledo with the responsibility of increasing the royal treasury and this indefatigable viceroy of Peru spent the first five years of his tenure traveling about the countryside consulting with experienced miners as to the best means of reviving the silver mining industry, particularly that of Potosí.

Through contact with his contemporary, the viceroy Velasco, Toledo was aware of the success of the Mexican patio amalgamation process perfected in 1555 at Pachuca by Bartolomé de Medina. After abundant mercury deposits were found in the Andes at Huancavelica in 1566, Toledo was determined to have the method introduced into his own viceroyalty. Earlier attempts by Potosí miners to amalgamate their ore had failed and they were skeptical, but Toledo sent for Mexican experts and arranged for a public demonstration in Cuzco using Potosí ore (Craig 1989). This time the results were successful and Toledo exulted he would soon "arrange the greatest marriage the world had ever seen between the mercury of Huancavelica and the silver of Potosí."[7]

In retrospect it seems the basic differences between patio amalgamation as practiced in Mexico, Peru, and Chile can be found in the Mexican use of *arrastras*[8] of various designs (Young 1970,70) to break up ore, versus the use of heavy duty stamp mills of Germanic design (Villalobos 1979, 265) needed to crush the harder Potosí ores. In Chile it was necessary to use the small, animal–powered *trapiche* (see Villalobos 1979, 263) in the Atacama Desert where water was nearly unobtainable; a photograph of a modern (1907) version used in Mexico, but called a "Chilean mill," is illustrated in Young (1970, plate 3).

It is not clear exactly who made the technological determination to forge ahead with a massive installation of waterpowered stamp mills at Potosí, but it was probably the viceroy Toledo himself who exerted great effort to stimulate and arrange financing for the construction of such stamp mills soon after the successful demonstration of 1572. Even more puzzling is the dearth of archival evidence for antecedent stamp mills in Alto Perú and the source for the corpus of artisans experienced in their construction. Such individuals seem to have sprung up out of the ground with the necessary knowledge of equipment design and exact dimensions adequate to immediately begin crushing ore. In other words, there seems to be little record of empirical experimentation in *ingenio* design once the Potosí community agreed to adopt this revolutionary change in beneficiation.

Ingenio Beneficiation at Potosí

By December 1574, at Toledo's urging, reservoirs, aqueducts, and a series of complex stamp mills powered by large waterwheels were installed and operating at full capacity in Potosí.[9] These new *ingenios* effectively supplanted the older battery of ten small–capacity mills built much earlier at the nearby village of Tarapaya (Craig 1993).

A remarkably detailed illustration of beneficiation by patio amalgamation as practiced at Potosí ca. 1620 is to be found in Arzáns (1965, 1:168). This naïve sketch (see fig. 2) shows the large *tapia*–walled compound surrounding a complex of stamp mills, sieves, flotation basins, amalgamation floors, roasting ovens, furnaces, warehouses, living quarters, offices, and chapel.[10] These self–contained mills were adapted to the special high–altitude conditions of Potosí and are more complicated than their medieval European counterparts. They reflect an abundance of available human labor (Indian *mita* draftees) and the large numbers of pack animals (llamas) necessary for transport of the ore down from the adjacent mines in Cerro Rico.

At "A" (fig. 2) hand–cobbed ore sorted at the *cancha* near the mine shaft arrives at the *ingenio* rear gate in leather bags of about 40 kg maximum weight. Llama pack trains usually consisted of gelded animals owned by Indians who specialized in providing transport service. To the right of this gate are the *galpones* "B" where the excess ore is stored until it can be passed through the system. On the left of gate "Q" are offices and bins "C" where copper, salt, lime, and other reagents are kept as well as a supply of *ucha* (dried llama dung fuel). From the *galpón,* laborers carried ore in bags to be dumped onto the boat–shaped stone *morteros* "D" then crushed by the descending iron shoes (*amaldanetas*) "E" attached to the vertical *mazos* (wooden shafts) "F." These banks of 4 to 7 stamps were actuated by elongated wooden axle and eccentric cams turned by the vertical over-

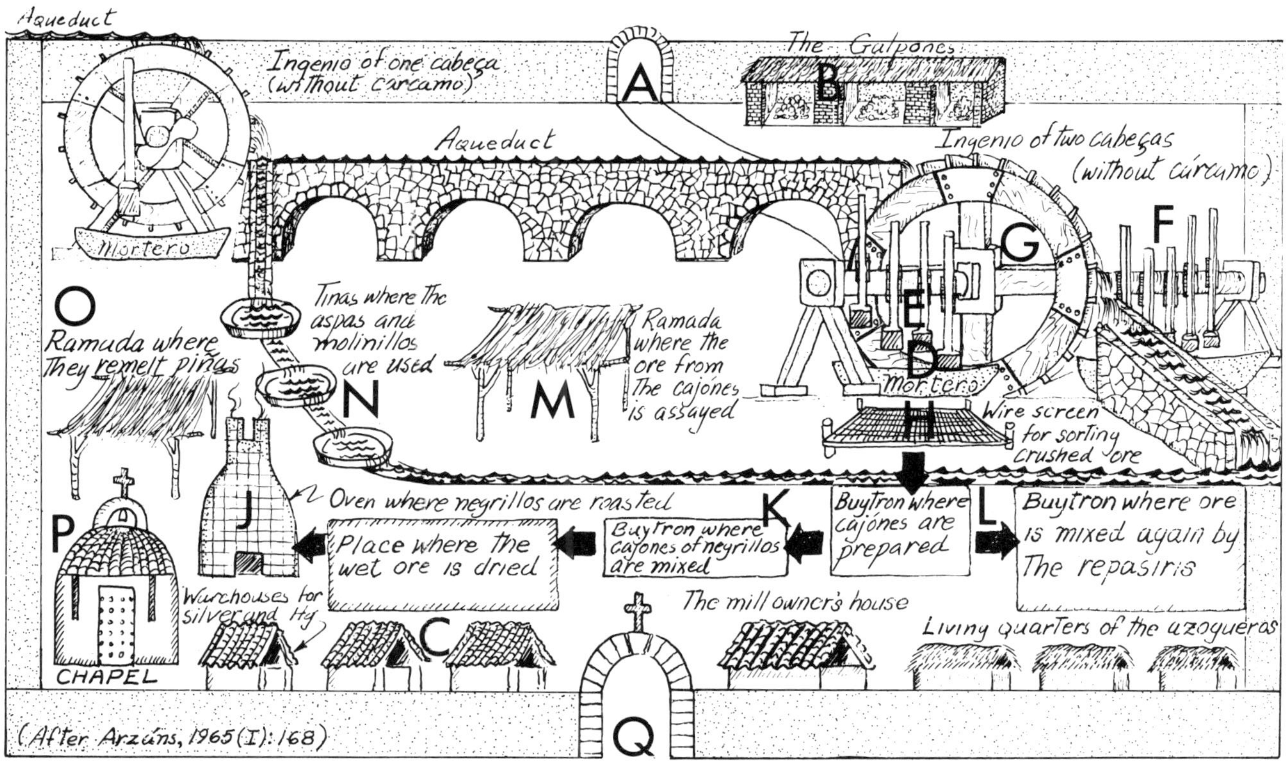

Figure 2 : A Potosí stampmill. After Arzáns 1965, 2:168.

shot waterwheel itself "G". To the left of this double–headed stamp mill are wire screens "H" used to sift the crushed ore which then may be sent to the *horno* "J" for roasting, if necessary, before being placed in a sunken enclosures (*cajones*) "K" in a standard amount and shape (also referred to as a *cajón*). If the ore was particularly "rebellious", it might be sent to another type of mixing surface "L" known as a *buitrón* (Villalobos 1979, 266;270) where it could be gently heated to improve the amalgamation rate. The progress of the amalgamation was periodically checked by an assayer located at a brush–covered shelter (*ramada*) "M" during the two–week soak time after the necessary amount of mercury had been squeezed through coarse Holland linen and sprinkled over the pulp in the *cajón* (Capoche 1959, 123).

When the assayer determined amalgamation was complete, the mixture was then carefully washed in large ceramic *tinas* (also called *cochas* by the Indian workers) "N" to remove the slimes and recover the *pella* of sponge amalgam.[11] In a separate *ramada* "O" excess mercury was recovered by placing the amalgam in linen bags and twisting them to squeeze out the mercury. The amalgam was then placed in a triangular wooden mold and pounded with a hammer to force out more mercury and finally heated in a crude retort to condense and recover the last drops of mercury. What remained were bright, shiny *piñas* of sponge silver to be melted in a small furnace and cast into fairly large ingots (20–35 kg) for transfer to the mint where they were taxed and assayed.[12]

This tight–knit community of as many as 40 Indian workers (two shifts), overseers, and perhaps even the mill owner and his family, all lived with their livestock within the compound walls where they could guard the valuable ore, reagents, sponge silver and supplies. No doubt they attended chapel "P" on Sundays.

Padre Barba is associated with many improvements in furnace construction and the patio amalgamation process in general. As the quality of silver ores at Potosí diminished, recoveries fell and mill operators cast about for any advantage that would improve the amalgamation scheme attributed to Andrea Carlos Corso in 1587 which merely required adding iron filings to the ground ore (pulp) mixture (Bakewell 1975). Copper, sulfur, salt and lead were also readily available and favorite reagents for experimentation.[13]

This basic system continued to be the norm for more than a hundred years but by the time the German expert Helms arrived (1797) he had nothing good to say about the antiquated beneficiation techniques he found there.[14] At this time the silver mines of Potosí were virtually exhausted and the remaining ores were mostly difficult sulfides mixed with antimony and tungsten compounds, traces of which may account for some of the problems we see in the coinage output from the mint.[15] In any event, the much admired European improvements to silver beneficiation were primarily in the form of labor–saving devices (mechanically stirred wooden barrels) that increased the amount of ore that could be processed during a normal working day. These "inventions" did not impress the rank and file of Potosí *ingenio* owners who continued to think in terms of cheap Indian labor even though the annual *mita* was reduced to a mere trickle of men before it was finally abolished.

Effects of Geology on Beneficiation

Early chroniclers had an imperfect understanding of silver mineralogy in general and of the Cerro Rico deposit itself. It was generally believed that the mountain was generating silver from within and that all veins must therefore increase in richness with depth (a most convenient point of view for those miners anxious to sell their exhausted claims to the flood of newcomers attracted to the legendary bonanza).

Neither have modern geologists been in agreement as to the basic structural and stratigraphic relationships present in the Potosí area (Ahlfeld 1956). However, the investigations of Rivas and Carrasco (1968) and Kelly and Turneaure (1970) have done much to elucidate the silver, tin, antimony, and tungsten mineralogy. In this regard, it is interesting to note that silver ores from nearby Porco which did not contain antimony or tungsten minerals could be smelted readily with a Castilian furnace and small bellows, whereas the more complex ores of Potosí evidently had melting points above the limits of hand–operation but were vulnerable to the forced draft of the indefatigable mountain breeze.

With the advent of satellite imagery (specifically LANDSAT 022345–098; 9 May, 1981), it is now possible to accurately identify the Cerro Rico deposit not as a volcanic neck as was originally proposed, but rather a highly mineralized igneous late–stage stock intimately related to the Kari–Kari batholith that was emplaced in early Tertiary time. Formations penetrated by this pluton have been severely faulted and deformed to the extent they now dip vertically along the east flank of Cerro Rico.

This subsequent tectonic activity that affected all of the Bolivian silver–tin district is probably responsible for the unusually complex nature of the Potosí silver ores and the injection of deep seated tungsten and antimony minerals that must have caused beneficiation problems by the time miners were into the sulfide zone.

From Silver Ore into Coinage

Figure 3 illustrates the complicated path that was developed during the colonial period in the viceroyalty of Peru in order to convert silver ore into coinage. This composite diagram combines several elements that in reali-

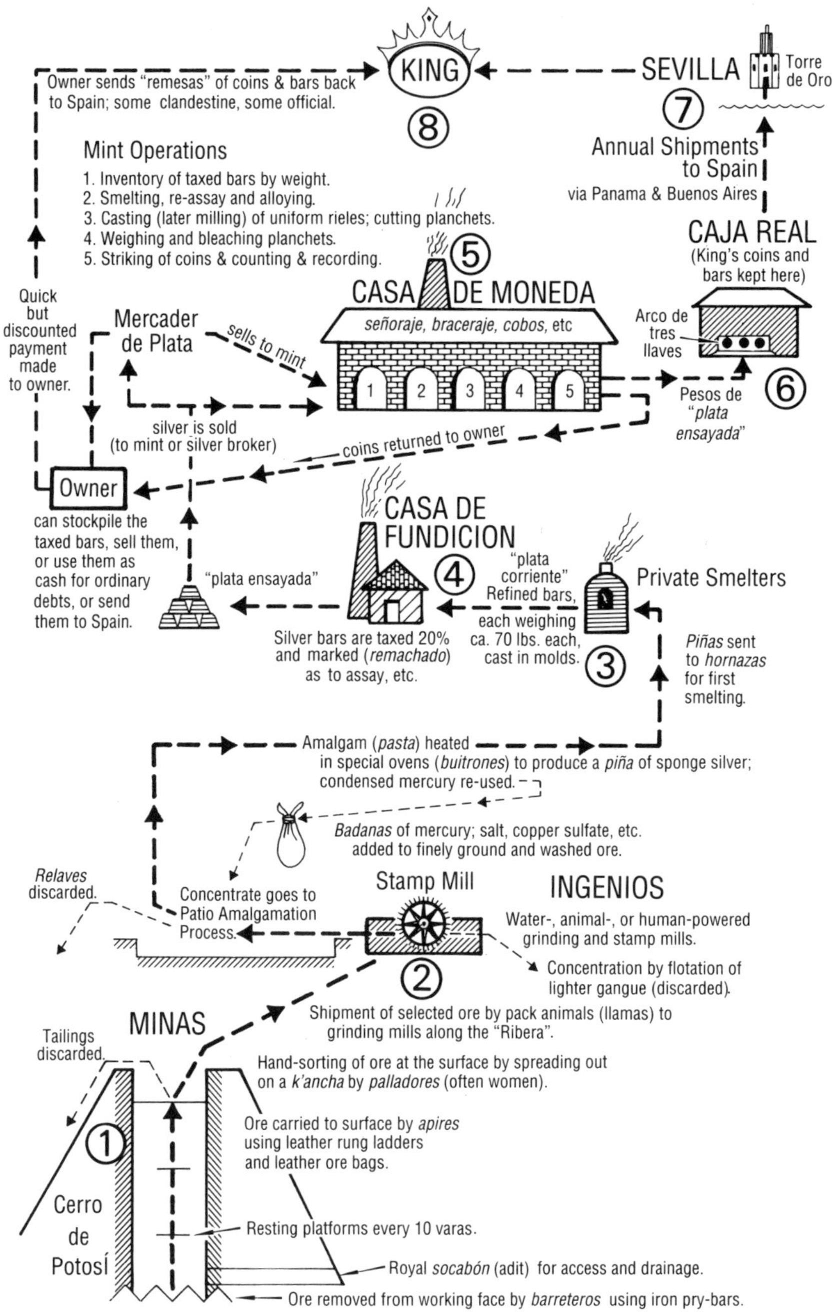

Figure 3 : From ore to coins to the King: the transfer of wealth from Alto Perú to Spain.

ty only existed at widely separated times. The journey begins in the many mines (1) where skilled Indian miners used long iron bars to pry out chunks of rock from silver veins deep within the mountain. After careful hand–sorting at the entrance, the better grade ore was sent down to the beneficiation mills (2) where it was accumulated and crushed for mixture with the reagents that were required for amalgamation (from 1572 onwards). Excess mercury was recovered at the mill using crude retorts. The amalgam was processed into nearly pure silver bars by smelters (3) that were originally operated separately from the mill but later became an integral part of the *ingenio* itself. Late in the colonial era the bars went from these private beneficiation facilities directly to the royal foundry (4) that at first was separate from the mint (5) but later became an integral part of this massive structure. The royal treasury (6) was also made a separate building at Potosí where both ingots and coins collected for taxes were held for safekeeping. By 1630, silver brokers (*mercaderes de plata*) were beginning to assume a quasi–official role and were actually allowed to be involved in mint operations. Some succumbed to temptation and indulged in multiple frauds that destroyed the reputation of the mint products. This caused the Crown to eventually take over all colonial mint operations from the franchisees in 1728. By this time the flow of New World specie being unloaded at the quayside customshouse (7) in Sevilla was greatly reduced from former days. Mysterious losses of gold and silver sometimes occurred between positions (7) and (8) for reasons beyond the scope of this paper.

Summary

For nearly thirty years after its discovery in 1545, rich silver ores from Cerro Rico at Potosí received little beneficiation beyond that developed prehistorically using the small clay *huayras* furnace with forced draft from nighttime katabatic wind. These ores were soon depleted and overall production slowly declined, much to the consternation of the Spanish king. He assigned viceroy Toledo the task of restoring his tax income from these mines. This he did in 1572 by successfully introducing the Mexican patio amalgamation process, using mercury from the nearby Huancavelica deposits. Fine grinding of low grade ores was essential to this technique and large European style hydraulic stamp mills (*ingenios*) were constructed to perform this task. They were supplied water from numerous reservoirs built by damming glacial tarns along the flank of the adjacent Kari–Kari mountains. All these physical advantages were in fortunate conjunction in what was otherwise a high–altitude location having an exceptionally harsh climate.

Toledo also organized the Indian *mita* corvée labor system that revitalized mining and supplied the necessary workers for hundreds of beneficiation mills that began to process the low grade discarded ores that blanketed the slopes of Cerro Rico. Their efforts resulted in silver production reaching a zenith in the 1590s, after which it slowly declined until the last water mill ceased operations in 1944.

The struggle to improve silver beneficiation techniques continued through the latter part of the colonial era as the deeper–seated ores from the chloride and sulfide zones were being mined. Their experiments to improve yields by amalgamation always revolved around the addition of cheap reagents that were locally abundant. The ore minerals involved presented problems of complex chemical reactions that were poorly understood by the owners of mines and mills.

Endnotes

1. The complete title of the Madrid edition of 1640 by Barba is: *Arte de los Metales en que se Enseña el Verdadero Beneficio de los de Oro, y Plata por Asogue; el Modo de Fundirlos Todos, y como se han de Refinar y Apartar unos de otros.* This edition was destroyed by the Inquisition and it is believed only three copies have survived. Other English editions appeared in 1670 and 1674, closely followed by translations in German, Italian, French and Dutch over the years until a facsimile edition was published at Potosí in 1967 under the editorship of Armando Alba.

2. Barba is perhaps best known for his advocacy of sliming (very fine grinding) of silver ores before amalgamation. He was also against the roasting of sulfide ores before amalgamation because he believed it created *vitrio* (copper sulfate), which he described as "...the mortal enemy of mercury." (Barba [1640] 1923, 104). In addition to his various concerns about the unnecessary loss of mercury itself, Barba insisted that the amalgam be carefully separated by hand washing at the end of the process. See endnote 9 for a probable source of slimes he may have purchased.

3. In Barba's (1640) treatise there is an often–reproduced wood cut illustration of a *huayra* that seems illogical, or even an imaginary construct of the artist. But a unique photograph of an exceedingly rare present–day example of *huayras* published by Peele (1893, 9) proves that both Barba and the illustrator have provided us with an accurate account of this remarkable prehistoric Andean invention.

4. According to Arzáns (1965, 1:142) only "plata blanca" and "plomo ronco" (argentiferous galena) were being smelted prior to the introduction of patio amalgamation because "*...no se mostraban rebeldes en las huayras,*" (i.e., they were tractable). This observation is interesting because it documents the continued use of prehistoric *huayras* technology for these two ores at least until 1576. Apparently the treatment of *negrillos* was not successful until long after the introduction of amalgamation.

5. The harsh climate of Potosí and adjacent mountainous areas never permitted more than a depauperate and rachitic plant community to develop prior to discovery of the silver mines. Stunted *quishuar* trees and the resinous, sponge–like *llareta* colonies were soon extirpated along with the *ichu* tussock grass for great distances in order to supply fuel for thousands of *huayras* smelters. Eventually (ca. 1590), it was necessary to import charcoal from lowland *carboneras* located as much as 80 leagues from Potosí. However, today this vegetation has gradually reestablished itself to the extent that even the slow–growing *llareta* can be

found on the upper slopes of Cerro Rico growing on discarded mine tailings from the early colonial era.

6. The archives contain many references to the scarcity of water at Potosí. Typical is the following: "...*este año* [1584] *ha sido muy falto de aguas y como sin ellas no pueden beneficiar los metales de plata*..." (AGI, Charcas 16, leg. 137). Evidently even though this was written ten years after construction of the lagoons was begun, there was still insufficient reservoir capacity to keep the water mills operating.

7. The Licenciado Ramírez of the Audiencia of Charcas, writing to the King on June 6, 1575 (AGI, Charcas 16, leg.61), said "The hill of Potosí so well known for the fame of its riches which from it have been taken out and may continue to be taken out, was failing and the Imperial Villa of Potosí might have been abandoned if Francisco de Toledo had not introduced and established the beneficiation of silver with mercury..." (translation mine).

8. See Prieto (1973, 44–45) for a view of an exceptionally large Mexican patio amalgamation site at the Hacienda de Proaño in Zacatecas, using ore from Fresnillo. This example is much more spread out than any developed under the restricted conditions that prevailed along the crowded banks of the stream flowing through Potosí.

9. Reading the documents describing these events, we sense a decided urgency in the preparations to install stamp mills and adopt amalgamation at Potosí. The herculean efforts of the Spanish to accomplish these tasks came about when they realized the slopes of Cerro Rico were literally covered with old discarded low–grade silver ore that was now waiting to be picked up and passed through the stamp mills. The dramatic results of their efforts can be seen in the surge of *quintos* shown in Craig (1993, table 1) that began to pour into Spain through the Torre del Oro customs house at Sevilla.

10. Prieto (1973) has published another detailed watercolor scene of a typical single–headed Potosí *ingenio* dated 1584 interesting for its view of the lifting cams on the 5–*mazo* stamp mill. This is perhaps the most valuable part of an otherwise often unreliable text.

11. From García de Llanos ([1609] 1983, 11) we discover there were actually *casas de lamas* in Potosí where poorly prepared slimes could be purchased to be roasted and re–amalgamated. This practice is not likely to have been common after the beginning of the seventeenth century when beneficiation techniques were improving while ore values were decreasing.

12. By 1733, when all colonial mints were under direct control of the Crown, the cost of refining silver ingots presented at a mint for assaying and stamping was fixed at 8 *maravedis per marco* (8 ounces) and charged to the owner. This calculation was based on the experience gained from assaying a test batch of 41,000 marcos. However, owners could have the refining done elsewhere if they chose (AGI Mexico 2794, leg.13, f.4v.). It is not clear whether this calculation was made in Spain by the Junta de la Moneda, established in 1730, or locally in Mexico.

13. Bargalló (1955) notes attempts by Gaspar Ortíz using brine (1586), the Corzo brothers using iron filings (1587), Juan Fernández Montano using copper sulfate and brine (1588), and so forth. Brine and sulfur were especially easy to obtain, but iron filings were relatively expensive.

14. Helms was a member of Baron Nordenflict's expedition to Potosí where he remained behind to try and improve what he considered to be outdated mining and beneficiation practices. By all accounts he was rather arrogant and openly contemptuous of the Potosinos. In 1798 his original book *Tagebuch einer Reiser durch Peru* was published in Dresden, and an English version appeared in 1806.

15. Butts and Coxe (1967, 310) claim the colonial silver alloys used to make coins at the Potosí mint did not successfully deoxidize the metal with the result that in the hand–striking of cobs we often see signs of embrittlement in the form of peripheral planchet cracks and even large splits. In other words, the silver alloy was not very ductile and the planchets were difficult to anneal.

References Consulted

Archival Sources

AGI — Archivo General de Indias [Sevilla.]
Charcas 16, leg. 61, leg. 137
Mexico 2794, leg. 13

Published and Other Materials

Agricola, G. [1556] 1950. *De Re Metallica.* Basileae: APVD Hieron Frobenivm et Nicolaven Episcopivm. Herbert C. Hoover and Louise H. Hoover, eds. and translators. New York: Dover Publications.

Ahlfeld, F. 1972. *Geología de Bolivia.* La Paz: Amigos del Libro.

Arduz Eguía, G. 1985. *Ensayos sobre la historia de la minería altoperuana.* Madrid.

Arzáns de Orsúa y Vela, B. 1965. *Historia de la Villa Imperial de Potosí.* Edited by Lewis Hanke and Gunnar Mendoza. 3 vols. Providence, Rhode Island: Brown University Press.

Bakewell, P. J. 1975. Registered silver production in the Potosí district, 1550–1735. *Jahrbuch für Geschichte von Staat, Wirtschaft und Gesellschaft Lateinamerikas* 12:67–103.

——. 1977. Technological change in Potosí: The silver boom of the 1570s. *Jahrbuch für Geschichte von Staat, Wirtschaft und Gesellschaft Lateinamerikas* 14:60–77.

——. 1984. *Miners of the Red Mountain: Indian labor in Potosí 1545–1650.* Albuquerque: University of New Mexico Press.

——. 1988. *Silver and entrepreneurship in seventeenth–century Potosí: The life and times of Antonio López de Quiroga.* Albuquerque: University of New Mexico Press.

Barba, A. A. [1640] 1923. *El Arte de los Metales.* Translation by Ross E. Douglas and E. P. Mathewson. New York: John Wiley & Sons.

Bargalló, M. 1955. *La minería y metalurgia en la América española durante la época colonial.* Mexico City: Fondo de Cultura Económica.

——. 1964. Sobre la introducción en el reino del Perú del beneficio de amalgamación de las minas de plata de Medina. In *Memorias del Primer Coloquio Mexicano de Historia de la Ciencia,* 143–167. Mexico.

——. 1967. El beneficio de amalgamación de patio: Originalidad, paternidad y primeras modalidades en México. *Ciencia* 25:177–184.

——. 1968. La guaira, horno de fundición del antiguo Perú: Estudio de las referencias de los cronistas. *Ciencia* 26:31–38. Also published in 1969 in *Minería* (91–92):43–49.

——. 1969 *La amalgamación de los minerales de plata en hispanoamérica colonial.* Mexico City: Cia. Fundidora de Fierro y Acero de Monterrey.

Beuchler, R. M. 1980. La compañía de azogueros y el Banco de Rescates de Potosí (1747–1779. *Boletín del Instituto de Historia Argentina y Americana Dr. Emilio Ravignani.* 26:67–116.

——. 1981. *Mining Society of Potosí, 1776–1816.* Ann Arbor, Mich.: University Microfilms International. [Published for the Department of Geography, Syracuse University, N.Y.]

Butts, A. and Charles D. Coxe, eds. 1967. *Silver: Economics, metallurgy and use.* Princeton: D. Van Nostrand Co., Inc.

Cañete y Domínguez, V. [1787] 1952. *Guía de la Provincia de Potosí.* Potosí: Editorial Potosí.

Capoche, L. 1959. *Relación general de la Villa Imperial de Potosí.* Vol. 122, Biblioteca de Autores Españoles. Madrid: Ediciones Atlas.

Cátedra de San Isidro. 1974. *La minería hispana e iberoamericana: Contribución a su investigación histórica: Estudios fuentes– bibliográficas.* Vol. 8. Tomo Primero: Documentos existentes en el Archivo General de Indias, Sección de Lima. León: Editorial Sever–Cuesta.

Craig, A. K. 1989. Mining ordenanzas and silver production at Potosí: The Toledo reforms. In *Precious metals, coinage, and the changes in monetary structures in Latin–America, Europe,*

Asia. Proceedings of the 1985 Stockton Conference on Monetary History, ed. Eddy van Cauwenberghe, 159–183. Leuven, Belgium: Leuven University Press.

——. 1993. The ingenious *ingenios*: Spanish colonial watermills at Potosí. *Culture, Form, and Place: Essays in Cultural and Historical Geography*, ed. Kent Mathewson, 125–156. Geoscience and Man, vol. 32. Baton Rouge, La.: Geoscience Publications, Department of Geography and Anthropology, Louisiana State University.

Elhuyar, F. 1789–90. Theorie der Amalgamation. *Bergbaukunde* 1:238–263; 2:200–298.

Evans, D. L. 1940. Structure and mineral zoning of the Pailaviri section, Potosí, Bolivia. *Economic Geology* 35(6): 737–750.

Greve, E. 1943. *Historia de la amalgamación de la plata*. Santiago de Chile: Imprenta Universitaria.

Helms, A. Z. 1806. *Travels from Buenos Ayres by Potosí to Lima; with notes by the translator, containing topographical descriptions of the Spanish Possessions in South America*. London: (printed for Richard Phillips, Bridgestreet) Black–Friars.

Hoover, H. and Hoover, L. H., eds. and trans. 1950. *De Re Metallica*, by G. Agricola, published 1556 at Basileae by APVD Hieron Frobenivm et Nicolaven Episcopivm. New York: Dover Publications.

Kelly, W. C., and Turneaure, F. S. 1970. Mineralogy, paragenesis and geothermometry of tin and tungsten deposits of eastern Bolivia. *Economic Geology* 65:609–680.

Llanos, G. de. [1609] 1983. *Diccionario y maneras de hablar que se usan en las minas y sus labores en los ingenios y beneficios de los metales*. La Paz: Museo Nacional de Etnografía y Folklore.

Niemeyer, H. F. 1981. Descripción de la hoya hidrográfica del río Copiapó. *Contribución Arqueológica No. 2*. Iquique: Museo Regional de Atacama.

Paoli, U. G. 1922. Il metallurgista spagnolo Alvaro Alonso Barba de Villa Lepe (1569–1662). *Archivio di storia della scienza* 3:150–168.

Peele, R., Jr. 1893. A primitive smelting furnace. *School of Mines Quarterly* 15:8–10.

Polo, J. T., 1911. *Reseña historica de la minería en el Perú*. Lima: Tipografía Peruana.

Prieto, C. 1973. *Mining in the New World*. New York: McGraw–Hill Book Co.

Rivas, S. and Carrasco, R. 1968. *Geología y yacimientos minerales de la región de Potosí*, 2 vols. Servicio Geológico de Bolivia GEOBOL, Boletín 11. La Paz: Ministerio de Minas.

Sánchez Gómez, J. 1982. La introducción del procedimiento de extracción de plata por medio del mercurio (procedimiento del patio) en España. *Actas II Congreso de la Sociedad Española de Historia de las Ciencias*. 3:293–302.

——. 1989. *De minería, metalúrgica y comercio de metales: La minería no férrica en el Reino de Castilla: 1450–1610*, 2 vols. Salamanca: Instituto Tecnológico de España, Colección Memorias.

Tandeter, E. 1993. Coercion and market: Silver mining in colonial Potosí, 1692–1826. Albuquerque, N.M.: University of New Mexico Press.

Tudela de la Orden, J. 1970. La minería y la metalurgia de la América española en los manuscritos de las bibliotecas de España. In *La minera española e iberoamericana*. Actas del primer coloquio español de historia de la minería. León.

Vázquez de Espinosa, A. [ca. 1629] 1948. *Compendio y Descripción de las Indias Occidentales*. The Smithsonian Institution Miscellaneous Collection 108. Washington, D.C.: Smithsonian Institution.

Villalobos R., S. 1979. *La economía de un desierto: Tarapacá durante la Colonia*. Santiago de Chile: Ediciones Nueva Universidad.

Young, O. E, Jr. 1970. *Western mining*. Norman, Oklahoma: University of Oklahoma Press.

The First Refining Mills in Potosí: Design and Construction

Peter Bakewell
Emory University, Atlanta, Georgia, U.S.A.

Abstract

Mechanical ore-crushing mills seem to have first been built in Potosí early in 1572 immediately after the introduction there of amalgamation as a refining method. Trituration on the large scale encouraged by the amalgamation method was possible only with the use of mechanical mills. Several types of machinery were tried over the succeeding decade. Among the first was a mill with five stamps, driven through gearing and a cam shaft by a horizontal water turbine. This mill was also equipped with a washing vat fitted with a powered rotating vane, driven off the cam shaft. Other energy sources used in the early and mid 1570s were human power (harnessed through treadmills), and horse (or mule) power (transmitted to millstones through a rotating beam, or to stamps through gearing). By 1577, however, Potosí had settled on a standard form of mill, which was to remain essentially unchanged from then on: a vertical water wheel, of 22-24 feet outside diameter, rotating on a massive axle, or cam shaft, and providing power to lift eight stamps arranged in one bank, or twelve stamps arranged in two (one on each side of the wheel). Building contracts provide detailed measurements of the major components of such mills. The contracts also show that there were present a small number of French and Italian millwrights (among a larger number of Spanish builders) in Potosí in the 1570s. This suggests that a variety of European mill-building designs and techniques may have been available there at this early date. Certainly, to judge by the similarity between the mills specified in contracts in the 1570s, and several devices illustrated by Agricola in *De Re Metallica*, this early machinery in Potosí was fully up to date in design.
Key words: Potosí, stamp mills, technology.

It was probably with the coming of refining by amalgamation to Potosí, an event for which the most likely date is 1571, that the first *ingenios*, or refining mills, appeared in the town and its environs (Bakewell 1977, 61-62). The documentary record, at least, does not show any earlier *ingenio*, prob-

In Quest of Mineral Wealth: Aboriginal and Colonial Mining and Metallurgy in Spanish America, edited by Alan K. Craig and Robert C. West, 1994. Geoscience and Man, vol. 33, pp. 287-302. Dept. of Geography & Anthropology, Louisiana State University, Baton Rouge, LA 70893-6010, U.S.A.

ably because the smelting techniques used in earlier years had not required dressing of large volumes of ore. What needed to be done could probably be managed with hand-held hammers, or under the traditional native *quimbalete* or *maray* — a large stone, made convex on its lower face, and to which a beam was lashed so that the whole could be rocked to and fro, breaking up ore set beneath it. Amalgamation's advantage over smelting was precisely that it could handle large batches of ore. But the ore had to be thoroughly pulverized before amalgamation, so that the maximum surface area of silver-bearing mineral was exposed to the mercury. Amalgamation on an industrial scale therefore was beyond the possibilities of crushing by hand; it demanded ore milling machinery.

The first contract for building mills *(ingenios)* that the notarial record in Potosí reveals is dated 9 January 1572.[1] (It would be no surprise to find *ingenio* construction the previous year, but the absence of notarial records in Potosí for 1571 deprives us of the most promising source of such information.) The contract is between Juan de Anguciana, *factor* (business agent) of the *Caja* (treasury office) of Potosí, and Hernán Martínez, described as an "*oficial y maestro de hacer ingenios de madera,*" and a resident of Potosí. The presence of such a *maestro* suggests that *ingenio* making was already in progress in the vicinity by January 1572. The contract was for two mills, to be built at the Chácara del Paya, three leagues from the town.[2] Each was to be capable of milling and sieving in a "*día natural*" (sunrise to sunset) 14 *quintales* (1,400 lbs.) of ore, using four stamps (*mazos*). And each was to be fitted with a washing vat (*tina de lavar*) and an apparatus called a *repasadero* — something in which the milled and sieved ores were to be *repasados*, or stirred up, during the amalgamation with mercury. This, of course, might in principle simply be a *patio*, or open yard. But an interesting little condition of the contract — that the *ingenios* did not have to be able to mill while washing and *repasando* — suggests that it was a mechanical device; that is, that whatever power source the mills used could be applied either to milling or to washing and the process of *repasar*. A remotely possible model was something on the lines of the mechanical amalgamating vessel recommended by Biringuccio in *De La Pirotechnia* ([1540] 1966): a millstone set horizontally and closely in a cylindrical wooden or stone bed.[3] In Biringuccio's illustrations of this device one man turns the millstone. The Potosí contract of January 1572 does not specify what energy source the mills were to have. Man-powered mills were certainly built from this time on; but working four stamps (unless they were very light) would have been beyond a single human's capacity. Hence, probably, waterpower or animals were to be used here. Each mill was priced at 650 pesos of *plata corriente.*

Agricola's *De Re Metallica* ([1556] 1950, 299) shows a machine that contains a device possibly similar to that of Biringuccio (fig. 1.) It is a mill

A—Water-wheel. B—Axle. C—Stamp. D—Hopper in the upper millstone. E—Opening passing through the centre. F—Lower millstone. G—Its round depression. H—Its outlet. I—Iron axle. K—Its crosspiece. L—Beam. M—Drum of rundles on the iron axle. N—Toothed drum of main axle. O—Tubs. P—The small planks. Q—Small upright axles. R—Enlarged part of one. S—Their paddles. T—Their drums which are made of rundles. V—Small horizontal axle set into the end of the main axle. X—Its toothed drums. Y—Three sluices. Z—Their small axles. AA—Spokes. BB—Paddles.

Fig. 1. Agricola's mill for amalgamating gold, with millstones and triple washing vats. ***Source*****: Agricola [1556] 1950, Book VII, p. 299.**

for crushing rock containing gold, and then amalgamating and washing it. The mill is driven by a vertical water wheel. On one side of the wheel is a stamp mill; on the other side, on a common shaft, an arrangement of horizontal millstones, driven via crownwheel and pinion, and a set of three powered vats, also driven off the shaft in the same way. The millstones and vats are laid out so that the ground slurry flows from the stones into the upper vat, and then sequentially into the other two. The difference from Biringuccio's arrangement is that mercury is placed in each vat. (Agricola [1556] 1950, 295-297).

There remains a far more informative mill-building contract dated six weeks later (the end of February 1572). This, in fact, may be the most highly detailed specification of a mill available for sixteenth- or seventeenth-century Potosí.[4] Quite quickly, perhaps, the design of mills became standardized and a matter of common knowledge. Building contracts certainly soon stopped giving precise instructions.

Here the power source is definitely water, driving a horizontal turbine wheel (*rodezno*) about 8 feet ("*de once a doce palmos*") in diameter, and 8 inches in section ("*lo alto un palmo*").[5] The wheel was to be mounted on a shaft (*guimbalete*) between 40 and 56 inches long ("*de cinco a siete palmos de alto*"), fitted at its upper end with a pinion of some 16 inches inside diameter ("*dos palmos de hueco*") and 21 inches length ("*dos y medio de largo*"). This in turn drove a crown wheel of one *vara* (ca. 33 inches) internal diameter. This was fitted to a horizontal shaft, 12 inches square ("*palmo y medio*") by between 11 and 12 feet long ("*de largo de dieciseis y dieciocho palmos*").[6] (See fig. 2) Twelve cams are specified, 40 to 48 inches long ("*de cinco a seis palmos de largo*"), to be attached to the shaft by *triángulos*. These triangles probably consisted of three boards fitted around the shaft to form an equilateral triangle, to one side of which the cam was secured. The surprising length specified for the cams is probably to be explained by supposing that one end of them projected from each side of the triangle, so that each revolution of the shaft produced two lifting motions from the cam. The shape of the cams is not defined; it is merely said that they should be chiselled to fit the triangles. But the well-known drawing of a Potosí *ingenio* in the *Atlas of Sea Charts*,[7] from the late sixteenth century, shows cams of a long-S shape (see fig. 3), and these of 1572 may have been similar. Agricola also shows cams of this design (fig. 4), and of roughly this length, for a stamp mill, though he has them set straight into the shaft rather than mounted on a structure attached to it. Interestingly enough, the Potosí contract also says that the cams may be fitted directly to the shaft if the builder prefers ("*y si le pareciere, asentarlas* [cams, *levas*] *en el propio eje*").[8]

The mill was to have five stamps. (Presumably, therefore, several of the twelve cams ordered were spares, since only one would be needed per stamp stem). Each of the stamp stems should be ca. 8 feet long ("*de doce pal-*

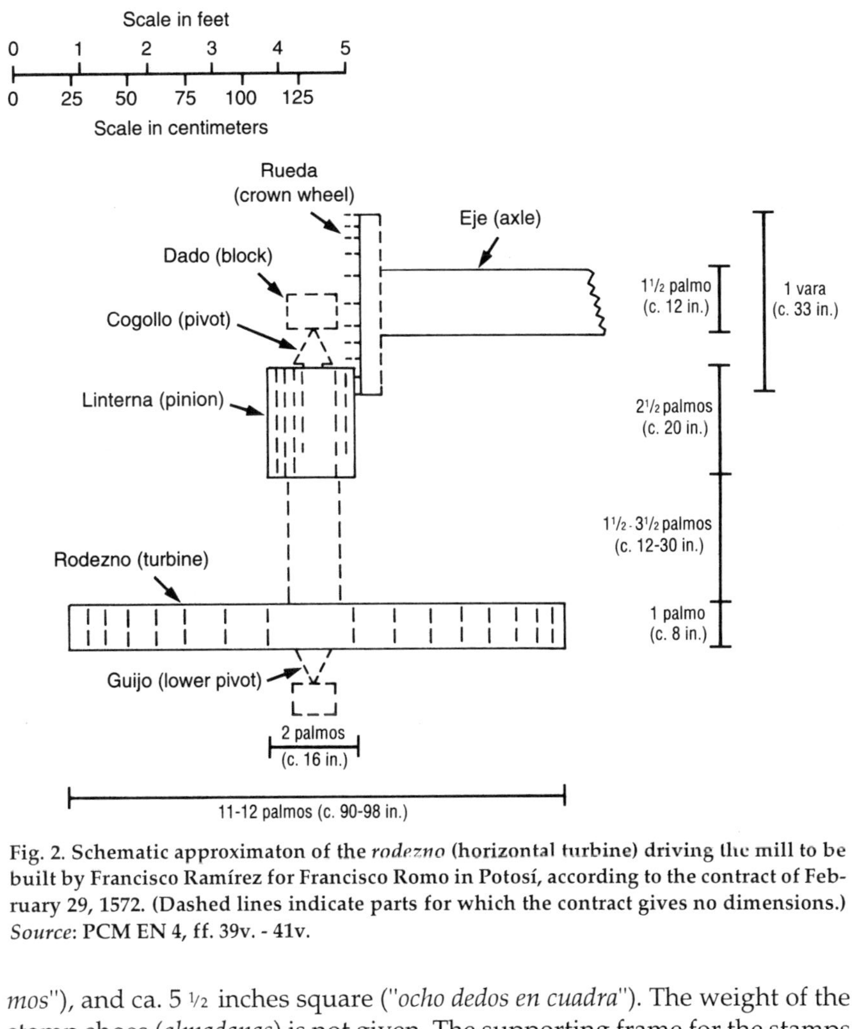

Fig. 2. Schematic approximaton of the *rodezno* (horizontal turbine) driving the mill to be built by Francisco Ramírez for Francisco Romo in Potosí, according to the contract of February 29, 1572. (Dashed lines indicate parts for which the contract gives no dimensions.) *Source*: **PCM EN 4, ff. 39v. - 41v.**

mos"), and ca. 5 1/2 inches square ("*ocho dedos en cuadra*"). The weight of the stamp shoes (*almadanas*) is not given. The supporting frame for the stamps should be 9 1/2-10 feet long ("*de catorce palmos o quince*").[9]

This mill was definitely to have a powered washing vat, driven off the main shaft by a small wheel (*rodezuela*) and pinion (*lanternilla*), with its spinning agitator (*molinete* or *rodeznillo*) seated on a bottom pivot (*guijo*) and bearing block (*dado*). Pivots and bearing blocks are, incidentally, specified also for the main turbine wheel and the shaft. The bearing material is not stated — presumably iron, or even steel, though hard wood is a possibility for the blocks. This mill, too, included a *repasador* (probably the same device as a *repasadero*), evidently powered, since it was to be mounted on the shaft.[10]

Fig. 3. Detail from a contemporary drawing of late 16th century ore-milling machinery in Potosí. *Source*: *Atlas of Sea Charts*, K3. By permission of the Hispanic Society of America, New York.

Ramírez, the builder, was to get 225 pesos of plata *corriente* for his labor. Romo, the owner, was to supply the necessary wood, uncut.

A later contract of the same year, from September 1572, suggests that designs were already becoming standardized: it calls simply for an *ingenio* "like that of Alonso Sierra."[11] This is a pity, since the agreement is for an *ingenio* de *grúa* (crane mill), of whose nature no description has yet come to light. Luis Capoche, in his *Relación* of Potosí of 1585, gives this as an early mill type, saying of the crane, "it is operated by Indians like a dock wheel."[12] This brings to mind a large wheel, wide enough across the rim to allow one or more men to stand inside it, and rotate it by walking forward.[13] From such a treadmill, power could easily have been transmitted to stamps through shafts and gears. Viceroy Toledo, in a letter of March 1573, comments on the large number of mills being built in and around Potosí for mercury processing "of water, stamps, and wheels that are called "*gruetas*" — which are perhaps the same thing as *grúas*.[14] In the mill of September 1572, a powered washing vat is in the contract. Márquez was to supply all needed nails, iron, and steel.[15] The mention of steel suggests its use in bearings.

A—AXLE. B—WATER-WHEEL. C—DRUM COMPOSED OF RUNDLES. D—OTHER AXLE. E—TOOTHED WHEEL. F—ITS SPOKES. G—ITS SEGMENTS. H—ITS TEETH. I—CAMS OF THE AXLE.

Fig. 4. Mill components. Note especially the cams, marked I. *Source*: Agricola [1556] 1950, Book IX, p. 372.

A particularly rich year in surviving contracts for *ingenio* building is 1577. By then, water-powered mills, with vertical, overshot wheels, were becoming standard in Potosí, as a result of their greater efficiency (Bakewell 1977, 73-75). (See fig. 5). Similarities of specifications in the contracts show that the design of these had settled within narrow limits. The wheels were usually of 22 feet inside diameter (24 feet outside). They drove either one bank (*cabeza*) of eight stamps, or two of six. In either case the wheel would be mounted directly on the main shaft. This may account for some of the superiority of this type of mill, since in it the frictional losses and fragility of wooden gearing were absent. The building time speci-

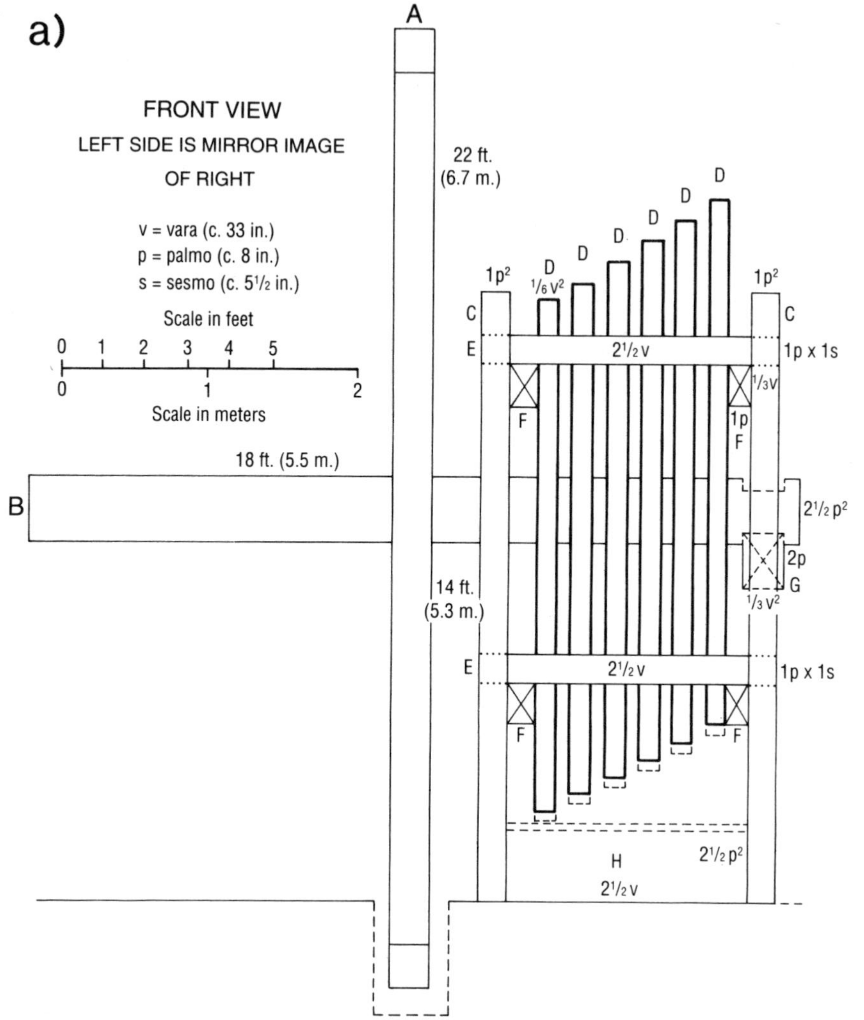

Fig. 5. a) front view and b) top view. Schematic approximation of an *ingenio de agua* (water-powered ore-crushing stamp mill) to be built near Potosí by Juan Martín and Diego Natera, *oficiales de carpintería*, for Francisco Logroño and Sebastián Gutiérrez, *vecinos y moradores*, according to a contract of February 21, 1577. Drawn from specifications in PCM EN 8, cuaderno 12, unfoliated.

This is a double-banked mill (*ingenio de dos cabezas*), with six stamps on each side of the wheel. Only one bank is drawn.

Moving parts are drawn in heavier line. Dotted lines show hidden parts of components. Dashed lines show dimensions that are uncertain, and parts added to the drawings that are not specified in the contract.

(continued on next page)

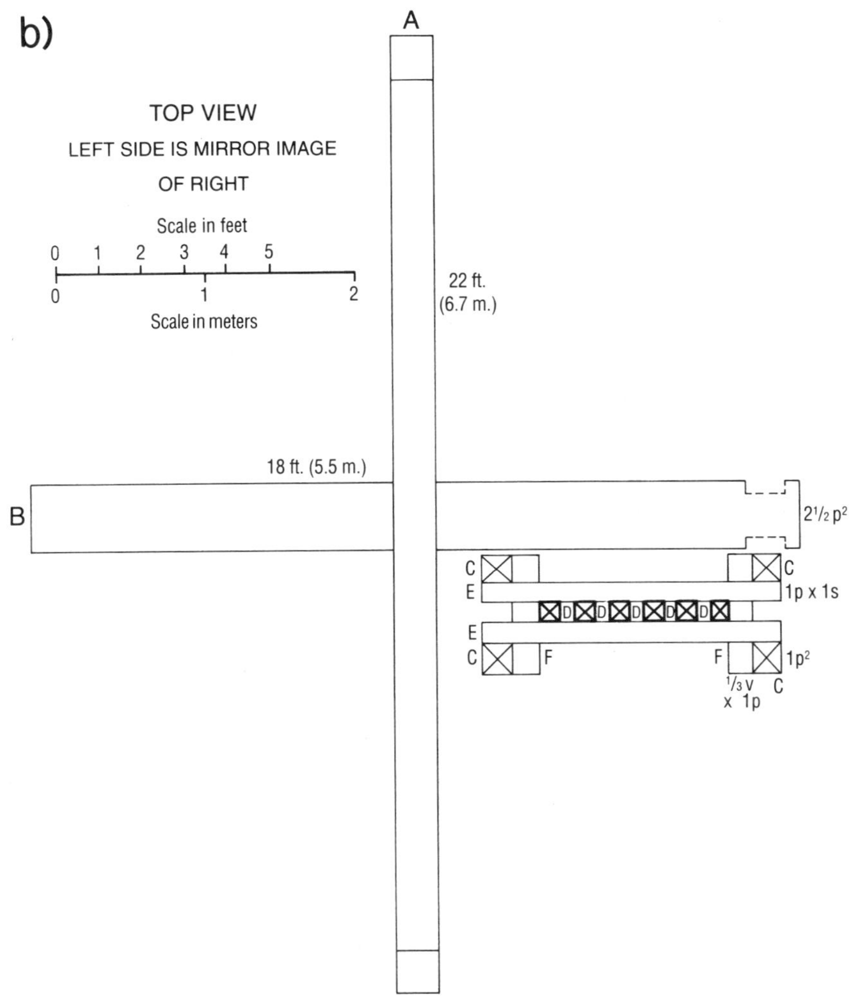

(Fig. 5 continued)

For simplicity's sake in making the drawings, 1 *vara* is taken here as 33 inches; 1 *palmo* as 8 inches; 1 *sesmo* [1/6 *vara*] as ca. 5 1/2 inches. In the drawings, v = *vara*; p = *palmo*; s = *sesmo*.

Key to components shown in the drawings:

A *Rueda*, or water wheel (overshot; 20 ft. inside diameter, 22 ft. outside; width, 1/3 *vara*)

B *Eje*, or axle (18 ft. long; 2 1/2 *palmos* square, or in diameter — unclear whether square or round)

C *Castillos*, or vertical posts of stamp mill (14 ft. high; 1 *palmo* square)

D *Mazos*, or stamp stems (12 ft. high; 1/6 *vara* square)

E *Cadenas*, or horizontal stringers of stamp mill (2 1/2 *varas* long; 1 *palmo* high; 1/6 *vara* wide)

(continued on next page)

Fig. 5 continued)

F *Palos para los castillos en que se asientan las cadenas*, taken to be horizontal crossmembers connecting the *castillos*, and on which the *cadenas* are seated (1 *vara* long; 1 - *palmo* wide; 1 *palmo* high)

G *Chumacera*, or pillow block, or shaft bearing (2 *palmos* high; ⅓ *vara* square)

H *Mortero*, or mortar block (2 *varas* long; 2 ½ *palmos* square)

N.B. These drawings are an attempt to "assemble" the *ingenio* from the list of parts given in the building contract, taking into account the dimensions specified for them. With one exception, the arrangement given here of the components seems likely to have been functional.

Bearings:

The exception is the mounting of the axle on its pillow blocks. Two of these *chumaceras* are called for by the contract: presumably one at each end of the axle. But the manner of mounting the blocks to the frame of the stamp-banks is not at all clear. The most likely position for them would seem to be between the two vertical posts, or *castillos*, at the outer ends of the banks, so that the axle itself would rotate between the *castillos*. But this — at least if other components are rightly placed here — would leave no room for the stamp stems and cam mechanisms. Possibly the *chumaceras* were fitted to separate posts resting on the ground; but no such posts are called for in the contract.

The nature of the bearings is also unclear. The contract contains no reference to iron, which may suggest that the bearings were wood-on-wood (liberally greased with tallow, no doubt). The right-hand bearing is drawn in that way here, with the axle cut down to form a circular journal; but this is pure speculation, unsupported by any part of the contract. It may be that, despite the absence of reference to metal, heavy bearing pegs of iron were what was actually used. They would be inserted into the ends of the axle, to run in the wooden *chumaceras*.

Stamps and stamp shoes:

A suspicion that the contract simply omitted mention of iron — perhaps because the two builders were purely carpenters — is reinforced by its lack of reference to metal stamp shoes and a metal slab to cap the mortar block. Iron shoes (*almadanetas*) were certainly in use in Potosí in 1577; stamps and mortar blocks lacking metal working surfaces would surely not last long. (See the contract for iron components, or *herramientas*, for an animal-powered *ingenio* made in Potosí on June 12, 1577, between Pedro Márquez and Melchior de Valencia. Márquez was to supply, among other things, eight *almadanetas* of 50 lbs. The weight of the iron shoes, as well as their resistance to wear, was crucial to the working of the mill. For this contract, see PCM EN 8, *cuaderno* 15, unfoliated.) In the front elevation of the mill given here, iron shoes and a mortar slab are drawn in with dashed lines.

The stamps are drawn here with a maximum lift of about 2 feet. This is again speculation, slightly informed by the *ingenio* illustration in the *Atlas of Sea Charts* (fig. 3). Similarly, the lifting order of the stamps is guesswork.

Cams:

No attempt is made to show the lifting mechanism of cams and tappets, since the contract is unrevealing on that subject. Forty cams (*levas*) are called for in it, to be made ⅙ *vara* square (length unspecified). Supposing that two cams were fitted to the axle for each of the twelve stamps, mortised into opposite sides of the shaft to produce two lifts per revolution, the substantial number of sixteen (forty less 2x12) were apparently intended as spares. Breakage, or rapid wear, was perhaps common. (The contract also specifies sixteen stamp stems — four being, apparently, spares.)

(continued on next page)

(Fig. 5 concluded)

Wheel:

The construction of the wheel is not obvious from the contract. Four *aspas* (spokes) of 22 feet length are specified, and this is shown in the drawing as the full diameter of the wheel, though it may have been greater. Certainly the inner diameter (*hueco*), excluding the rim of water boxes, is given as 20 feet. The number and size of the boxes are not described, though they were, presumably, at least 2 feet deep. Neither is the depth of the wheel-pit stated. Hence the height of the axle above ground is also uncertain. The *aspas* (22 feet x 1/3 *vara* x 1/6 *vara*) of the wheel were to be reinforced by eight *contra-aspas* (8 feet x 1/3 *vara* x 1/6 *vara*). The arrangement of these is not described.

Other parts:

The contract requires various other beams and pieces of wood to be supplied: four *grillos* (clamps?) for the *cadenas*; eight "*palos para trabajones de cadenas y castillos*"; "other things necessary for the *ingenio*." For lack of indication of their position, these have not been included in the drawings here.

Wood:

The woods called for are: for the wheel, *cedro*; for the axle and stamp stems, *tipa*; for the *chumaceras*, *soto*. In the contracts of the time, *tipa* is the standard material for axles and stems, being apparently valued for its hardness. *Cedro* is the preferred wood for components in contact with water: wheels and washing vats. It has not yet proved possible to find the botanical names of these woods. For the other components, no particular wood is named.

Price:

The price of the mill (wood and labor) was 1,350 *pesos ensayados*. The buyers were also to supply five Indian workers for two months at their own cost, though the builders must feed them. The *ingenio* should be finished within two months of the arrival of the wood on site.

---

fied in contracts was usually about seven months. Building cost had also standardized: 900 *pesos ensayados* for a mill of eight stamps, and 1,350 for the double-banked type with twelve.[16]

Luis Capoche, in the *Relación* of 1585, lists, though with all too little description, the types of mills used in Potosí up to then. He gives the three types (*de grúa*, *rodezno*, and vertical water-wheel) described here from building contracts. He uses the term *rodezno de alavés* instead of the plain *rodezno*, which presumably links the type of turbine in question to Alava in Spain, although the nature of the connection is not explained. Alava, one of the three Basque provinces, was, like the Basque country generally, a mining region, with lead workings (Sánchez Gómez 1989, 264). Possibly Alava had produced a distinctive type of milling device; but if so, at the present state of knowledge, its form is beyond even speculation.

Capoche also refers to other types for which no contracts have appeared (Capoche [1583] 1959, 117). The first type of *artificio* used in Potosí, he says, was foot driven by so-called *mozos de pies* (foot servants). Mills of this type operated in the town itself. They were clearly distinct from the *grúa* machines (also possibly foot-powered, as suggested above), which he gives separately. Then there were others operated by hand (*de mano*). Yet others were horse-powered.[17] Some of these were like gypsum mills, he says. This was possibly the type, common later in Mexico, in which a round millstone, attached to a central pivot by a beam, was pulled around a circular path, rolling on its circumference, by the animal. Other horse-driven mills had stamps. These undoubtedly were of the same general design as the machines driven by a turbine or a vertical wheel, with the horses or mules simply substituting for the wheel. The power would be carried to the main shaft through a crown wheel and pinion. Of all these types, Capoche notes, only stamp mills driven by animals or by a vertical waterwheel were still found by the mid 1580s, "being the most suitable thing for milling" (Capoche [1583] 1950, 117). Another source, of 1581, mentions only *ingenios de agua* and *ingenios de caballo* for that year: 55-60 of the first in and around Potosí, and 25-30 of the second.[18] Refiners continued to prefer those two types. In fact, as the building of dams above Potosí proceeded in the closing decades of the sixteenth century, finally providing enough water for year-round milling when rainfall was normal,[19] animal-powered mills were ever fewer, and the vertical water wheel became all but universal.[20]

Still, it was in the 1570s — the most remarkable decade in Potosí's entire mining history, with the installation of both amalgamation and the fully organized *mita* — that the essential winnowing-out of mill types took place. Rather little changed in the design of *ingenios* after then, partly because, no doubt, such sophisticated devices and mechanisms as the powered washing vat were present from the start, in 1572. Their early appearance raises questions of technology transfer, on which a little speculation may not be out of place (although little more than speculation can be offered).

The best hints come from the building agreements. Hernán Martínez, *oficial y maestro de hacer ingenios de madera*, was the builder in the contract of January, 1572. Where would such a master of making wooden machinery come from: from elsewhere in Peru, where he might have been making flour mills; or possibly from Mexico, where ore-crushing mills had existed since the mid-1550s, when amalgamation started to be used there? It is impossible to say. But at least Hernán Martínez is a classic Spanish name, so that his knowledge and skills were probably acquired somewhere in the empire.

Most interestingly, though, foreign carpenters appear in Potosí in the 1570s. The first name to emerge is Jorge de Candía, one of builders of the *grúa* specified in the contract of September 1572. "Candía" was an old Spanish word for Crete. So this man, like the famous conqueror of Peru, Pedro de Candía, may have been a Greek — a people much skilled in making artillery and relied upon by the Spanish for that purpose in the Mediterranean (Lockhart 1972, 129-132). In 1577 contracts there appear builders named Juan Siguer [?] de Niza [Nice], resident in Potosí;[21] Juan de Nápoles, resident[22]; and Nicolás Ginovés, *carpintero y maestro de hacer ingenios*.[23] This suggests a French and Italian presence, also, among the machine makers of the 1570s, possibly bringing in techniques and designs from their respective regions along the north Mediterranean shore. The foreigners, it is true, are far outnumbered by men with good Spanish names.

Another tantalizing possibility is the influence of foreign texts. A very hypothetical connection was raised earlier, in reference to the *repasadero*, with Biringuccio; but there is really no cause to think that the *Pirotechnia* (Biringuccio [1540] 1966) was known in Potosí in the sixteenth century.[24] Agricola's *De Re Metallica* may be a different matter. Research to date has raised no allusion to it in Potosí in the 1570s, or indeed at any time before Alvaro Alonso Barba's references in the *Arte de metales*, a book composed in and around Potosí in the early seventeenth century (Bargalló 1955, 225-228). But *De Re Metallica* was definitely present in Spain in 1557, a single year after its publication in Basle (Sánchez Gómez 1989, 355-356). So it, or its content, might easily have reached America by the time *ingenio* building began in Potosí. The designs appearing in the town in the 1570s notably resemble those in Agricola's Book VIII. The specification of the stamp mill called for in the contract of February 29, 1572, for example, could well have been written from Agricola's illustrations of such apparatus, except for requiring five stamps where he shows four; similarly, the specifications of washing vats (Agricola [1556] 1950, 284-287). Whether this indicates direct influence by Agricola, or whether it is a matter of drawing on common sources cannot be said. But the resemblances are striking. And, at the very least, they show that in milling, if possibly not in methods of ore extraction, Potosí brought itself up to date with an impressive rapidity in the 1570s, in response to the demand for high-volume processing made by amalgamation.

Endnotes

1. PCM EN 4, ff. 12-14. Potosí, 9 January 1572, *concierto* of Hernán Martínez, *oficial y maestro de hacer ingenios de agua, residente* of Potosí, with the *Factor* Juan de Anguciana, to build two *ingenios de moler metales de plata* at the Chácara del Paya. The first recorded cession of a *sitio* for an *ingenio* in Potosí itself is recorded in BAN CPELA Minas Catalogue No. 126, *acuerdo* of the cabildo of Potosí, 17 March 1572. The Pedro Hernández to whom the site was granted is very possibly Pedro Hernández de Córdoba, the man usually credited with introducing amalgamation to Potosí.

2. This "Paya farm" is not to be found on modern maps. Many of the early mills of Potosí were built at Tarapaya, on the river of the same name, roughly ten miles north-west of the town. Possibly, then, a connection exists between the names Paya and Tarapaya. Three leagues equal about ten miles.

3. Vannocchio Biringuccio [1540] 1966, IX:11: "Make a large bricked pan of stone or wood and arrange a millstone inside which rotates in the hollow of this pan like that of a flour

mill. In the hollow of the pan put some of your material containing gold after it has been crushed well in a mortar and then washed and dried. Grind it in the said mill, moistening it meanwhile with vinegar or water in which sublimate, verdigris, and common salt have been dissolved. Then enough mercury to cover these materials is poured over them. They are agitated inside for an hour or two, by turning the mill either by hand or with a horse, depending on the device you have arranged....Finally, thus prepared, the mercury is separated from the earthiness with a sieve or by washing and thus the mercury is retrieved. Then, make it rise with a little cucurbit or pass it through a purse and there will remain at the bottom the gold, silver, copper, or whatever metal you placed in the pan under the millstone to be ground."

4. However, I confess that some of the detail may have been lost, or muddied, by my inability to read the script with certainty. See PCM EN 4, ff. 39v.-41v. Potosí, 29 February 1572, agreement between Francisco Ramírez, *oficial de carpintero*, and Francisco Romo, for the building of an *ingenio de madera* to Romo's plan. The site was outside Potosí, between the town and Pora (? - possibly Porco), below the flour(?) mill (*molino*) of Susporas (?).

Francisco Romo was one of the men who had been experimenting with amalgamation in Potosí in the 1560s. "Y despues vio [el testigo] a cabo de cuatro o cinco años que Gonzalo Santos y Juan Carrasco y fulano Concha y *Francisco Romo* todos trabajaron mucho con cierto azogue que les venia en hacer ensayes y experiencias y todo salia perdido y no se salia con ninguna cosa, y antes de esto ciertas personas que llamaban los valencianos los cuales no vio este testigo mas de ser publico y notorio habian hecho muchas experiencias y nunca salieron con ello, sino todos huyendo y perdidos, y asi en tiempo del dicho Damian de la Bandera cuando se intento el dicho beneficio de los azogues le tenian por cosa perdida y disparada, en tal manera que un Pedro Hernandez, persona que vino a casa de este testigo a asistir en la labor del socavon que llaman de Nicolas de Benino, que este testigo le tenia asalariado para la labor del dicho socavon, estando en la dicha labor, hizo sus ensayes con hasta ocho libras de azogue que tenia y saliendo con ello pidio socorro al dicho Damian de la Bandera como tal corregidor, el cual le envio a su excelencia para tratase [*sic*] lo susodicho y de alli el susodicho le escribio y dio cuenta de lo que pasaba en esta dicha Villa ..." Testimony of Gonzalo Durán in "Probanza de los méritos y servicios de Damian de la Bandera," 1586 (AGI Charcas 42; Bargalló 1969, 178-179).

5. The *palmo* is taken here to be a quarter of a *vara*, or about 21 cms (8.2 inches). The *dedo*, a measurement appearing below, was a twelfth of a *palmo*, or 1.74 cms (0.68 inches).

6. The shaft was to be fitted with a "work of triangle, quadrangle, and eighths" ("labor de triangulo y cuadrangulo y octavas"). This is confusing, if, as the wording implies, all three were to be applied to this one shaft. Triangle and quadrangle would seem to indicate wooden structures of three or four sides surrounding the shaft, serving as anchors/platforms for the cams lifting the stamps themselves. "Eighths" might mean cutting the shaft to produce eight equal facets around its diameter, which would also serve as bases for cams. Georgius Agricola ([1556] 1950, Book VIII) shows such faceted shafts in trituration mills in various illustrations.

7. Hispanic Society of America (New York), Ms. K3.

8. Agricola's S-cams at first sight seem to be single pieces of wood passing through the shaft (see fig. 4). Obviously enough, such construction would pose problems of carpentry. But closer observation of the illustration shows halves of the S-cams (marked `i' like the cams already in the shaft) lying on the ground. Thus these S-cams, though apparently continuous, are evidently two separate pieces mortised into the shaft. Agricola's text ([1556] 1950, 373) seems to confirm this form of construction. Clearly this arrangement would have been less strong than a single-piece cam, which perhaps is why the Potosí contract calls for single-pieces to be attached to a mounting on the outside of the shaft.

9. This is actually the specified length of the *cadenas*, or horizontal stringers connecting the main vertical posts of the mill and holding the stamp stems vertical. The *cadenas* were to be of square section, of about 1 *palmo*.

10. The builder "ha de armar un repasador en el eje."

11. PCM EN 4, ff. 46-47. Potosí, 1 September 1572. *Concierto* of Jorge de Candía and Gaspar de la Irúa (?), with Hernando Márquez; all *estantes* in Potosí.

12. "Otros [ingenios], de grua, que la traen indios como rueda de muelles..." Luis Capoche, 1959, 117, col. 2.

13. For a Spanish treadmill-driven crane of these same years, see George Kubler, 1982, illustration 55. From the dimensions of the frame of this crane given in this picture, it seems that the wheel had an inside diameter of about 9 ft.

14. AGI Lima 29, vol. 1. f. 26, Potosí, 20 March 1573, Viceroy Toledo to crown ("Gobierno," para. 14). *Gruetas* are mentioned again ca. 1576 in a manuscript listing the early owners of *ingenios* in Potosí, and referring to "otros muchos ingenios ansi de agua como de cavallos, mazos, e gruetas e otros artificios hasta en cantidad de dozientos de cada genero poco mas o menos que al presente estan hechos y cada dia se van aumentando" (Bargalló, 1969, 207, citing BN Madrid, MS 3040, ff. 403-410, para. 13).

15. The materials of the *tina* are specified — "tablas de cedro de dos dedos de grueso, y las tablas de suelo de cuatro dedos de grueso, del tamaño que hicieron [?hicieren] otra para Alonso Sierra." The axle of the vat should be at least a *tercia de cuadra* (1/3 *vara* square) in cross section. The contract also specifies a *zangarilla*, or hut, "para labrar los dichos metales en la dicha tina." The *tina* and *grúa* are to be made in Potosí; though it is not clear from the wording whether the mill is to be in Potosí. Márquez was apparently to provide (or at least freight) all the necessary wood, supply two Indian workers, and all *clavazón* (nails), *hierro* and *acero*. Candía and Irúa (?) were to get 440 *pesos de plata corriente* for their work.

16. See various mill contracts in PCM EN 8, particularly in *cuaderno* 11 (unfoliated), from January and February 1577. Contracts are also to be found in *cuadernos* 12, 13, 14 of the same book.

17. The term of the time was *ingenio de caballo*, though it is probable that the animal more commonly used was the mule.

18. Spain, Biblioteca Nacional [BN], ms 3040, ff. 46-50v., Don Diego Cabeza de Vaca to "excelentisimo señor" [Viceroy Enríquez?], Potosí, 8 April 1581.

19. The first reservoir, or *laguna*, was finished in 1576. One Pedro Sande is described as the supervisor of dam building above Potosí in the late sixteenth century. His origins are unknown. But, according to Capoche, Enrique Sandi, a partner of Nicolás del Benino in the construction of a famous adit (*socavón*) in the Cerro of Potosí in the 1550s and 60s, was a native of London. Benino was himself a Florentine. Here, then, is further evidence of foreign technical expertise and effort in early Potosí to add to that of the foreign carpenters given below. For the *lagunas*, see Bargalló; for Sandi and Benino, Capoche, 1950, 106.

20. Except, in all probability, in the small mills known as *trapiches* that were used in the seventeenth and eighteenth centuries by Indian refiners to process ores stolen from mines, or taken from them as part-payment of wages. What sort of crushing devices were used there is an open question.

21. PCM EN 8, cuaderno 12 n.f., contract of Potosí, February 21, 1577.

22. PCM EN 8, cuaderno 11, ff. 520-521, contract of Potosí, January 2, 1577; cuaderno 11, n.f., Potosí, contract of February 2, 1577; and cuaderno 13, n.f., contract of Potosí, March 5, 1577.

23. PCM EN 8, cuaderno 12, n.f., contract of Potosí, March 2, 1577.

24. Though the possibility certainly exists that knowledge of Biringuccio's *De La Pirotechnia* (1540), or of the matters that Biringuccio treats, may have reached Peru via Mexico, to which a number of German miners travelled from the mid-1530s onward.

Acknowledgements

I should like to record here the help of Judith Hope Reynolds in reading and transcribing several of the manuscripts referred to in the endnotes.

References

Archival Sources

AGI	—	Archivo General de Indias [Sevilla, Spain] Lima 29 Charcas 42
BAN CPELA	—	Bolivia, Archivo Nacional, Cabildo de Potosí, Libros de Acuerdo [Sucre]
BN	—	Biblioteca Nacional [Madrid, Spain]
PCM EN	—	Potosí, archive of the Casa de Moneda, Escrituras notariales [Bolivia]

Published Materials

Agricola, G. [1556] 1950. *De Re Metallica.* New York: Dover Publications.

Bakewell, P. 1977. Technological change in Potosí: The silver boom of the 1570s. *Jahrbuch für Geschichte von Staat, Wirtschaft und Gesellschaft Lateinamerikas* 14: 57-77.

Bargalló, M. *La minería y la metalurgia en la América española durante la época colonial.* Mexico City: Fondo de Cultura Económica, 1955.

___. 1969. *La amalgamación de los minerales de plata.* Mexico City: Compañía Fundidora de Fierro y Acero de Monterrey S.A.

Biringuccio, V. [1540] 1966. *The Pirotechnia.* Cambridge, Mass.: The M.I.T. Press.

Capoche, L. [1585] 1959. *Relación general de la Villa Imperial de Potosí.* Biblioteca de Autores Españoles, vol. 122. Madrid: Ediciones Atlas.

Kubler, G. 1982. *Building the Escorial.* Princeton: Princeton University Press.

Lockhart, J. M. 1972. *The men of Cajamarca. A social and biographical study of the first conquerors of Peru.* Austin: University of Texas Press.

Sánchez Gómez, J. 1989. *De minería, metalurgia y comercio de metales. La minería no férrica en el Reino de Castilla, 1450-1610.* Salamanca: Universidad de Salamanca.

Silver Mining at Huantajaya, Viceroyalty of Peru

Kendall W. Brown, Brigham Young University, Provo, Utah, U.S.A.

Alan K. Craig, Florida Atlantic University, Boca Raton, Florida, U.S.A.

Abstract

Although bonanza silver lodes at Huantajaya may have been the richest in South America, hyperaridity of the Atacama Desert made production difficult. Subsurface geology of this extensive deposit was complicated by multiple faulting. Chronic scarcity of water, food, and workers added to refining and transportation difficulties, resulting in relatively low profits compared to the more famous mines at Potosí. Unusually egalitarian relations arose between colonial mine owners and voluntary workers from nearby villages who were often paid in water, ore and staples. Ore from Huantajaya often was carried more than 1,500 kilometers inland to Potosí for beneficiation by the nearest waterpowered stamp mills. The peripheral location and isolation of Huantajaya discouraged effective colonial administration. These conditions tempted mine owners to evade taxes and smuggle silver.
Key words: Huantajaya, silver, Atacama Desert, Tarapacá, Potosí, isolation, hyperaridity.

Introduction

Mining of precious metals is a subject neglected by economic geographers who seldom discuss it in the modern literature. Reasons for this neglect include a general perception that gold and silver mining is an arcane, complicated and ephemeral activity, largely relegated to the past. Historical geographers have ignored some important themes of mining that evolved within the Spanish colonial empire. This paper is an effort to combine the researches of an historical archivist and a field geologist to focus on a little-known silver deposit having abundant documentation surviving from the sixteenth century to the present.

In Quest of Mineral Wealth: Aboriginal and Colonial Mining and Metallurgy in Spanish America, edited by Alan K. Craig and Robert C. West, 1994. Geoscience and Man, vol. 33, pp. 303-327. Dept. of Geography & Anthropology, Louisiana State University, Baton Rouge, LA 70893-6010, U.S.A.

Huantajaya (figs. 1 and 2) is a remarkable example of human ingenuity in a setting of extreme environmental constraints. In this case study we show how the mining center survived, in spite of every adversity, to become the largest desert mining complex in South America during the colonial period. Although Huantajaya was located within sight of the Pacific Ocean and the primitive port of Iquique (fig. 3), transportation of supplies and ore was mostly carried out over inland routes across some of the most

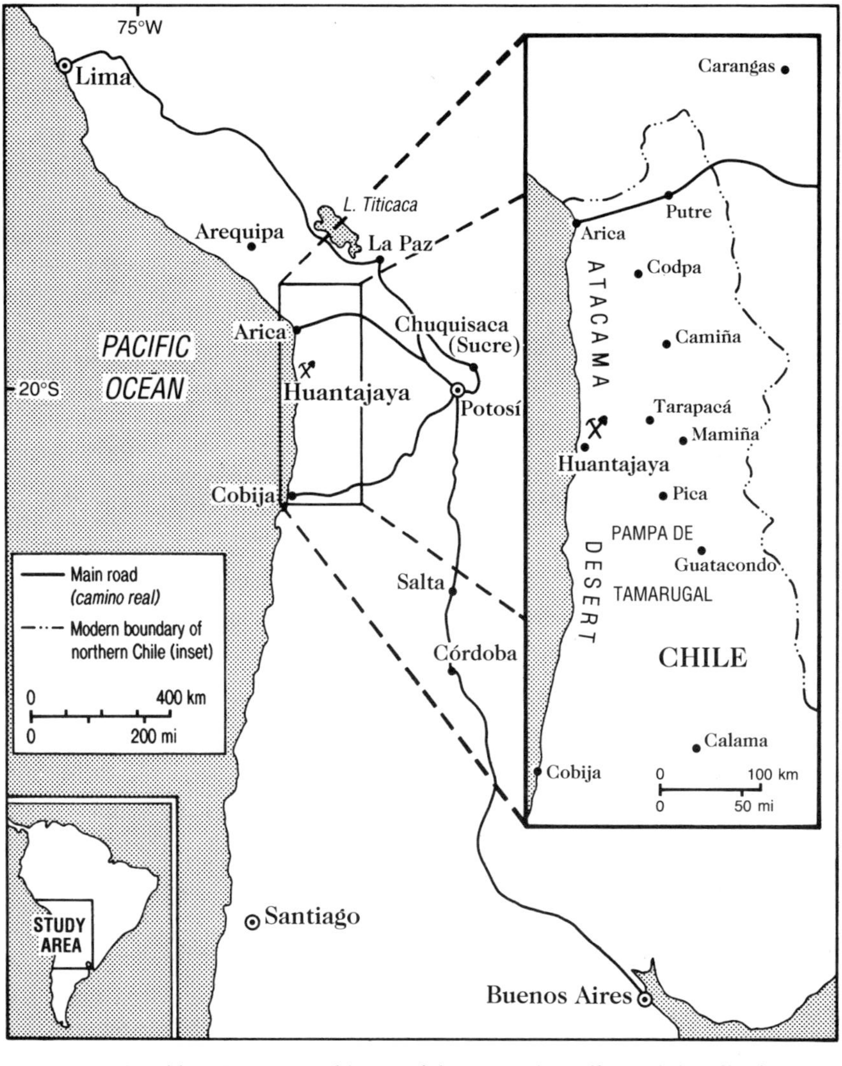

Fig. 1. Regional location map and insert of the Huantajaya silver mining district.

Fig. 2. East flank of Cerro Huantajaya showing numerous old *cateos* (prospecting pits), *canchas* (ore-sorting platforms), and entrance shafts. Modern prospectors are using the company bulldozer to scrape outcrop surfaces for signs of mineralization.

arduous terrain in the South American continent. This resulted in excessive production costs because the only large-scale refining facilities were located at Potosí where numerous hydraulic stamp mills were in operation after 1576. Mine owners at Huantajaya never resolved the problem of acquiring enough water to construct their own mills needed to crush and concentrate all but the richest native silver ores.

In addition to the problems caused by distance, we must add those caused by the inland location of regional viceregal administrative centers, which was extremely disadvantageous for Huantajaya mine operators, not to mention the often corrupt nature of the officials stationed there who were responsible for collecting taxes on specie produced from the mines.

Very few mineral deposits are of sufficient volume and richness to allow truly long-term development necessary for the evolution of complex social institutions. Potosí and Huantajaya are two important exceptions. Their contrasting geographic elements make an interesting comparison: the former, located near a highland Indian population, had massive forced labor problems, while the latter, in an uninhabited desert, attracted volunteer workers through special incentives. Potosí was plagued by disastrous floods that repeatedly destroyed beneficiation facilities, while Huantajaya

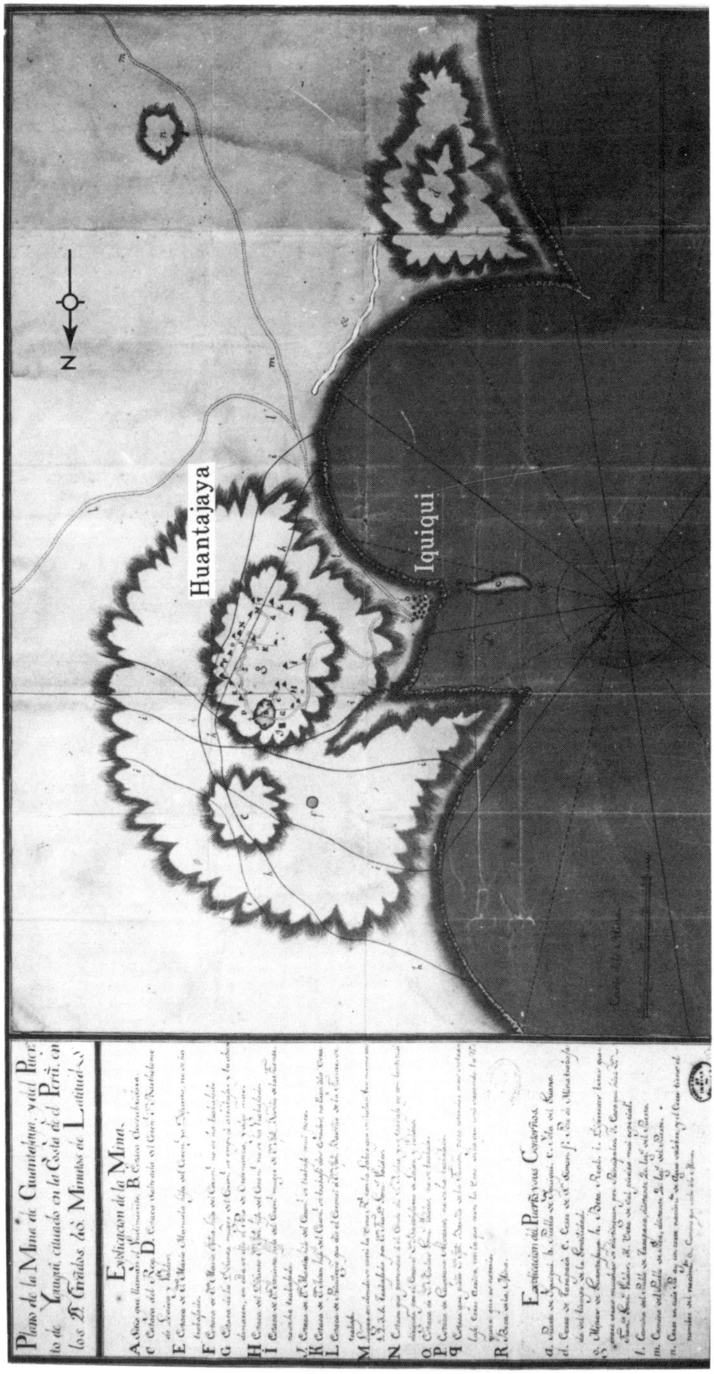

Fig. 3. Plano de la Mina de Guantajaya y del Puerto de Yquiqui cituado en la Costa de el Perú... A late eighteenth century sketch map showing locations of the discovery claim, the King's claim, the hundamiento and important mines. AGI, Mapas y Planos, Py Chile 43. Published with permission.

languished in a total absence of rain. Many schemes were devised to overcome the spatial handicaps of both mining centers, but in the end, Huantajaya's problems were found to be simply too formidable. As high grade surface deposits dwindled, profits fell precipitously; isolated from the mainstreams of the colonial economy, mine owners turned increasingly to illegal means of reducing expenses.

As more modern archival work has been published on aspects of Potosí (Bakewell 1983; Tandeter 1993) and adjacent areas that provided logistical support, including Arequipa (Brown 1986), it has become evident that several other regional mining centers made substantial contributions of silver that added to the fame enjoyed by the viceroyalty of Alto Perú. Of these, Huantajaya offers a vivid example of human response and perseverance under the most extreme environmental conditions imposed on the miners who staked their claims in the heart of the Atacama Desert.[1]

Discovery and Early History

Although some authors suggest San Agustín de Huantajaya was not exploited before 1556 (Mendiburu 1874-1890, 8:164), others state categorically that Inkas prehistorically mined its silver (Wormald 1968; Officer 1912, 14) but they offer no firm evidence for their views, such as that found by Craig on a site visit in 1992 (fig. 4).[2] Still other chroniclers suggest the actual discovery was made by Portuguese prospectors,[3] but are unable to agree on the exact date. We do know, however, that Diego de Almagro and his ragged band struggled past these outcrops in 1537 on their retreat from Chile back to Peru by way of Caleta Icaiza (Iquique). Other Spaniards may have discovered Huantajaya as early as 1540, thus antedating the great strike at Potosí five years later.[4]

In 1540 a pioneer settler of Arequipa, Lucas Martínez Vegazo, held an *encomienda* (tribute grant) which included the Indians of Tarapacá district. Prospectors hired by Martínez Vegazo soon began mining operations at Huantajaya, using his *encomienda* Indians as workers (Trelles Arestegui 1982). Unfortunately, there was no eloquent witness to chronicle these early days when lucky miners simply gouged out chunks of native silver from the conspicuous surface outcrops. Some bonanzas of nearly pure silver weighed as much as one hundred pounds (Pizarro [1571] 1969, 390-394) and Lucas Martínez Vegazo soon became one of the richest men in southern Peru (Trelles Arestegui 1982, 42).

When civil war broke out in 1544, Martínez Vegazo made the unfortunate mistake of supporting the unsuccessful rebellion headed by Gonzalo Pizarro and was soon stripped of his *encomienda* and mines. Through a series of appeals these were partially restored to him in 1557 and we know his mining operations continued at Huantajaya as late as 1560 (Trelles Ar-

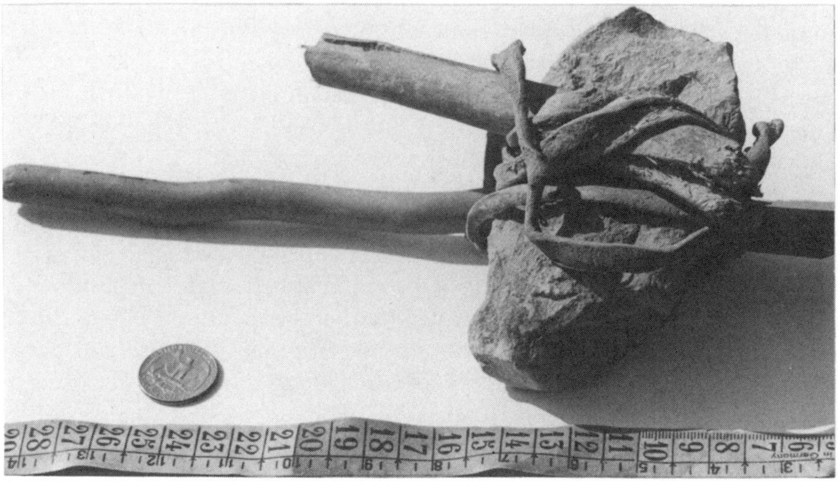

Fig. 4. Stone hammer found with skeletons of two prehistoric Indian miners trapped by a cave-in of early workings at Huantajaya, shown in figure 11. Photo by Ralph Cané.

estegui 1982, 103-112). He died in 1567 and it is unclear whether his young bride became involved in the mines. Apparently his Portuguese associates carried on for another decade and in 1572, Alonso Muñoz, a royal mining inspector, sent ore samples to the king and marvelled that the mines were "a thing so rich."[5]

Following the Muñoz report, Huantajaya disappears from the historical record for nearly a century. We can only speculate about the reasons for this decline. The mines fade from view at a time when there was a great upsurge in mining activity at Potosí, discovered in 1545 and in decline by 1560, but revitalized by the Toledo reforms and patio amalgamation in the 1570s (Paz Soldan 1980, 264-265; Craig 1989). Demographic factors no doubt contributed to the abandonment of Huantajaya by the few remaining Indians who, as they became increasingly hispanicized, were attracted to Potosí and its nearest entrepot, Arica.

Nothing in the early records suggests that experienced miners were ever present at Huantajaya. When the easily removed surface ores were exhausted, it was soon apparent the subsurface geology presented a number of problems caused by complex faulting. Miners discovered they could not follow a vein or ore body for any considerable distance before it disappeared, only to reappear again at some unpredictable distance with both vertical and horizontal displacements.

Geological Conditions

As early as 1766 an accurate subsurface map of the Huantajaya silver mines, associated veins, and surface claims was produced by Col. Antonio O'Brien acting on orders of Viceroy Amat in Lima.[6] Modern geological information did not become available until the present century when a consulting geologist hired by an English mining syndicate made a detailed study of the underground workings (fig. 5) where serious brine flooding had occurred (Officer 1922, 570). The silver deposits described in his report were distributed over an elliptical area roughly 4 x 2 km located 12 km inland from the port of Iquique. Since then, work by government geologists mapping the Iquique and Caleta Molle quadrangles makes the regional geological relationships quite clear (Novoa 1970).

The subdued coastal range of mountains in which Huantajaya is found represents an uplifted horst-like feature caught between the gigantic coastal escarpment (Paskoff 1978) and the famous nitrate beds that fill an elongated graben-like depression extending northward along the foothills of the Andes.

Fig. 5. A typical inclined colonial shaft opening, reenforced (note small wooden bents) by the English mining company and further strengthened (large bags of tailings) by the present operators who are using it for access to and ventilation of some lower levels. All these unstable workings are highly dangerous to enter in this seismically active area.

Silver ore bodies at Huantajaya were emplaced in Oxfordian (Jurassic) limestones. Analysis shows highly mineralized hydrothermal solutions were repeatedly injected into these steeply dipping beds, often near metamorphosed contacts with younger intrusive dacite porphyry dikes and sills.

According to plate tectonics theory, the silver itself may have originated in ocean floor deposits far to the west in the Pacific Ocean at some point along a South Pacific spreading center. As this new material was shoved eastward, it eventually became subducted under the overriding mass of the South American continent and was converted into a molten state deep beneath the surface. Finally it was forced by great heat and pressure into overlying sedimentary beds along planes of crustal weakness caused by fractures.

From the geological cross-section (fig. 6) it is evident that colonial miners at Huantajaya faced some daunting problems caused by a complex series of fault displacements. At least four distinct faulting events can be observed, and measurements of the dominant system shows throws of nearly 200 meters. An interval of reverse faulting further complicates the patterns of simple gravity faults in such a way that makes it extremely difficult for unskilled miners to follow faulted ore bodies.

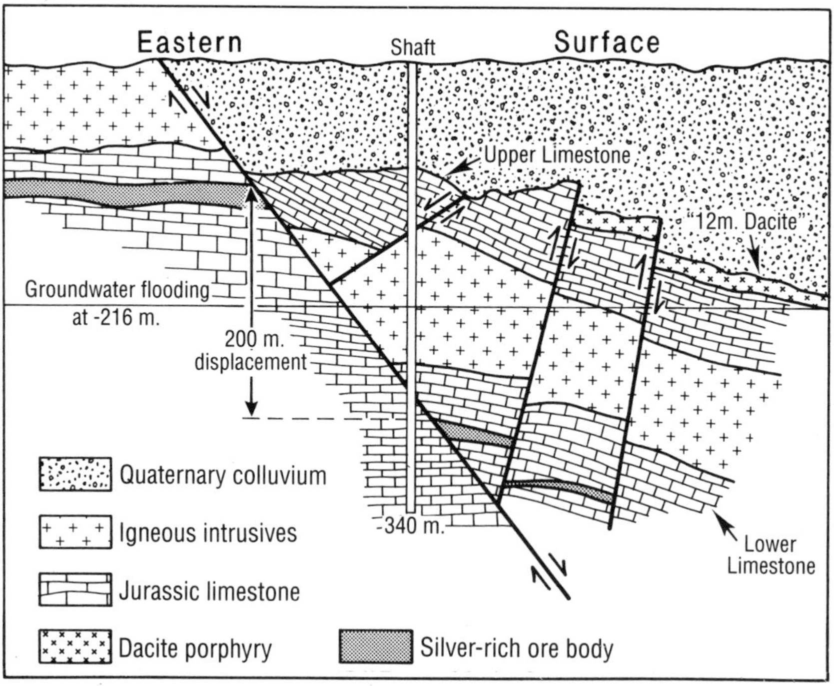

Fig. 6. Schematic subsurface geology of the Huantajaya silver deposits. After Officer, 1922.

In the western half of the mineralized district, some veins cropping out on the surface originally contained native silver as the dominant mineral. Archival records indicate it continued to occur at shallow depths along with argentite, proustite, and other silver sulfides. Ore shoots did not always have well defined boundaries, but they were essentially stratabound within the steeply dipping limestone beds. Mineralization was strongest along zones of multiple fractures; where there was an intersection of two such zones, some splendid bonanzas were found. These glory holes continued to add to the reputation of Huantajaya throughout the colonial era.

Huantajaya in the Eighteenth Century

According to folk tradition, the mines of the Huantajaya district were not rediscovered until 1680 when Juan de Loayza briefly attempted to reopen the workings. He had no notable success and the mines again lay idle until 1718 when his son, Bartolomé, began to work the San Simón and Tajito veins. This initiated nearly a century of relatively constant mining, and the period when the mines are best known to historians.

Bartolomé Loayza enjoyed considerable success as his increasingly experienced men discovered a series of profitable old and new veins. When his miners encountered a single silver mass weighing 3,200 pounds that produced 34,000 pesos when smelted in Potosí, others were attracted to the scene and filed numerous claims. In 1746 Loayza sent samples of these massive nuggets to Madrid where they were displayed in the royal museum of natural history and gained fame for their "incomparable richness."[7]

In the late eighteenth century the Spanish Minister of the Indies, José de Gálvez ordered Peruvian authorities to send representative ore samples from new discoveries. On receiving some ore from the Huantajaya district, Gálvez chastised the intendant for sending refined silver instead of ore. However, when the metallurgist Francisco Chabaneau assayed it, he declared it to be "virgin silver."[8] Such bonanzas spoiled some mine owners to such an extent that they despised ores that would have been considered rich at other Andean mines. But the chronic lack of water, food and fuel at Huantajaya made processing very expensive in comparison with other sites. For these reasons the mines remained comparatively small scale operations throughout their useful life.

When Col. O'Brien inspected the mines in the 1760s there were only seven miners operating fourteen claims.[9] By 1799 there were twelve mine owners, but only seven mines were active.[10] In 1807, a visiting artist made an evocative sketch (fig. 7) of the remaining buildings. The general desolation is in sharp contrast to the optimistic accompanying narrative. Today

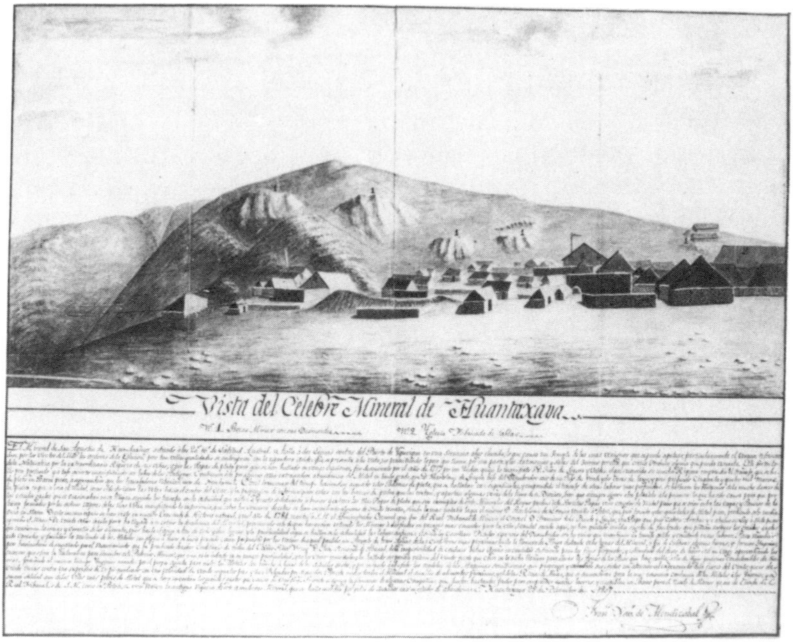

Fig. 7. View of Huantajaya in 1807 from an aquarela by Francisco Navarro de Mendizabal. AGI, Mapas y Planos Py Chile 16. Published with permission.

(fig. 8) these last vestiges of the colonial era have been destroyed by the present operators.

Working Conditions

Labor in the mines of Huantajaya was almost entirely voluntary, in contrast with Potosí where mine owners had grown accustomed to depend in large part on the *mita* (corvée labor by Indians).[11] At Huantajaya the workers came voluntarily from the small villages scattered along the Andean foothills springline to the east of the Atacama Desert where seeps provided just enough water for subsistence crops and livestock. Unlike the workers at virtually all other Peruvian mines, there were few Indians among them. Most were mestizos, with a few mulattoes and *zambos* (mixed black and Indian) from the province of Tarapacá.

Mine owners acquired workers by advancing them money and goods and then forcing them to pay off the debt by working in the mine. This form of debt peonage became entrenched so that by the 1760s miners were receiving three-fourths of a peso per day for the heavy work of a *barretero* (who used a large iron pry bar to dislodge ore). In addition he received two rations of water and two pounds of *pallacos* (small bits of ore). *Apiris*

Fig. 8. Administration building and assay laboratory of Cía. Minera Fabian Ltda. Their installations have been erected on the bulldozed foundations of the colonial settlement shown in fig. 7.

who carried the ore out of the mine in leather bags made a half peso per day, plus water as well as *pallacos*. But the wages went to pay off the previously contracted debt, so they survived by trading their *pallacos* for food, coca leaves, and other necessities. The records suggest many workers contracted simultaneous debts with several mine owners who then had great difficulty in collecting the labor owed them. This suggests workers were scarce at Huantajaya and that the operators were unable to coerce labor to any great extent. We see an extension of this situation in the fact that wages rose to a peso a day for *barreteros* in the 1790s and three-fourths of a peso for *apiris*.[12] These wages were considerably higher than at other regional mines at this time (Brown 1986, 65).

In addition to debt peonage, some workers established informal partnerships with individual mine owners and collected the *pallacos* from abandoned stopes. These *pallaquiris* or *zanapacos* shared with the owner whatever they found (Barriga 1941, 110). In the early years these *pallaquiris* produced substantial amounts of silver from ores the owners disdained. As high-graders, however, they did not have the means for processing their *pallacos* and consequently sold them to dealers (*rescatores*), merchants, or other mine owners who became involved in refining. The number of *pallaquiris* varied with the season and with fluctuations in mining activity. In the 1760s there were about 200 such workers out of a total work force of 350 (Bermúdez 1975, 34).

The typical work day began before daylight as miners assembled and were taken to the workings by the owner or his assistant in order to start at sunrise. Wooden doors were installed at tunnel entrances and locked at night to prevent theft of silver ore. As workers filed past these doors, *porteros* (gatekeepers) recorded their names. Miners then went to the working face where there was little or no supervision and stayed there until 11:00 a.m., when they came out for lunch. At the doors they were searched by the *porteros* to make sure they were not stealing ore. After group prayers before a crucifix, they received a ration of water and a pound of *pallacos* and then had lunch. For those whose wives were not at the mine to cook for them, peddlers offered food for sale. About 1:00 p.m. work began again and lasted into mid-afternoon when there was another distribution of water and *pallacos*, followed by prayer. On Sundays the miners received religious instruction, attended mass, and then the operators distributed weekly allotments of dried meat, maize, and coca leaves. In the afternoon, a few miners cleaned and picked over the accumulated ore that was sent to Tarapacá, Pica, or Guarasina for small-scale beneficiation and artisanal refining.

It would have been far less costly to process the ore immediately at Huantajaya, but the extreme aridity of the site made this impractical. Ore had to be crushed by water-driven grinding or stamp mills (Craig 1993) before smelting which required great amounts of wood or charcoal for fuel. Refining by mercury amalgamation involved considerable preparatory washing. Since wood, charcoal, and water were unavailable at Huantajaya, it was necessary to import such items or take the ore elsewhere. The owners clearly found it cheaper to haul the ore to more practical surroundings. In at least one instance, Bartolomé Loayza seriously considered a proposal to ship his ore all the way to Lima for processing.[13] Under these circumstances, whatever option was chosen, operating costs were extremely high. Mine owners could only process very rich ore and abandoned any that yielded less than fifteen *marcos* of silver per *cajón* (7.5 pounds of silver per 2.3 metric tons), which would have been considered valuable ore at Potosí (Cañete y Domínguez 1952, 28). At Huantajaya mine owners calculated they could not profitably process ore yielding less than 25 *marcos* per *cajón* whereas some Peruvian mines were operating at this time with as little as four *marcos* per *cajón*. In 1807, tailings at Huantajaya were hand sorted and reportedly averaged thirteen *marcos* per *cajón*, for a total in place estimated at 512,000 *cajónes*.[14]

Mining Ventures

Given Tarapacá's desolate location, the inhabitants had little capital to invest in the mines at Huantajaya. Operators there consequently had to rely on their own resources or seek investments from outside the region. This was true of even the most important mine owners. For example, in 1747 Bartolomé de Loayza formed a joint venture company with three investors from Lima. His partners were Pedro Gutiérrez Cosio, Felipe Gonzáles Cosio, and Captain Francisco Montestruque. They agreed to supply the necessary capital, with Loayza and Gonzáles Cosio sharing management of the mines. The contract was to last twelve years but within three years the Limeños accused Loayza of mismanagement. Now the mines were seized by government decree and handed over to González Cosio while Loayza and his family were banished from Huantajaya in 1751. Suddenly, all three Lima investors died within a five-week period from December 1752 to January 1753, and Loayza regained control of his mines. He then successfully dissolved the company so that heirs of the deceased partners could not claim interest in the properties.[15]

However, by this time the real power at Huantajaya was in the hands of José Basilio de la Fuente to whom Loayza owed 137,000 pesos. Fortunately for Loayza, José Basilio was his son-in-law. By 1772 de la Fuente was the only active mine operator in Huantajaya and owned nearly all the claims (Villalobos 1979). His death two years later marked the end of an era.[16] Later, a son of de la Fuente's named Francisco, whose children were too young to carry on operations, continued mining, but with little prosperity. Francisco reported that between 1803 and 1812, his mines generated 160,311 pesos, but costs reduced the net profit to only 26,222 pesos (Wibel 1975, 158). After independence, the family struggled on, trying to work the mines through a large joint-stock company (Wibel 1975, 375).

Royal Administration of Huantajaya

The rich silver mines at Huantajaya remained outside viceregal control until the mid-1700s, when competition between the royal *cajas* (treasuries) of Carangas, Potosí, and Lima began to impinge upon Huantajaya's independence. Prior to 1750, Huantajaya was theoretically subject to the *caja* of Carangas located some 300 miles away over the most wild and rugged terrain imaginable.[17] From this distance, treasury officials were unable to exercise firm control and venality infected what little contact existed.[18] Mine owners and silver merchants registered very little Huantajaya output with the Carangas *caja*. The lack of official complaint implies Carangas officials were probably being paid off by the owners in return for a relatively free hand at the mines.[19]

All this came to an end in the 1750s, when Ventura de Santelices y Venero became *corregidor* of Potosí and pressured the Huantajaya operators to register their silver there.[20] He evidently hoped increased mining revenues at the Potosí *caja* would create the impression his administration had breathed new life into that mining center. Surprisingly, the Huantajaya refiners accepted this plan, although their cooperation must have been laced with more than a little self-interest. Meanwhile, the royal officials at Potosí, which was even more distant that Carangas, had little hope of exerting effective control over Huantajaya. The operators there recognized this fact and knew they would only have to register as much silver as they pleased. Still, it was probably easier to transport silver to Potosí than to Carangas because it could then be sent by ship to Arica and from there by pack animals along a well established route to Potosí. Silver dispatched to Potosí could be exchanged for an equivalent amount of coins produced there at the mint, whereas at Carangas, one of the fiscal backwaters of the empire, sufficient coins were seldom available. Potosí officials also reportedly gave a better price for silver than at Carangas, whose officials allegedly cheated the Huantajaya mine owners by claiming their silver was lower grade than it actually was.

The Carangas treasury officials viewed Santelices as a kind of bureaucratic claim jumper. They immediately complained to the viceroy in Lima and accused the Huantajaya refiners of smuggling. The treasurer and comptroller at Carangas were desperate to hold on to the little income and jurisdiction their *caja* possessed and offered to take turns visiting Huantajaya to register its bullion. In 1756 Santelices in Potosí backed down and offered to provide the Carangas *caja* with 12,000 pesos to buy Huantajaya silver. By this time the obdurate Carangas officials and the rebellious mine owners refused to compromise; the officials confiscated some silver bars and initiated criminal suits against alleged smugglers.[21]

Meanwhile, reports of these controversies came to the attention of Spanish ministers with a result that on September 2, 1761, Charles III ordered an investigation[22] and the viceroy sent Col. O'Brien to Huantajaya where he compiled the first detailed report on geology and mining operations and the map to which we have referred earlier (endnote 6).

The province of Tarapacá was created from the southern part of Arica, both of which belonged to the viceroyalty of Peru. However, Carangas lay within the boundaries of the new viceroyalty of Río de la Plata, created in 1776. A dispute over Huantajaya soon developed between Lima and Buenos Aires, both of which wanted to claim its silver for taxation. Meanwhile, the mine owners registered their silver anywhere they wished, if, indeed, they did so at all. But Lima finally prevailed and established a royal assay office in the easily accessible *caja* located at Tacna in 1780.

Silver Production

Lax fiscal control and widespread corruption make it impossible to know exactly how much silver was produced at Huantajaya. In 1826 it was calculated that over the preceding one hundred years it had yielded an average of 750,000 pesos of silver per year (Bollaert 1851, 108). If so, the mines must have been tremendously productive around mid-century. Oscar Bermúdez (1975, 33) speculates Huantajaya may have experienced its "most important phase" during the 1760s, but that seems unlikely since it reportedly produced 340,000 to 425,000 pesos of silver each year during that decade.[23] But the narrow veins and complex faulting subjected the district to an unusual number of booms and busts. By 1777 production had climbed to about 500,000 pesos per year.[24]

Analysis of the treasury accounts from Carangas and Arica from 1770 to 1819 gives some idea of the size of Huantajaya productions as well as the level of smuggling before 1780 (table 1). In the decade before establishment of the assay office at Tacna, Carangas registered an average of 182,696 pesos annually (table 1). This included silver from nearby mines as well. But when Carangas finally lost the right to tax Huantajaya silver in 1780, its registry declined to an average of 116,511 pesos. In other words, Huantajaya was only accounting for one third of the silver taxed by Carangas (some 65,000 pesos per year). Yet when Arica began taxing the output (table 2), the average for the 1780s was 284,424 pesos. Judging from the Arica tax accounts, Huantajaya production held at essentially the same level from 1780 to 1799, at approximately 280,000 pesos per year. During the period 1800-1809, production dropped by nearly 50 percent, to 155,000 pesos, and in the following decade to only 55,000 pesos per year. Therefore, if Huantajaya did in fact average 750,000 pesos per year between 1726 and 1826, its most productive period was evidently the first half of that period.

The Mercury Problem

Aside from ore quality and the arid environment, mercury was probably the factor that most influenced silver output at Huantajaya. This was true even though mine owners continued to smelt significant portions of their ore. From 1782 to 1784, for example, the viceroyalty of Peru used 8,312 *quintales* of mercury to refine 884,601 *marcos* of silver for a ratio of 106.4 *marcos* per *quintal*. This coincided with the general assumption that a refiner ought to produce about 100 *marcos* for each *quintal* of quicksilver used. Yet at Huantajaya during those years, the ratio was 483 *marcos*, nearly five times the viceregal average. While this reflected the generally richer ore, it also meant the Tarapacá refiners continued to smelt part of the mines' output. Since they consumed no mercury in the process, the *correspondencia*

Table 1. Value of silver registered at Carangas Caja, 1718-1790 (in pesos de ocho)

Year	Silver	Year	Silver
1718-19*	53,812		
1719-20	40,612	1757-58	186,722
1720-21	35,413	1758-59	143,198
1721-22	38,969	1759-60	187,850
1722-23	37,070		
1723-24	43,398	1764-65	125,908
1724-25	63,400	1765-66	136,020
1725-26	57,174	1766-67	145,776
1726-27	53,640	1767-68	99,868
1727-28	64,770	1768-69	99,645
1728-29	65,144	5-12/	137,675
1729-30	64,871	1770	97,645
1730-31	64,490	1771	66,650
1731-32	48,138	1772	202,162
1732-33	38,215	1773	170,974
		1774	201,289
1735-36	84,080	1775	324,730
1736-37	139,036	1776	280,802
		1777	256,916
1743-44	84,264	1778	210,132
		1779	15,656
1745-46	54,000		
1746-47	88,449	1784	154,793
1747-48	71,480	1785	113,260
		1786	54,925
1749-50	52,546	1787	92,670
1750-51	57,515	1788	46,934
1751-52	57,515	1789	157,674
		1790-	195,322
1754-55	75,524		

*Prior to 1770, most fiscal years for the *caja* ran from May 1 to April 30.

Source: Calculated from the fiscal data contained in TePaske and Klein (1982, 2:29-42).

Table 2. Silver registered at the Arica Caja, 1780-1819 (in *pesos de ocho*)

Year	Silver	Year	Silver
1780	762,837	1800	269,604
		1801	182,608
1782	237,463		
1783	113,128	1803	122,819
1784	168,520	1804	133,912
1785	200,070	1805	153,789
1786	306,159	1806	130,590
1787	243,583	1807	136,511
1788	274,862	1808	95,471
1789	253,197	1809	81,797
1790	149,374		
		1811	90,687
1792	249,060	1812	140,000
1793	234,264		
1794	276,326	1815	19,868
1795	239,207		
1796	305,251	1817	8,846
1797	330,925	1818	20,855
1798	501,789	1819	46,912

Source: TePaske and Klein (1982, 2:8-24).

(ratio) became inflated. But lower grade ores, of which there was an increasing amount, required amalgamation with mercury.

After the Tacna *caja* acquired fiscal responsibility for Huantajaya in 1780, it also supplied the miners with mercury. In Table 3 we see the amount used fluctuated significantly. In part, this was due to the vagaries of mining, but also it was caused by the uncertain nature of the mercury supply. For example, in 1803 the Tacna *caja* had no mercury at all to sell because the supplies from Spain had been cut off by war. Although mercury arrived in 1802 and 1803, Tacna did not receive a new supply until 1804. Meanwhile, Huantajaya had to smelt or do without. Other shortfalls are known for brief periods in 1785, 1786 and 1788. This scarcity of mercury was a chronic source of anxiety and seriously hindered production.

Table 3. Mercury purchased from and marks of silver registerd at Arica Caja, 1780-1809

Year	Quintales of Mercury	Marcos	Year	Quintales of Mercury	Marcos
1780	98	88,352	1795	240	27,768
1781	?	23,793	1796	244	35,432
1782	35	22,773	1797	250	38,456
1783	30	13,241	1798	252	60,136
1784	50	19,556	1799	268	31,408
1785	227	23,261	1800	247	?
1786	250	35,591	1801	206	21,152
1787	297	28,312	1802	49	26,932
1788	353	31,947	1803	0	14,233
1789	355	29,429	1804	96	15,518
1790	41	32,175	1805	93	17,812
1791	53	24,553	1806	181	8,519
1792	98	28,120	1807	127	4,427
1793	171	27,256	1808	175	11,070
1794	176	32,952	1809	86	9,474

Sources: AGI, Charcas 1301A, 1301B; AGI, Lima 632-634; Escobedo to Sonora, 5 February 1787, no., 22, AGI, Lima 1340; TePaske and Klein (1982, 2:8-21).

For the period 1780-1809, the correlation between the amount of mercury sold by the Arica *caja* and the *marcos* of silver registered the following year is 0.643 (P=.007). This means that the mercury sold the previous year statistically explains 41 percent of the variance in the quantity of silver produced.[25] This surprisingly strong statistical influence is noteworthy.

Final Decline of Huantajaya

Despite several private and governmental attempts to solve production problems, by the early 1800s Huantajaya was clearly in distress. Lack of capital was a perennial problem among mine operators, and it was to provide working funds that a *Sociedad Mineralógica* was formed in Arequipa in April 1792. Essentially a joint-stock company, the Society acquired capital through the sale of shares on the pledge that monies would be invested in mining. Although the amount sought was 50,000 pesos, by late 1794 only about 22,000 had been raised which the directors decided to invest at

Huantajaya where relatively small amounts could sometimes produce good returns. In this way, an operator could hire a few workers to search for a displaced vein and then expand the work force if it was found (Brown 1986, 159-160). Unfortunately, we find no evidence that the Society's efforts enjoyed any special success at Huantajaya. As mining production dwindled in the early 1800s, acquiring new capital became an increasing problem.

Lack of coins to pay the workers was another difficulty. Miners often had to accept their wages in *pallacos* because the operators had little or no cash.[26] Officials had long recognized that Huantajaya needed a *banco de rescate* with sufficient coins to purchase silver in ingot form from the mine owners.[27] Only in 1792 was such a service authorized and in the following year the bank opened in Tarapacá with an operating fund of 50,000 pesos (even though the Mining Tribunal in Lima had recommended beginning with a reserve of 200,000 pesos). The continued exchange operations of this bank required replenishment of specie. By 1794 the officials were complaining that the Arequipa treasury did not have sufficient coins to redeem the bars of silver which the Tarapacá bank had purchased (Brown 1986, 160-161). The bank seems to have functioned only fitfully thereafter.

The general decline was exacerbated by problems in acquiring a steady supply of mercury. Wars in Europe interrupted the major shipments that came from Spanish mines at Almadén. Peruvian mercury from the mines at Huancavelica was reduced to a trickle by mismanagement and the struggles associated with the wars of independence.

Like mercury, gunpowder was another government monopoly which affected mining operations. Blasting greatly accelerated ore production over the traditional use of pry bars, particularly when veins were contained in hard rock that was not easily fractured. Tarapacá had unlimited quantities of saltpeter, sulphur was available from nearby volcanoes, and good charcoal could be made from the mesquite-like *tamarugo* trees. As a result, mine operators claimed they could manufacture their own blasting powder for one *real* per pound, whereas the government monopoly sold it at three and one-half to four *reales*.[28]

The greatest difficulty at Huantajaya continued to be the high cost of processing ore as compared to other Andean mines. Everyone searched for ways to cut production costs by overcoming the desert's handicaps. Despite Huantajaya's isolation, operators there were well informed about the attempts by the Crown and the Mining Tribunal of Peru to introduce new technology.[29] For example, in 1788 one of the mine owners appealed to the viceroy for permission to have some ore samples analyzed by José Coquet, a metallurgist whom the Crown had sent to Peru from Mexico. When the Nordenflicht expedition arrived in Potosí to introduce the barrel system of amalgamation and otherwise upgrade Peruvian mining and beneficiation techniques, Huantajaya mine owners took large samples of their ore to Potosí for analysis. They came away disappointed, however, because the higher costs of the barrel system made the method unprofitable. Another aborted scheme involved shipping ore to Lima for analysis by Nordenflicht in the laboratory he intended to establish.[30] In 1793 permission was requested to ship 1,200 pounds of selected tailings to Spain where it was hoped experts would devise a means of profitably extracting silver from these lower grade ores.[31]

The final colonial effort to overcome the lack of water and high costs of production at Huantajaya occurred in 1807, when Matias Gonzáles Cosio, a mine owner, proposed the construction of a canal to bring water down out of the Andes. Viceroy Abascal named an engineer to conduct a technical study and in that same year the project was declared impractical. There was too little water available in the highlands to be channeled westward across the desert to Huantajaya. A careful survey and mapping showed aqueducts would be needed to carry water across various depressions, including one 300 yards long and 60 yards high. Costs of the system were calculated at 754,780 pesos with a possibility of recovering 160,000 pesos from the sale of irrigation water, but this still would have left a deficit of nearly 600,000 pesos.[32]

Operations at Huantajaya struggled on, but the mines never recovered their former splendor. In the years immediately after independence various attempts were made to utilize the accumulated colonial tailings (see fig. 9 for a modern example), but none was successful. When the deep underground workings were abandoned, brine seeped in during the nineteenth century, until today all the lower levels have become filled with corrosive mine waters. It is ironic that the very element that was lacking for so long when rich ore was present, is now abundant while the mines are exhausted.

Conclusion

This multidisciplinary study of a Spanish colonial silver mining center provides a detailed reconstruction of the effects the geographic factors of site and situation had on development of an isolated mining community. The human response to extraordinarily harsh conditions at Huantajaya is surprising. Of various alternatives available to them, mine owners often chose a course of action that seems difficult and unpredictable viewed by today's standards.

Social aspects of life at Huantajaya are also surprising. At no time in the long history of this community do we find miners engaged in the debauchery that prevailed at nearby Potosí. Relationships between owners and workers were unusually egalitarian for the times. Even so, much later under foreign ownership, workers and their wives were buried apart from their overseers (fig. 10).

Fig. 9. Lixivation vats (capacity 500 tons/day) of operators processing relatively low grade colonial tailings. With silver worth $1.47/oz. in 1990, operations were shut down.

We conclude that a greater degree of cooperative effort, ingenuity, and overall tenacity were requirements for survival in the extremes of the Atacama Desert. If nature had endowed Huantajaya with ample water, it is conceivable the richness of its ores would have gained it far greater fame, and profoundly changed the political economy of this isolated corner of the viceroyalty.

Endnotes

1. In many parts of the central Atacama Desert vegetation is totally absent and it is accurate to say that most of the biomass consists of buried wind-blown dormant seeds and invisible bacteria. In 1835 when Charles Darwin passed within sight of the ruins of Huantajaya, he remarked that the only vegetation to be seen was yellow lichens growing on the bleached bones of dead animals.

2. On a recent site visit in 1992 some very compelling evidence for prehistoric mining activity was obtained by Craig from mine guards who had just discovered two skeletons of aboriginal miners in a collapsed section (fig. 11) of the earliest workings. A hafted stone hammer (see fig. 4), small wooden wedges, and a guanaco skin ore bag accompanied the remains. The hammer has been donated to the Museo de Arte Pre-Colombino in Santiago. Since the remains were found near the western zone of copper mineralization and removed from context before my arrival, it is not possible to say with certainty whether silver or copper was the ore being extracted.

Fig. 10. Remains of the worker's cemetery established by the English mine owners ca. 1900.

3. Juan Valdivia (1847, 175): "The geographers find themselves divided about the time in which these mines were discovered." See also AGI, Lima 1357, Jorge Escobedo's opinion attached to Viceroy Abascal to Minister of Hacienda, January 23, 1807.

4. AGI, Lima 1530, Relación de las minas de oro y plata que ahora se entiende ay descubiertas en este reino del Perú (no date). Since conspicuous dike-like surface veins crisscross the western side of Huantajaya that faces the shoreline and the great escarpment rising up from the sea (see fig. 2), any European sailing along this coast looking for mines would be tempted to investigate, just as did the prehistoric prospectors mentioned in note 2.

5. AGI, Lima 93, Alonzo Muñoz To Felipe II, 1572.

6. The original map is held by the Archivo Nacional de Chile (Mapoteca No. 485 "Mineral de San Augustín de Guantajaya, Arica." dated 1766). It contains extensive notes in the "Explicación."

7. AGI, Lima 1353, Report by Jorge Escobedo attached to "Expediente sobre la representación de Pedro Lafite," 1793.

8. AGI, Lima 1351, The pertinent correspondence and reports are contained in "Expediente sobre el descubrimiento de una mina de plata en el Mineral de Santa Rosa, del partido de Tarapacá," 1786-1789.

9. AGI, Lima 1351, Capitulo 41, Antonio O'Brien, "Mineral de Guantajaya: descripcion de sus minas y de las demas del Partido de Tarapacá," 1765.

10. Biblioteca Nacional del Perú, F55, "Matrícula general de los mineros de este virreinato del Perú," 1790; AGI, Lima 659, "Estado general y actual del real y importante Cuerpo de Minería del Perú," 1799.

11. The Count of Superunda (Viceroy of Peru 1745-1761) did grant two *mitas* of 50 Indians each to the principal operators at Huantajaya, perhaps as a result of the rich discoveries

Fig. 11. View of the *hundamiento* (collapse feature) caused by uncontrolled removal of ore-bearing pillars from many old stopes and galleries. A small, possibly prehistoric, tunnel has been exposed at the upper-left corner while below to the right at the deepest part of the collapse feature is the opening in which the prehistoric miners were discovered. A conspicuous fault seen on the right has caused typical downthrown displacement of the ore body.

made there, but the number of workers was insignificant compared to the annual *mita* at Potosí (see Bakewell 1983 for details).

12. Barriga 1941, Memorias I, 110, n. 6.

13. AGI, Lima 427, Cap. 11, "Expediente sobre una compañía de minas del cerro de Huantajaya," 1756.

14. AGI, Lima 1357, Viceroy Abascal to Minister of Hacienda, January 23, 1807.

15. Note 13, *passim*.

16. José Basilio de la Fuente's death is generally given as 1774 (Villalobos 1979, 134), but his son dated it in 1775 in AGI, Lima 1620, "Oja de servicios de Francisco de la Fuente," 1798.

17. For a vivid description see "Informe del Contador General de Indias Tomás Ortíz de Landázuri," AGI Lima 1351, June 22, 1769.

18. Treasury officials at Potosí had so little confidence in their counterparts at Carangas that they feared for the security of Crown shipments of mercury dispatched there. See Ventura de Santelices to the Marqués de la Enseñada, AGI Lima 643, October 9, 1753.

19. Antonio de Ulloa reported that an inspector sent to the Carangas Caja had discovered treasury officials there smelted silver bars presented by refiners without collecting any royal tax at all. The implication is that these officials did so for a share in the illicit profits. Memorial of Antonio de Ulloa to Crown (Biblioteca Nacional, Madrid, MS 19568, September 14, 1771).

20. For details of Santelices' long career at Potosí and his concern about Carangas and Huantajaya, see Enrique Tandeter (1980, 34-47).

21. AGI, Lima 868, Informe de los oficiales reales de Carangas, January 1, 1768.

22. AGI, Lima 1351, Informe del Contador General de Indias Tomás Ortíz de Landázuri, July 23, 1766.

23. AGI, Lima 1351, Landázuri, and ANP Derecho Indígena 14, Cuaderno 335, folio 5.

24. AGI, Buenos Aires 59, folio 12, Viceroy of Buenos Aires to Gálvez, No. 101, February 5, 1779.

25. r = .0643; therefore r = .413, with a proportionate reduction in error estimate of 41.3 percent.

26. AGI, Lima 1353, Escobedo to Gardoqui, March 6, 1793, but attached to Pedro Lafite to Crown, February 26, 1793.

27. For example, see Guirior to Gálvez, No. 506, AGI Lima 659, October 5, 1779.

28. Mining Tribunal of Peru to Crown, No. 6 (AGI, Lima 1354, April 15, 1794) and "Expediente sobre el laboratorio químico metalúrgico mandado hacer bajo de la dirección del Barón Nordenflicht," (AGI, Lima 703, 1790-1792).

29. De la Fuente to Croix (AGI, Lima 1446, May 31, 1788); also see Fisher, 1977, especially chapters 2-4.

30. Representation of the Mining Deputation of Huantajaya, December 24, 1791, in "Expediente", note 28 above.

31. AGI, Lima 1353, Lafite to Crown, February 26, 1793.

32. AGI, Lima 625, Expediente concerning Francisco Mendizábal, Sarjento-Major, Brigada de Ingenieros, 1808.

References

Archival sources

AGI — Archivo General De Indias [Sevilla]

Lima	93	(1572)	1340	(1787)
	427	(1756)	1352	(1765-1780)
	625	(1808)	1353	(1793)
	632-624		1344	(1794)
	643	(1753)	1357	(1807-1809)
	659	(1779)	1446	(1788)
	703	(1790-92)	1530	(n.d.)
	868	(1768)	1620	(1798)

Buenos Aires 59 (1779)

Charcas 1301A, 1301B

Mapas y Planos, Py Chile 43

ANC — Archivo Nacional de Chile, Santiago
Mapateca no. 485 (1776)

ANP — Archivo Nacional de Perú, Lima

Biblioteca Nacional, Madrid. MS 19568 (1771)

Published and Other Materials

Bakewell, P. 1983. *Miners of the Red Mountain: Indian labor in Potosí, 1545-1650.* Albuquerque: University of New Mexico Press.

Barriga, V. M., ed. 1941-1952. *Memorias para la historia de Arequipa.* 4 vols., Arequipa: Editorial Portugal.

Bermúdez, O. 1975. *Estudios de Antonio O'Brien sobre Tarapacá: Cartografía y labores administrativas, 1763-1771.* Antofagasta: Universidad del Norte.

Bollaert, W. 1851. Observations on the geography of southern Peru, including a survey of the Province of Tarapacá, and route to Chile by the coast of the desert of Atacama. *Journal of the Royal Geographical Society of London* 21:99-130.

Brown, K. W. 1986. *Bourbons and brandy: Imperial reform in eighteenth century Arequipa.* Albuquerque: University of New Mexico Press.

Cañete y Domínguez, P. V. [1787] 1952. *Guía histórica, geográfica, física, política, civil, y legal del gobierno e intendencia de la Provincia de Potosí.* Potosí: Editorial "Potosí."

Craig, A. K. 1989. Mining ordenanzas and silver production at Potosí: The Toledo reforms. In *Precious metal, coinage, and the changes of monetary structure in Latin-America, Europe, and Asia: 1500-1800,* Eddy Van Cauwenberghe, ed. Proceedings of the 1985 Stockton Conference on Monetary History. (Also in *Studies in Social and Economic History* 2:159-183.) Leuven, Belgium: Leuven University Press.

—. 1993. The ingenious *ingenios*: Spanish colonial watermills at Potosí. In *Culture, form, and place: Essays in cultural and historical geography,* Kent Mathewson, ed., 125-156. Geoscience and Man vol. 32. Baton Rouge, LA: Geoscience Publications, Department of Geography and Anthropology, Louisiana State University.

Echeverría, F. X. 1952. Memoria de la Santa Iglesia de Arequipa. In *Memorias para la historia de Arequipa,* vol. 4, V. M. Barriga, ed., 1-261. Arequipa: Imprenta Portugal.

Fisher, J. 1977. *Minas y mineros en el Perú Colonial, 1776-1824.* Lima: Instituto de Estudios Peruanos.

Mendiburu, M. de. 1874-1890. *Diccionario histórico biográfico del Perú.* 8 vols. Lima: Imprenta de Torres Aquirre.

Novoa, A. T. 1970. *Carta geológica de Chile: Cuadrángulos Iquique y Caleta Molle, Provincia de Tarapacá.* Escala 1: 50,000. Santiago: Instituto de Investigaciónes Geológicas.

Officer, H. G. 1922. Geology of the Huantajaya Silver District. *Engineering and Mining Journal-Press* 113:568-572.

Paskoff R. 1978. Sur l'évolution géomorphologique du gran escarpement côtier de désert chilien. *Géographie Physique et Quaternaire* 32(4): 351-360.

Paz Soldan, M. M. 1980. *La moneda colonial en el Perú: Capítulos de su historia.* Lima: Banco Central de Reserva del Perú.

Pizarro, P. [1571] 1969. *Relation of the discovery and conquest of the kingdoms of Peru.* New York: Kraus Reprint.

Tandeter, E. 1980. La rente comme rapport de production et comme rapport de distribution: Le cas de l'industrie minière de Potosí 1750-1826. Thèse de 3e cycle en historie, École Practique des Haute Études, Paris.

—. 1993. *Coacción y mercado: La minería de la plata en el Potosí colonial 1629-1826.* Archivos de Historia Andina 15. Cusco: Centro de Estudios Regionales Andinos.

TePaske, J. J. and Klein, H. S. 1982. *The Royal Treasuries of the Spanish empire in America.* 3 vols. Durham: Duke University Press.

Trelles Arestegui, E. 1982. *Lucas Martínez Vegazo: Funcionamiento de una encomienda peruana inicial.* Lima: Pontífica Universidad Católica del Perú.

Ureta y Peralta, P. 1792. Descripción de la ciudad de Arica y su vasta jurisdicción correspondiente a la Intendencia de Arequipa en el Perú. *Mercurio Peruano* 6:191.

Valdivia, J. 1847. *Fragmentos para la historia de Arequipa.* Arequipa: Imprenta de Mariano H. Moreno & Cía.

Villalobos R. S. 1979. *La economía de un desierto: Tarapacá durante la colonia.* Santiago: Ediciones Nueva Universidad.

Wibel, J. F. 1975. The evolution of a regional community within Spain, empire, and Peruvian nation: Arequipa, 1780-1845. Ph.D. diss., Stanford University.

Wormald, A. 1968. *Frontera Norte.* Santiago: Editorial Orbe.

Attempted Technological Innovation in the Late Colonial Peruvian Mining Industry, 1776-1824

J. R. Fisher, University of Liverpool
Liverpool, England

Abstract

The Spanish Crown sent European scientists to modernize the technology of the mining industry in Peru during the period between the transfer of Upper Peru to the viceroyalty of the Río de la Plata and the establishment of Peruvian independence. In 1776 the viceroyalty of Peru lost control of 63 percent of its silver output, as Potosí and Oruro were transferred to the control of bureaucrats in Buenos Aires, who made strenuous efforts to encourage the flow of Upper Peruvian silver away from Lima towards the Atlantic ports of the new viceroyalty. In the same year the commissioning by the Crown of a general inspection of Peru inaugurated a prolonged period of administrative reorganization which laid the foundations for the establishment in 1787 of a Mining Guild and Mining Tribunal, and the dispatch to Peru in 1788 of a team of German mining specialists, led by the Swedish scientist Thaddeus von Nordenflicht, which was entrusted with the modernization of the process employed for the refining of silver ore.

In 1777 registered silver production in the truncated viceroyalty of Peru stood at 2.1 million pesos; by 1799 it had increased by 250 percent to 5.4 million. It fell back slightly in 1800-1809 to an average of 5 million, and more severely in 1810-1819 to an average of 3.2 million. Its decline was arrested in 1820, when production reached 4 million pesos, as a result of the installation of steam-driven pumping engines at the viceroyalty's most important mining center, Cerro de Pasco. Although this recovery was short-lived, it confirmed that local entrepreneurs were capable of financing successful technological improvements, notwithstanding political instability. By contrast, there was little relationship between technological innovations promoted by the Crown and the increase in output in the late colonial period.
Key words: Cerro de Pasco; mercury; Potosí; silver; mining technology; Trevithick, Richard.

In Quest of Mineral Wealth: Aboriginal and Colonial Mining and Metallurgy in Spanish America, edited by Alan K. Craig and Robert C. West, 1994. Geoscience and Man, vol. 33, pp. 329-342. Dept. of Geography & Anthropology, Louisiana State University, Baton Rouge, LA 70893-6010, U.S.A.

Introduction : Mining and the Late Colonial Economy of Peru

Notwithstanding the efforts of the Bourbon monarchy to diversify the economic and commercial structures of the Hispanic world in the eighteenth century, the production and export to Spain of silver remained the most significant and conspicuous feature of Spanish America's economic life until the collapse of imperialism in the second decade of the nineteenth century. Between 1700 and 1800 the registered output of imperial silver multiplied fourfold from 8.3 million to 33 million pesos, and, despite the rapid economic growth of non-mining regions such as Cuba and Venezuela during the last quarter of the century, shipments of treasure (primarily silver) represented in this period almost 60 percent of the value of American exports to Spain (Fisher 1985, 66-71). In 1700, 53 percent of total silver output had been concentrated in the viceroyalty of Peru, particularly in Upper Peru (modern Bolivia), where legendary Potosí was the principal producer; by 1800, largely as a consequence of developments in central and northern Peru, the combined output of Lower and Upper Peru had increased by 250 percent to 10 million pesos, although primacy had passed to Mexico, which registered an average of 22.9 million pesos a year in the period 1790-1810 (Fisher 1992, 225-235).

Although eclipsed by Mexican output by the end of the colonial period, silver remained the key to the economic structures of Peru, oiling the wheels of overlapping trade networks (international, inter-American, and internal), and determining in considerable measure the development of domestic agriculture and industry (Assadourian 1982). Given this context, the decision of the Crown in 1776 to transfer Upper Peru, which accounted for 63 percent of the silver output of the viceroyalty of Peru, to the newly-created viceroyalty of the Río de la Plata, threatened to undermine the long-established economic and commercial relationships in the region, not least because the authorities in Buenos Aires (capital of the new viceroyalty) sought immediately to curb the export to Lower Peru of unminted silver from Potosí and Oruro in exchange for agricultural goods and manufactures (Céspedes 1947, 120-121). This administrative action confirmed the already-existing, albeit strictly illegal, trend for Peruvian silver to seep overland to the Atlantic ports of Buenos Aires and Montevideo for export to Europe, although until 1778, when the Regulation of Free Trade formally opened these and other ports to direct trade with Spain, all international trade was supposed to be conducted through the Peruvian port of Callao (Villalobos 1965). The pessimism induced among Lima's mercantile community by the loss of overt control of Upper Peru was compounded in the early 1780s by a series of further problems: they included the abolition of the forced distribution of merchandise to Indian communities (the *reparto*), provisionally in 1780 as a response to the indigenous Rebellion of Túpac Amaru and definitively in 1784; rigorous fiscal reform; and the virtual cessation of trade with Europe in 1780-1783 as a consequence of Spain's entry into the War of American Independence (Fisher 1970, 5-28). Even the recovery of international trade in 1785-1787, when goods worth 24 million pesos were landed in Callao, provoked complaints from the merchant guild that the flood of imports, if not curbed, would soon drain the viceroyalty of its circulating capital (Fisher 1970, 131).

These and subsequent representations from the Lima merchant guild (Vargas Ugarte 1956) provided the basis for the long-standing historical myth that the viceroyalty of Peru was the economic casualty of the Bourbon programme of imperial reform in South America (Anna 1979, 3-8). The reality is quite different: the former monopolists in Lima-Callao certainly suffered in the 1780s, as control of trade passed to smaller firms and recently-arrived immigrants from Spain (Brown 1986), but the Peruvian economy as a whole grew substantially from the mid-1780s, and remained relatively buoyant until well into the second decade of the nineteenth century. This is clearly indicated by recent studies of not only the mining industry (Fisher 1976; Deustua 1986) but also international and internal trade (Chocano 1982; Fisher 1988), manufacturing (Salas de Coloma 1986), and agriculture (Haitin 1986). The whole edifice of economic expansion was underpinned by the substantial growth of silver production in Lower Peru, which, as table 1 (Fisher 1976, 124-127; Fisher 1992, 221) shows, doubled its output in the twenty years after the separation of Upper Peru; production in Upper Peru, by contrast, stagnated at or about the 1776 level until the end of the century, and then declined sharply (Buechler 1981, 24,

Table 1. Registered silver production in Lower Peru, 1770-1824 (millions of pesos, by quinquennia)

Quinquennium	Production	Index
1770-1774	10.4	100
1775-1779	11.6	112
1780-1784	13.6	127
1785-1789	14.9	143
1790-1794	20.4	196
1795-1799	23.7	228
1800-1804	22.6	217
1805-1809	19.9	191
1810-1814	15.8	152
1815-1819	11.5	111
1820-1824	6.8	65

31). Indeed, by 1804 Potosí was producing less silver than Cerro de Pasco in central Peru (Fisher 1976, 112). It is impossible, unfortunately, to establish exactly what proportions of total output came from particular mining centers, for the figures to which we have access are derived from the registration process (i.e. the delivery of silver to exchequer offices for assaying, casting into bars and payment of taxes) and are distorted to some degree by the tendency of merchants to take crude silver from a variety of mining camps to Lima for registration. However, it is clear (fig. 1) that by far the most important center was Cerro de Pasco — in 1771-1824, 40 percent of total Peruvian production was registered in the nearby town of Pasco; it was followed by the northern center of Hualgayoc — the nearby Trujillo treasury registered 16 percent; and Huarochirí, which accounted for much but not all of the 21 percent of output registered in Lima. The three remaining treasuries — in Arequipa, Huamanga and Puno — accounted for 12 percent, virtually all the residue (fig. 2). What relationship, if any, was there between this impressive growth of mining output in central and northern Peru in the late colonial period and the attempts of the Spanish Crown and its representatives in the viceroyalty to promote technological innovation?

The Nordenflicht Mining Mission (1788-1810)

In 1788, a thirteen-strong team of European scientists, engineers and artisans landed in Buenos Aires, commissioned by the Spanish Crown to spend ten years in Peru; their primary task would be to instruct the local mining community in modern European methods of refining silver, al-

Fig. 1. A view of Cerro de Pasco in 1827. *Source*: Rivero y Ustáriz 1857.

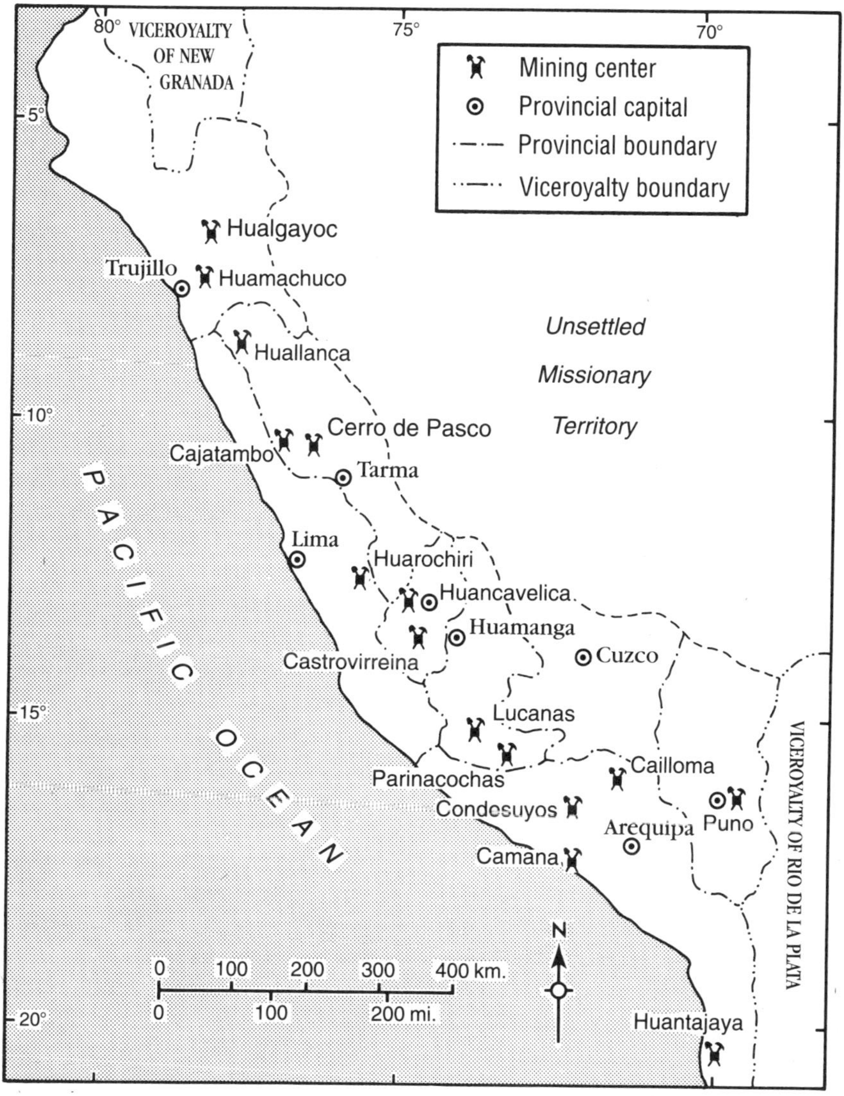

Fig. 2. Major mining centers in late colonial Peru.

though it was hoped that they might also improve the extraction of ore from the primitive Peruvian pits. Only one of the team, Isidro de Abarca, was a Spanish speaker, and he was recalled to Spain for obscure reasons while still in Buenos Aires (Fisher 1976, 56-57); eleven were Germans, recruited mainly in Saxony, and the thirteenth member, entrusted with the direction of the mission was a Swedish scientist, Thaddeus von Nordenflicht, who had been recommended to the Ministry of the Indies by Fausto de Elhuyar, with whom he had studied at the Freiberg metallurgical institute. By 1810, when the Council of Regency formally ended the mission, only four of its members — Nordenflicht, two mining engineers and one artisan — remained in Peru, the others having died, deserted, or secured licences to return to Europe (Fisher 1976, 66-67).

The background to the dispatch of the mission is relatively straightforward. At one level it reflected the culmination of a long process of inspection and reorganization of the mining industry in Peru, which had begun in 1777 with the arrival in Lima of the general inspector (*visitador general*) Antonio de Areche, whose comprehensive instructions included a somewhat vague requirement to stimulate mining; his principal practical measure was a 30 percent reduction in 1779 in the price at which the Crown monopoly supplied mercury to miners, although his attempts to increase production of this essential amalgamating agent at Huancavelica were unsuccessful, with the result that mercury shortage was held to be the principal reason for a fall in silver production in Peru in the early 1780s, a period when, of course, shipments from Spain were suspended because of international hostilities (Fisher 1976, 21). Areche's more able successor, Jorge Escobedo, who served in Lima from 1782 until 1787 as both general inspector (until 1785) and superintendent of the treasury (1784-1787), adopted a more rounded policy. It embraced the establishment of the Mining Tribunal and the Mining Guild in 1787, and the drawing up of plans, shelved indefinitely following his recall to Spain later that year, for the establishment in Lima of a mining college, capable of instructing the viceroyalty's ignorant miners in modern technology (Molina Martínez 1986, 309-337; Fisher 1976, 21-26; Fernández Alonso 1991, 159-204). This same broad aspiration, although with a more precise focus, lay behind the Nordenflicht mission, which grew out of the Crown's decision in 1781 to establish a Royal Metallurgical School in the Basque town of Vergara, Spain. As early as 1783 the Minister of the Indies, José de Gálvez, had commissioned one of its researchers, Juan José de Elhuyar to go to New Granada in an attempt to persuade its miners to abandon the patio process of amalgamation in favour of smelting, and the minister's enthusiasm for innovation in this sphere intensified in 1786 when he received reports from Juan José's brother, Fausto de Elhuyar, of the invention by the Austrian scientist, Inigo von Born, of a new method of amalgamation which apparently combined the benefits of the traditional patio process with modern technology (Montgomery Keelan 1990; Whitaker 1951). Essentially, Born's "new method of amalgamation," as he and many of his contemporaries described it involved no new principles, but its use of efficient, modern machinery had proved capable at Schemnitz (Hungary) of compressing into three or four days a process which traditionally had taken a month or more, and in substantially reducing mercury consumption. His machinery — detailed diagrams of which were sent to Madrid in 1788 by Nordenflicht (fig. 3) — comprised up to ten copper boilers, each capable of holding two hundredweight of ground ore and one hundredweight of mercury, together with smaller quantities of water, pyrites and salt, which were heated as they were rotated by a sliding rake attached to a water wheel (Fisher 1976, 55). Having been instructed by Gálvez to recruit mineralogists willing to go to America to disseminate this innovation, Fausto recommended the appointment of Nordenflicht for the Peruvian mission early in 1788; less than a year later, with their arrival in Potosí on 24 January 1789, the proud Swede and his humourless Saxon colleagues came face-to-face with the difficulties of persuading conservative entrepreneurs to abandon time-tested, if slow and primitive, technology in favour of expensive new machinery.

The process employed at Potosí, as in all other American mining centers, for the refining of all but the highest-grade silver ores (which were smelted), was unchanged since the 1570s, and involved simply the mixing of crushed ore, mercury, and the other ingredients mentioned above in an

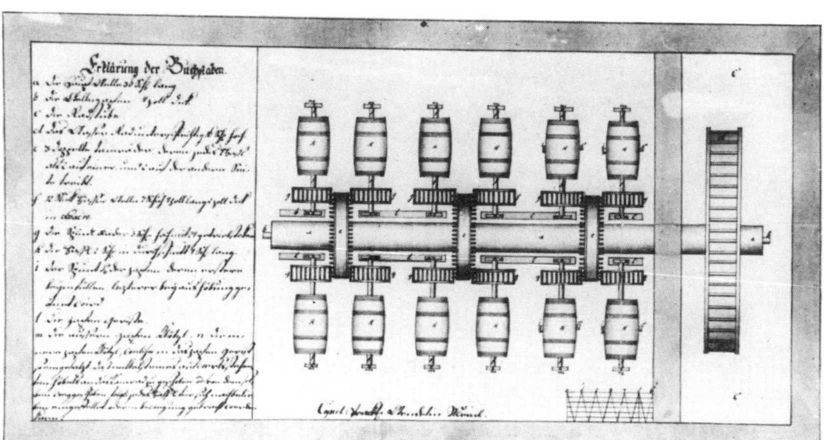

Fig. 3. A 1788 plan, prepared by Nordenflicht, of Born's amalgamation machinery. ***Source*****: AGI, Sevilla.**

open-air patio, followed eventually by the separation of the silver-mercury amalgam by washing, and the recovery from it of some of the mercury by sieving and heating in a condenser (fig. 4). The process was slow, inefficient, and labour-intensive, but it required little investment in equipment or the consumption of fuel, which was both scarce and expensive in most of the mining centers of Peru, which tended to be established at altitudes well in excess of 10,000 feet (Hualgayoc stood at 13,340 feet, and Cerro de Pasco even higher at 14,206 feet). European writers of the eighteenth century, whose romantic engravings of Potosí often showed palm trees and scantily-clad Indians, failed to appreciate the harshness of physical conditions in the treeless high Andes, and their ignorance was probably shared by officials in Madrid.

Nordenflicht's experiences in Potosí need not detain us on this occasion, except insofar as they established a pattern which would be repeated when he moved on in 1790 to Peru. Altogether he constructed four sets of barrel machinery at Potosí, three for individual miners and a fourth for the local mining guild (Buechler 1981, 65-108). An investment in the latter of 40,000 pesos and a series of thirty-one tests, details of which were reported in Lima by the Mercurio Peruano, succeeded in producing silver yields only marginally better than those which would have been expected from the patio process. Although seven members of his team had been in Peru since January 1790, Nordenflicht's first real opportunity to demonstrate the new process there was delayed until 1793, following an inspection of the royal mercury mine at Huancavelica, and the somewhat uncoordinated dispersal of several of his colleagues on a variety of tasks at other mining centers, including Cerro de Pasco and Hualgayoc.

The second set of experiments began at the end of 1793 in a new laboratory in Lima, the construction of which had consumed 35,000 pesos of the mining tribunal's funds, derived from a levy on registered silver. They took the form of parallel tests, using identical consignments of ore, which Nordenflicht and his German assistants refined in their new machinery, alongside Peruvian specialists, recruited by the mining tribunal, who employed the "native method." Amidst increasing acrimony between the rival teams, three tests were run, confirming the significant time saving secured by the barrel machinery and a better silver yield, but producing inconsistent results in terms of mercury consumption; in May 1794, overriding Nordenflicht's protests, the viceroy cancelled a fourth comparative test, and endorsed the mining tribunal's argument that Nordenflicht had failed to demonstrate that the new process could bring about significant improvements in refining (Fisher 1976, 62-63). Whereas the members of the parallel mission to Mexico readily accepted that the Born process was inappropriate for American conditions and forged a new role for themselves in the mining college opened in 1792 by Fausto de Elhuyar, the un-

Fig. 4. A late-eighteenth century depiction of the patio process, in use at Hualgayoc. *Source*: Domínguez Bordona 1936.

fortunate Nordenflicht and his men were left stranded in Lima, virtually forgotten by the Crown. They were literally idle for over two years (1795-1797), much to the exasperation of the mining tribunal which was required to pay their salaries, amounting to 18,000 pesos a year, and, although granted several commissions thereafter — including a further survey of the royal mine at Huancavelica (fig. 5) and the construction there of a new warehouse for storing mercury — made no contribution whatsoever to the rising curve of silver production in Peru in the 1790s.

It might be argued that the presence of the Nordenflicht mission actually hampered the development of Peruvian mining in the 1790s, by diverting for its support funds that might have been used for more direct investment in the industry. This is not mere idle speculation, for it is clear that one of the principal reasons for the rapid growth in output at Cerro de Pasco in this period was the decision of the mining tribunal, taken in 1796, to invest its surplus funds — which grew, of course, as silver registration increased, since they came from a levy of one real on each mark of silver registered — in supporting the digging of a major draining tunnel, the Yanacancha *socavón*, begun in 1794 with local capital. Between 1797 and 1821, when the wars of independence brought activity to a halt, the total invested in this project and in a deeper adit, the Quiulacocha *socavón*, reached a total of 247,000 pesos, much to the chagrin of miners in some of the smaller Peruvian deputations, such as Huamanga and Puno, who complained bitterly of having to pay the levy to support their richer colleagues elsewhere in the viceroyalty (Fisher 1976, 44). In other major centers, such as Hualgayoc, increases in production can be explained

Fig. 5. Huancavelica in 1827. Source: Rivero y Ustáriz 1857.

primarily by local investment in drainage tunnels and, at a different level, by the greater availability of mercury, as a consequence of both the stabilization of production at Huancavelica, which satisfied about 50 percent of viceregal demand, and the shipment from Cádiz of large consignments of European mercury (most of it from the Spanish mine of Almadén), particularly in the periods 1788-1797 and 1802-1805 (Fisher 1976, 80).

Richard Trevithick

If Peru had secured independence from Spain in 1810, one might have concluded without qualification that there was absolutely no relationship between attempted technical innovation and silver production; in fact, the viceroyalty's continued loyalty to Spain until 1824 requires us to modify this suggestion. Flooding of pits was the constant bugbear of Peruvian miners, except at Huantajaya, where a total lack of water created a different but equally grave problem. There were various ways of dealing with flooding (fig. 6), the most important of which was the cutting of a sloping adit beneath the pits in the hope that water would drain into it. Apart from the seemingly interminable disputes between miners that these necessarily collaborative ventures provoked, their principal drawback was that their very success in enabling pits to be deepened meant that they provided only short-term palliatives as increased activity soon brought the miners to the new water levels. This is precisely what occurred at Cerro de Pasco in 1812, as many pits reached the level of the Yanacancha adit, causing an abrupt fall of 68 percent in output at Peru's principal mining center, and a fall in total viceregal production to the lowest level recorded since 1786 (Fisher 1976, 114-116). As noted, the mining tribunal took immediate steps to subsidise the cutting of a deeper adit, but this made slow progress, partly because of basic engineering errors, and remained uncompleted at independence. While it was being started, a group of Lima capitalists took the bold initiative of dispatching an agent to England, along with 30,000 pesos in capital, in the hope of persuading the inventor Richard Trevithick to supply high-pressure steam engines, which would be used to pump water up to the level of existing adits, thereby enabling surrounding shafts to be deepened by a proposed 110 feet. The first consignment of equipment, accompanied by Cornish craftsmen and an engineer, was landed in Callao in 1815, followed two years later by Trevithick himself, who supervised its installation at Cerro de Pasco and solved the fuel problem by suggesting the working of nearby coal seams. By the end of 1819 three engines were working at Cerro de Pasco, and in the following year, 1820, silver registration there soared by 350 percent to 2.7 million pesos, the highest level recorded since 1804, as new drainage pits, from which water was pumped by the new engines, allowed the miners to deepen their pits and obtain ac-

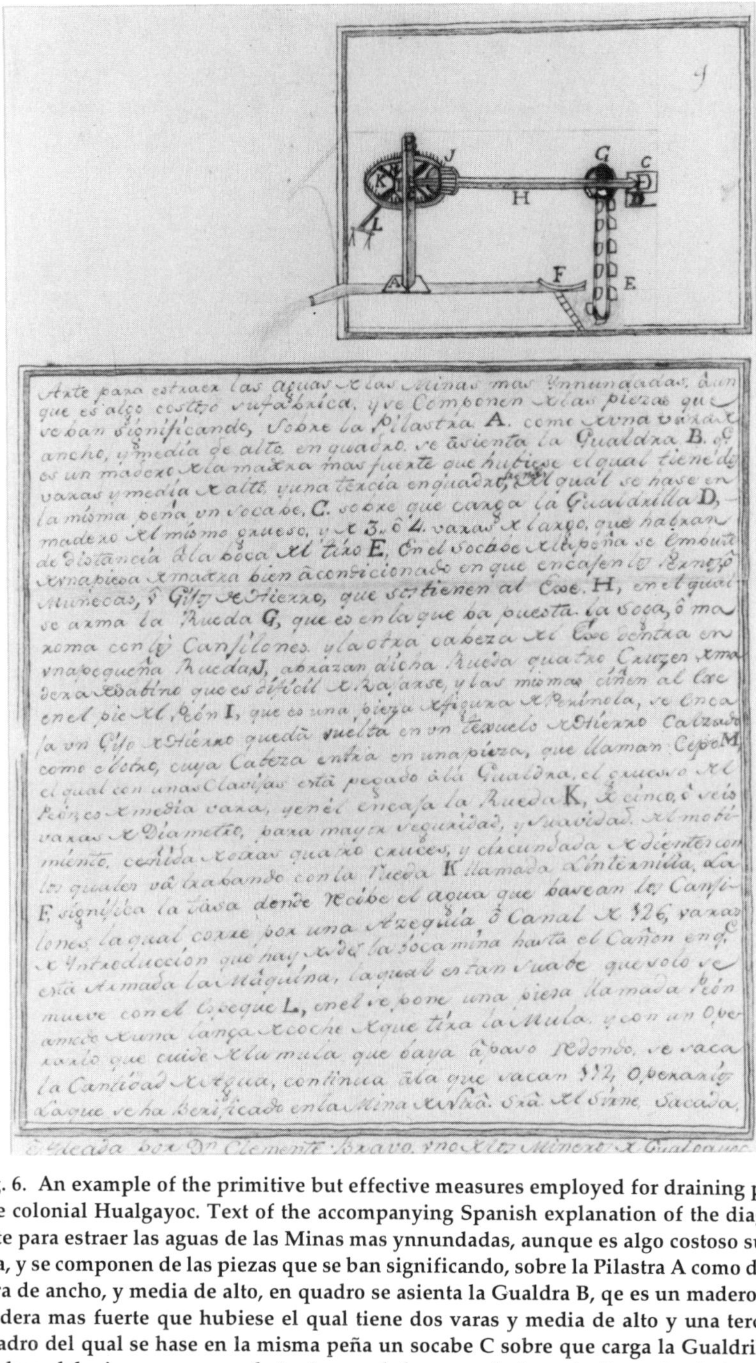

Fig. 6. An example of the primitive but effective measures employed for draining pits at late colonial Hualgayoc. Text of the accompanying Spanish explanation of the diagram: Arte para estraer las aguas de las Minas mas ynnundadas, aunque es algo costoso su fabrica, y se componen de las piezas que se ban significando, sobre la Pilastra A como de una vara de ancho, y media de alto, en quadro se asienta la Gualdra B, qe es un madero de la madera mas fuerte que hubiese el qual tiene dos varas y media de alto y una tercia en quadro del qual se hase en la misma peña un socabe C sobre que carga la Gualdrilla D, madero del mismo grueso y de 3 o 4 varas de largo que habran de distancia a la boca del

(Fig. 6 concluded)
tiro E. En el socabe de la peña se embute de una pieza de madera bien acondicionado en que encajen Los Pernos o Muñecas. o Gijos de Hierro, que sostienen al Exe H. en el qual se arma la Rueda G, que es en la que ba puesta la soga, o maroma con los canjilones y la otra cabeza del Eje dentra en una pequeña Rueda J, abrazan dicha Rueda quatro cruzes de madera de sabino que es dificil de Rajarse, y las mismas ciñen al exe el el pie del Peón I, que es una pieza de figura de Perinola, se encaja un Gijo de Hierro que da vuelta en un texuelo de Hierro calzado como el otro, cuya Cabeza entra en una pieza, que llaman cepo M, el qual con unas clavijas está pegado al la Gualdra el grueso del Peón es de media vara, y en el encaja la Rueda K, de cinco o seis varas de diametro, para mayor seguridad, y suavidad del mobimiento, ceñida de otras quatro cruces, y circundada de dientes con los quales va trabando con la rueda K llamada linternilla, la F, significa la tasa donde recibe el agua que basean los Canjilones la qual corre por una Azequia o Canal de 126 varas de yntroduccion que hay desde la bocamina hasta el cañon en qe está armada la maquina, la qual es tan suabe que solo de mueve con el espeque L, en el se pone una pieza llamada Peon a modo de una lanza de coche de que tira la Mula y con un operario que cuide de la mula que baya a paso redondo se saca la cantidad de Agua continua a la que sacan 112 operarios la que se ha benificado en la Mina de Nra. Sra. del Sirne. Sacada e ydeada por Dn Clemente Bravo uno de los mineros de Gualgayoc. *Source*: Author's collection.

cess to the rich ores lying beneath the natural water level. Unfortunately for the investors, whose 1812 contract with the local miners provided for them to be rewarded with variable shares (15-50 percent) of the ores dug from specific locations, the promise of a prolonged bonanza was almost immediately dashed by the occupation of Cerro de Pasco by a patriot army in November 1820, and subsequent fighting in the region over the next four years, which led to the dispersal of workers and the destruction of some of the new machinery. Nevertheless, the setback, although serious, was temporary, and production resumed in 1825, albeit at a low level. Recovery after independence was relatively slow, and production did not surpass the 1820 figure until 1841 (Deustua 1986, 244); on the other hand, once re-established it was substantial, and, as in the late colonial period, underpinned a complex network of economic activity in central Peru throughout the 1830s and 1840s.

The concluding moral, perhaps, is that in the last resort Peruvian investors and miners had a clearer idea than the Spanish monarchy of what was good (and bad) for them in terms of technological innovation.

References

Archival Source

AGI — Archivo General de Indias [Sevilla]
Mapas y Planos, Minas 28.

Published and Other Materials

Anna, T. E. 1979. *The fall of the royal government in Peru*. Lincoln and London: University of Nebraska Press.

Assadourian, C. S. 1982. *El sistema de la economía colonial: mercado interno, regiones y espacio económico.* Lima: Instituto de Estudios Peruanos.

Brown, K. W. 1986. *Bourbons and brandy: Imperial reform in eighteenth-century Arequipa.* Albuquerque: University of New Mexico Press.

Buechler, R. M. 1981. *The mining society of Potosí 1776-1810.* Ann Arbor: University Microfilms International.

Céspedes del Castillo, G. 1947. *Lima y Buenos Aires.* Sevilla: Escuela de Estudios Hispano-Americanos.

Chocano M., M. 1982. *Comercio en Cerro de Pasco a fines de la época colonial.* Lima: Universidad Nacional Mayor de San Marcos.

Deustua, J. 1986. *La minería peruana y la iniciación de la república, 1820-1840.* Lima: Instituto de Estudios Peruanos.

Domínguez Bordona , J. 1936. *Trujillo del Perú a fines de siglo XVIII: Dibujos y acuarelas que mandó hacer el obispo D. Baltasar Jaime Martínez Compañon.* Madrid: Talleres Gráficos C. Bermejo.

Fernández Alonso, S. 1991. *Presencia de Jaén en América: La visita general de Jorge Escobedo y Alarcón al virreinato del Perú en el siglo XVIII (1782-1788).* Jaén: Diputación Provincial.

Fisher, J. R. 1970. *Government and society in colonial Peru.* London: Athlone Press.

——. 1976. *Silver mines and silver miners in colonial Peru, 1776-1824.* Liverpool: Centre for Latin American Studies, University of Liverpool.

——. 1985. *Commercial relations between Spain and Spanish America in the era of free trade, 1778-1796.* Liverpool: Centre for Latin American Studies, University of Liverpool.

——. 1988. El impacto del comercio libre en el Perú. *Revista de Indias* 48:401-420.

——. 1992. *Relaciones económicas entre España y América hasta la independencia.* Madrid: Editorial Mapfre.

Haitin, M. 1986. Urban market and agrarian hinterland: Lima in the late colonial period. In *The economies of Mexico and Peru during the late colonial period, 1760-1810,* ed. N. Jacobsen and H.-J. Puhle, 281-298. Berlin: Colloquium Verlag.

Molina Martínez, M. 1986. *El real tribunal de minería de Lima (1785-1821).* Sevilla: Diputación Provincial.

Montgomery Keelan, S. 1990. The Bourbon mining reform in New Granada, 1784-1796. In *Reform and insurrection in Bourbon New Granada and Peru,* eds. J. R. Fisher, A. J. Kuethe, and A. McFarlane, 41-53. Baton Rouge and London: Louisiana State University Press.

Rivero y Ustáriz, M. E. 1857. *Collección de memorias científicas, agrícolas e industriales.* Brussels: H. Goemaer.

Salas de Coloma, M. 1986. Los obrajes huamanginos y sus interconexiones con otros sectores económicos en el centro-sur peruano a fines del siglo XVIII. In *The economies of Mexico and Peru during the late colonial period, 1760-1810,* eds. N. Jacobsen and H.-J. Puhle, 203-232. Berlin: Colloquium Verlag.

Vargas Ugarte, R., ed. 1956. Informe del tribunal del consulado de Lima, 1790. *Revista Historica* 22:266-310.

Villalobos R., S. 1965. *Comercio y contrabando en el Río de la Plata y Chile.* Buenos Aires: Editorial Universitaria de Buenos Aires.

Whitaker, A. P. 1951. The Elhuyar mining missions and the Enlightenment. *Hispanic American Historical Review* 31:557-585.

Early Petroleum Extraction and Tar-boiling in Coastal Ecuador

**Karen E. Stothert, The University of Texas,
San Antonio, Texas, U.S.A.**

Abstract

Since early in the historic period petroleum has been extracted from hand- dug wells on the Santa Elena Peninsula of Ecuador and on the north coast of Peru. It was refined by boiling and the resultant tar widely used during the colonial era as a naval store, sealant, and medicine. This study of the primitive petroleum industry is based on historical sources, present day informants and archaeological evidence from the Santa Elena Peninsula.
Key Words: Santa Elena, petroleum, tar-boiling, furnaces, Ecuador, Amotape (Peru).

Introduction

This paper describes the evolution of the primitive petroleum industry in coastal Ecuador using multiple lines of evidence. Organization of the early tar works and archaic tar-boiling technologies is reconstructed based on historical, ethnographic and archaeological information. Although tar produced in some modern refineries is a relatively unimportant commodity, this research demonstrates that tar was the first product of the primitive petroleum industry in Ecuador, where its uses and production technology evolved during the early historic period.

Where petroleum emerges at the surface, natural "tar seeps" form as happens at many locations in the New World, including portions of the adjacent coasts of Ecuador and Peru. On the Santa Elena Peninsula of Ecuador (fig. 1), viscous, low gravity petroleum passes through formations and is extruded onto the ground. This crude oil may saturate the earth, lose its volatile fractions, form pools of natural asphalt, or even flow downslope into the sea. These seeps have been prominent features of the landscape since the late Pleistocene, and some contain fossilized remains

---

In Quest of Mineral Wealth: Aboriginal and Colonial Mining and Metallurgy in Spanish America, edited by Alan K. Craig and Robert C. West, 1994. Geoscience and Man, vol. 33, pp. 343-354. Dept. of Geography & Anthropology, Louisiana State University, Baton Rouge, LA 70893-6010, U.S.A.

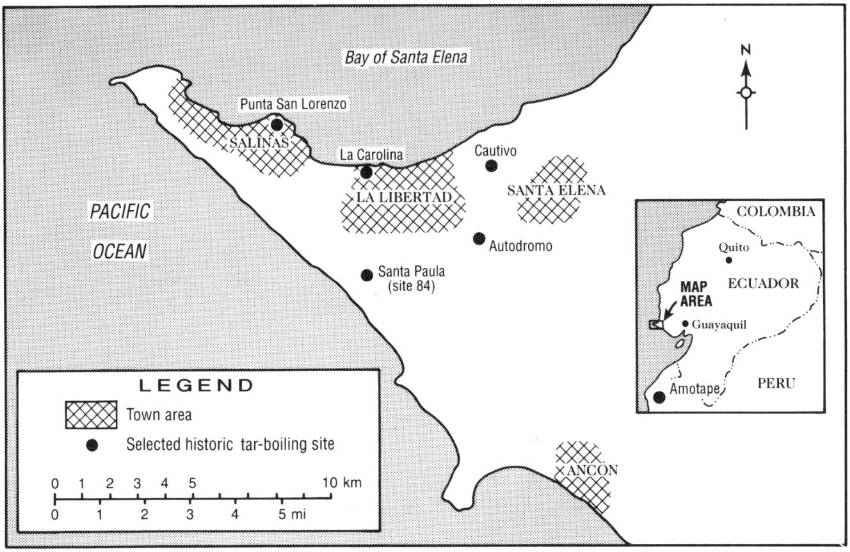

Fig. 1. The Peninsula of Santa Elena.

of Pleistocene animals (Edmund 1965; Hoffstetter 1952; Lemon and Churcher 1961; Spillman 1931, 1948). Today, earth around the seeps of Santa Elena is mined for *tierra brea* — bitumen-soaked earth, used to pave roads and driveways. Modern oil wells which are drilled into deeper formations dot the landscape, but the natural asphalt has been exploited by people for more than four millennia.

Early Petroleum Exploitation in Ecuador and Northern Peru

Prehistoric Valdivia people brought pitch (an indiscriminate synonym for bitumen or asphalt) to their habitation sites in makeshift seashell plates. Guangala and Manteño-Huancavilca people used this substance to manufacture various kinds of shell artifacts. Olaf Holm (pers. comm.) suggests it may have been used to make prehistoric watercraft seaworthy.

After European entrance into Ecuador and Peru, refined tar was produced as a naval store used in construction and maintenance of sailing ships. In the early 1530s, Blas de Atienza reported that Spanish sailors stopped at Santa Elena to take on water and get pitch from the seeps (Szászdi and Léon Borja 1978, 209 and note 8). For the next two centuries ships bound from Lima to Panama frequently stopped at Santa Elena for provisions and repairs. Here Indians were engaged in mining and delivering pitch to the port (Anonymous [1605] 1973, 65). According to González (1988, 253), during the sixteenth century some natives of

Colonche on the Santa Elena Peninsula supplied pitch as their tribute payment. This suggests they were in possession of the tar pits and involved in tar-boiling as were others in northern Peru (Espinoza Claudio 1981, 106-7; Szászdi and Léon Borja 1978, 209).

Most colonial sources for Ecuador dwell on the medicinal applications of *brea* (pitch or tar) and its importance to the shipping industry. Cobo ([1890-] 1956, I:109) remarks on the quantity from Santa Elena, describing its color, odor, and especially the medicinal properties. However, its principal use was to *alquitranar las jarcias* (waterproof the rigging) of ships (Lizárraga [1605] 1968, 55). A secondary use for tar was in the colonial wine and brandy industry of the south coast of Peru where it was used to seal the ceramic bottles (Rice and Smith 1989; Holm 1971, 272).

Eighteenth century *brea* production in Ecuador and Peru was characterized by political and economic fluctuations. The English pirate William Dampier (1906, 58), who raided Santa Elena in 1709, noted bituminous material on the bay shore seeping out just above high-water mark. It was gathered by Indian fishermen who kept it in jars until Spaniards came to use it on their ships. However, by 1741 Alsedo y Herrera's ([1741] 1938, 63) description of the resource indicates that provisioning ships in the area had become more important than tar-boiling. After 1768 the high price paid in Lima for *brea* as a sealant revitalized the industry. Rivalry between Santa Elena and the tar works at Amotape in northern Peru developed in the late 1700s until excessive production again caused the price in Lima to drop dramatically (Macera 1963, 27-28; Laviana Cuetos 1987, 332-34).

The long established government monopoly, poorly administered from Lima, caused commercial problems and the mines were shut down (Lavinia Cuetos 1987, 341-343). In 1787 the mines were reopened with José Antonio Rocafuerte, a Guayaquil merchant, as lessor of both Amotape and Santa Elena. Struggles between private operators and the state continued until the war of Independence in 1821. Macera (1963, 19-23) concludes that the labor situation in northern Peru affected the development of the tar-boiling industry since there were very few Indians in Amotape and few available wage laborers. Requena ([1774] 1984, 69-70) lamented the scarcity of entrepreneurs, but admitted there was considerable risk and that lack of fuel in the arid lands of Santa Elena made the process expensive.

Historical Archaeology of Tar-boiling

Several colonial tar-boiling sites were formerly present on the western Santa Elena Peninsula and in petroleum concessions abutting Santa Paula where sherds of large earthenware vessels were present on the surface until their recent destruction. The only relatively intact tar-boiling locale today is OGSE-84 in the Santa Paula concession, where petroleum was extracted from hand-dug pits until the 1940s.

Descriptions of the early twentieth century industry by present-day informants indicate a pit about 5 meters in diameter was excavated in the surface sand to a depth of 1-2 meters. Once stable strata were reached, a pit 1.5 meters square was dug to a depth of 4-5 meters. Petroleum, dripping like honey, would enter the pit from the walls, forming a pool at the bottom. This was collected in buckets and brought to the surface by means of a giant tripod (fig. 2). When pits became dangerous because of cave-ins, they were abandoned and a new location was selected between old pits. At Santa Paula many of these old pits are now obscured by dunes and erosion. Petroleum spillage around these excavations has stabilized and hardened the surface so that, earlier in this century, it was possible to roll in wooden barrels to fill with crude oil and use donkeys to drag them to the collection station. Sidewalls of these old pits and the backdirt surrounding them contain the remains of industrial rubbish left in still earlier centuries.

One such industrial midden, Site 84, about 50 m in diameter, was selected for archaeological study. Exploration trenches were extended back from the edges of several of the old, circular hand-dug pits. Material excavated included fragments of large tar-boiling jars with carbonized petroleum adhering to the interiors, charcoal and wood ash, fragments of fire-reddened adobe bricks and lumps of unbaked adobe, all indicating the nearby presence of a kiln or furnace.

The excavated charcoal was always in the form of sticks less than 3 cm in diameter. This use of rather small sticks for fuel may reflect the serious lack of more suitable material noted by Requena ([1774] 1984) in his eval-

Fig. 2. A tar pit at Santa Paula photographed in 1917 showing the owner, Sr. Lecaro, on the horse, and workers gathered around a tripod used to remove crude oil from the deep, hand-excavated well.

uation of the industry in the 1770s. During the ninteenth century, southwestern Ecuador experienced a long series of dry years further impacting the rachitic coastal vegetation.

Fragments of earthenware tar-boiling vessels dominate the midden at Site 84, many being discarded inside the abandoned furnaces. Frequently these fragments were filled with carbonized petroleum, likely produced when the crude oil was overcooked. A morphometric study of the ceramic assemblage shows these vessels varied in detail, but conformed closely to certain standard dimensions in order for them to be used successfully on open furnaces. The external ridges may have facilitated transport if they were lashed to poles as shown in a painting (fig. 3) made in the 1780s by the Bishop of Trujillo (Peru), Baltazán Martínez Compañón. The painting depicts a tar-boiling operation near Amotape in northern Peru and illustrates various steps in the production of tar that were probably common to the operation in Santa Elena.

Seven degenerated adobe furnaces (fig. 4) were found in close association with the industrial midden, and five were excavated thus allowing a reconstruction of the structure and its function. The furnaces were arranged in two lines, one group having the firebox oriented in the direction of prevailing winds. The largest and best preserved furnace originally measured 3 m long and 1.7 m wide with preserved portions about 0.7 m above the foundation. When it was abandoned, broken vessels were thrown into the fireboxes and the interior filled with melted adobe, dune sand and fallen bricks. Subsequently this debris became cemented into a matrix.

Figure 5 shows the back and front walls constructed of unfired (but partially fire-reddened) adobe bricks with a standard size of 40x20x5 cm. Compact white ash layers within the walls indicated firebox locations, although their entrance structure can not be reconstructed accurately. Furnace interiors were divided into compartments by several transverse walls or buttresses projecting toward the center from both front and back of the structure. This created 3 or 4 fireboxes open to each other along the furnace mid-line (fig. 6). Since the firebox width corresponds to the outside diameter of the ceramic vessels, it is likely they (or the adjacent walls) were intended to support these containers over the fire. This design, which allowed much heat to escape into the air, seemed wasteful to one of the colonial inspectors who was very critical of the efficiency of the tar-boiling furnaces (Requena [1774] 1984).

In Macera's description of the furnaces at Amotape, tar-boiling jars were mounted over trenches lined with bricks, which reads as if the furnaces were dug into the ground. This method of insulating furnaces extends into the prehistoric of northern Peru (Shimada, Elera and Chang 1991) and is a strategy still used in rural areas in coastal Ecuador.

Fig. 3. A drawing by Bishop Baltazán Martínez Compañón entitled "La Saca y Beneficio de la Brea del Mineral de Amotope," made ca. 1740. *Source*: Martínez Compañón 1985, II. Reproduction from the manuscript in the Biblioteca del Palacio de Madrid.

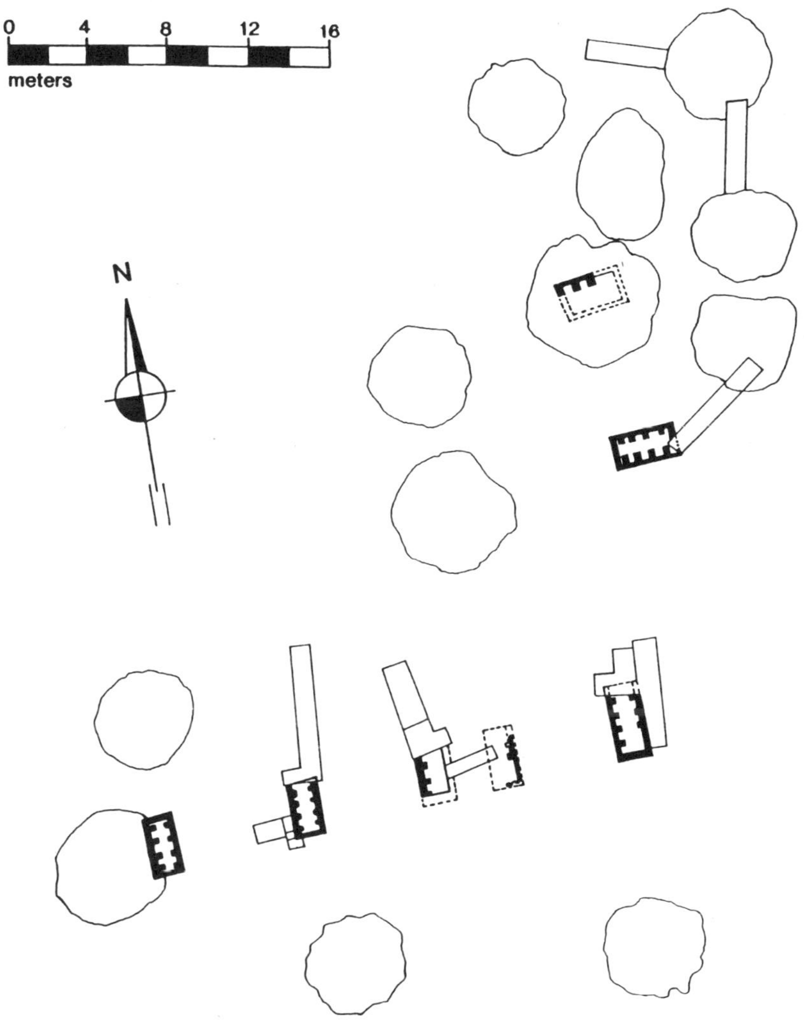

Fig. 4. Plan of Section A of Site 84, showing the excavations (rectangles); seven partially-preserved adobe furnaces (in heavy black); and large circular pits excavated early in the twentieth century to extract petroleum.

Dating Site 84

Petroleum contamination prevents radiocarbon dating of the construction and use of these furnaces at Santa Elena and as yet there is no possibility of cross-dating with the miscellaneous historic artifacts associated with the site. However, similar adobe structures and industrial midden have been studied near the tar seeps at Talara, northern Peru where it is gener-

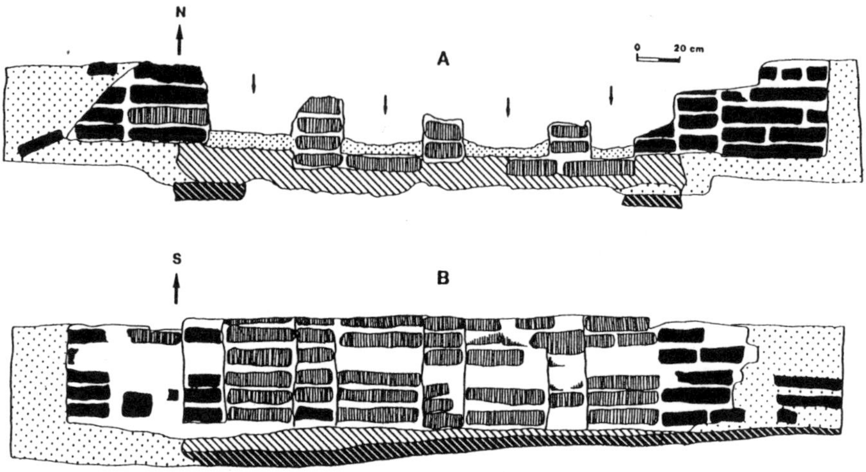

Fig. 5. Drawing of north and south walls of excavated Feature 7, an adobe brick furnace. A. The incomplete south wall showing four entrances into the firebox with concentrations of compact white ash (dotted) inside the apertures; the layers of sand and tar (black hatched) beneath the foundation; areas where the sand has been reddened by heat (light hatched); and the contrast between the dark bricks which are relatively unaltered by fire and the cross-hatched ones which are reddened by exposure to high heat. B. The exterior of the north wall showing the distribution of unbaked (black) and reddened (cross-hatched) adobe bricks, and the sand and tar layers below the structure.

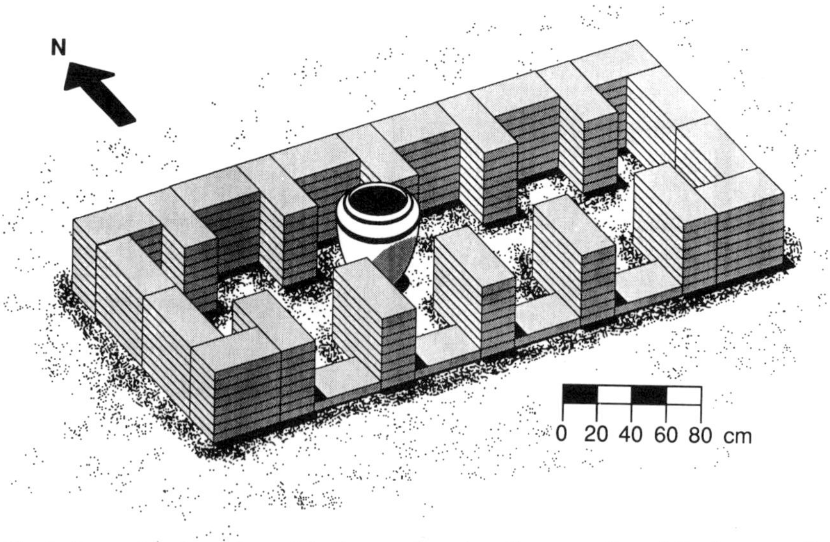

Fig. 6. Reconstruction of an adobe furnace based on the excavation of the lower portion of the walls whose original height is unknown. The tar boiling vessels may actually have rested on the buttresses that jut into the firebox of the furnace.

ally believed they are of the late eighteenth century (Richardson and Décima 1977). Although little comparative study has been done, the ceramic vessels used for boiling the tar at sites in northern Peru and on the Santa Elena Peninsula are very similar (Petersen 1963, fig. 60). Tar works at Santa Paula can be estimated to be late eighteenth century since they are like the ones described by Requena and the furnaces are similar to the ones pictured by Bishop Martínez Compañón in the 1780s (fig. 3).

Since ceramic jars were abandoned for tar-boiling before the 1870s in Ecuador, and perhaps as early as the 1850s in Peru, it seems likely Santa Paula tar works were in use during the first half of the nineteenth century. These furnaces may have been part of the operation developed by Rocafuerte who is reported to have had 300 men working for him at the beginning of the nineteenth century when he also had the concession for tar resources in the Amotape-Talara "mines" of northern Peru. This may not have been the height of the early tar-boiling industry, but Macera (1963) indicates huge quantities of tar were exported from Santa Elena in 1800, 1807 and 1813.

Oral Histories

By interviewing elderly workers who had participated in tar-boiling as young men, we accumulated another body of descriptive data. According to informants who worked for the Carolina Oil Company in the 1920s and 1930s, tar was made on the peninsula before 1931 in the following manner. The *maestro* in charge would fill a 400 gallon metal tank with petroleum, preferably old, well-oxidized *petroleo grueso* and set fire to the surface (see figs. 2 and 7, both dated 1917). After about three hours of burning, the product was tested by dipping out a small quantity with a very long handled spoon. This sample was dropped into a water filled hole in the ground where it congealed into a ball. The tar was judged sufficiently cooked if material from the broken ball was glassy. If it was still gummy, more firing was needed. When the batch was ready, it needed to be extinguished quickly. This was done by going up-wind from the flames and pouring water from a large wooden barrel onto the surface of the vat (one informant remembered a metal lid that could be slid onto the tank to extinguish the fire). Later the cooled tar was packed for transport to Guayaquil in wooden cases which originally contained lard.

Another refining technique involved digging a hole in the ground and mounting the boiling vessel on top. By installing a gas burner in the hole and a distillation device on top, low octane gasoline was produced in small amounts. This technology is similar to that used today in salt works and sugar-making by rural families in the Caribbean and South America. There seems to have been considerable variability in firing techniques,

Fig. 7. Santa Paula in 1917, showing the owner, Sr. Lecaro, in the white hat. The person in charge of tar-boiling is holding a long "spoon" and stands near the burning petroleum. The large barrels hold water for extinguishing the blaze.

showing the "frontier ingenuity" of people with limited resources who undertake industrial processes.

Summary and Conclusions

Artisanal petroleum exploitation in coastal Ecuador and northern Peru during the historic period has been reconstructed by using historical documents, collecting oral histories, and analysis of excavated materials from Site 84 where tar-boiling was conducted in the late eighteenth and early nineteenth centuries. The traditional technology involved boiling crude oil in ceramic vessels mounted on adobe furnaces. Iron vessels replaced ceramic jars in the mid-nineteenth century and by 1917 huge iron vats were in use. In the early twentieth century, "refining" consisted of directly igniting the surface of the petroleum, thus eliminating the need for furnaces and scarce firewood to stoke them, as well as reducing labor costs.

Ownership of the crude oil resource remained in native hands during the early colonial period until it was declared a monopoly of the Spanish Crown. Thereafter it was exploited by a series of white and creole lessors until more recent times when foreign companies as well as Ecuadorian entrepreneurs have gained concessions from the government to operate the "mines" of Santa Elena.

While tar was the chief product of the primitive petroleum industry during the colonial era, it became a relatively unimportant product in this century. Small quantities produced in a modern oil refinery continued to be exported from Santa Elena until the 1940s.

Acknowledgements

This research was sponsored by the Museo Antropológico, Dr. Olaf Holm, Director, Banco Central del Ecuador, Guayaquil. Other individuals to whom I am indebted for their assistance include Dr. Jorge Marcos, Alexandra Izyak, Roberto Lindao Quimi, Dr. James B. Richardson III, Dr. James H. Burton, Ing. Iván Abad, Jorge Racines Sarmiento, Dr. Julio Estrada, Carlos Lecaro, Stanley Moss, Robert Norris , Manuel Lainez, and Guillermo Pareja Cucalón.

References

Alsedo y Herrera, D. de. [1741] 1938. *Compendio histórico de la Provincia de Guayaquil*. Guayaquil: Imprenta Gutenberg de Elicio A. Uzcategui.

Anonymous. [1605] 1973. Descripción de la gobernación de Guayaquil (1605). *Revista del Archivo Histórico del Guayas*. 4:55-94.

Cobo, B. [1890-93; Ms. completed 1653] 1956. *Historia del Nuevo Mundo*. Biblioteca de Autores Españoles, 2 vols. (91 and 92). Madrid: Ediciones Atlas.

Dampier W. 1906. *The Voyages of Captain William Dampier*. John Mansefield, ed. 2 vols. London: E. Grant Richards.

Edmund, A. G. 1965. *A late Pleistocene fauna from the Santa Elena Peninsula, Ecuador*. Life Sciences Contribution, No. 6. Toronto: Royal Ontario Museum, University of Toronto.

Hoffstetter, R. 1952. *Les mammifères pleistocene de la République de L'Equateur*. Paris: Mémoires de la Société Géologique de France, 66:1-391.

Holm, O. 1971. La cerámica colonial del Ecuador. *Boletín de la Academia Nacional de Historia*, No. 116 (julio-diciembre de 1970).

Laviana Cuetos, M. L. 1987. *Guayaquil en el siglo XVIII: Recursos naturales y desarrollo económico*. Sevilla: Escuela de Estudios Hispano-Americanos.

Lemon, R. R. H. and Churcher, C. S. 1961. Pleistocene geology and paleontology of the Talara Region, Northwest Peru. *American Journal of Science*, 259:410-429.

Lizárraga, Fray R. de. [1605] 1968. *Descripción breve de toda la tierra del Perú, Tucuman, Río de la Plata, y Chile*. Biblioteca de Autores Españoles, vol. 216. Madrid: Ediciones Atlas.

Macera Dall'Orso, P. 1963. *Historia del petróleo peruano, I, Las breas coloniales del siglo XVIII*. Lima.

Petersen G., G. 1963. La industria petrolera del Perú en su primer centenario, 1863-1963: Exploración y afloramiento de petróleo. *Boletin de la Empresa Petrolera Fiscal* 6:1-198.

Requena, F. [1774] 1984. La descripción de Guayaquil. In *Francisco Requena y su descripción de Guayaquil*, M. Laviana Cuetos , ed. 25-123. Sevilla: Escuela de Estudios Hispano-Americanos.

Rice, P. M. and Smith, G. C. 1989. The Spanish colonial wineries of Moquegua, Peru. *Historical Archaeology* 23(2): 41-49.

Richardson III, J. B. and Décima Zamecnik, E. 1977. The Talara Tar Pits: 11,000 Years of Human Utilization. Paper presented at the Society for American Archaeology, New Orleans.

Shimada, I., Elera, C. G. and Chang, V. 1991. Cupisnique ceramic kilns and production: excavation and experimental data. Paper presented at the 56th Annual Meeting of the Society for American Archaeology, New Orleans.

Spillman, F. 1931. *Die Säugetiere Ecuadors im Wandel der Zeit.* Quito: Universidad Central.

——. 1948. Beiträge zur Kenntnis einer gravigraden Reisensteppentieres (*Eremotherium carolinense* gen. et sp. nov.), seines Lebensräumes und seiner Lebensweise. *Paleobiologica* 8:231-279.

Szászdi, A. and León Borja, D. 1978. Los recursos y el desarrollo económico de Guayaquil, 1535-1605. In *Wirtsschaftskräfte und Wirtsschaftswege*, vol. 4 : Übersee und allgemeine Wirtsschaftsgeschichte, J. Schneider et al., eds. , 201-217. Nurnberg: Kommission bei Klett-Cotta.

FROM GEOSCIENCE PUBLICATIONS

GEOSCIENCE AND MAN

Vol. 5: **Man and Cultural Heritage: Papers in Honor of Fred B. Kniffen**, edited by H. J. Walker and William G. Haag, 1974, 236 pages. $12.50.

Vol. 6: **Ostracoda: The Henry V. Howe Memorial Volume**, edited by W. A. van den Bold, 1974, 182 pages. $15.00.

Vol. 8: **Aspects of Trinity Division Geology**, edited by Bob F. Perkins, 1974, 228 pages. $11.50.

Vol. 10: **Grasslands Ecology: A Symposium**, edited by R. H. Kesel, 1975, 50 pages. $9.50.

Vol. 12: **Coastal Resources**, edited by H. J. Walker, 1973, 127 pages. $6.00.

Vol. 16: **The Mississippi River Delta: Legal Geomorphologic Evaluation of Historic Shoreline Changes**, by David Joel Morgan, 196 pages. $15.00.

Vol. 17: **Poverty Point Culture**, by Clarence H. Webb, 1982, 2 ed. rev., 86 pages. $25.00.

Vol. 18: **Research Techniques in Coastal Environments**, edited by H. J. Walker, 1977, 320 pages. $18.00.

Vol. 19: **Man and Environment in the Lower Mississippi Valley**, edited by Sam B. Hilliard, 1978, 165 pages. $12.00.

Vol. 20: **Phytoplankton of the Gulf of Mexico: Taxonomy of Calcareous Nannoplankton**, edited by R. W. Pierce and G. F. Hart, 1979, 97 pages. $12.00.

Vol. 21: **Historical Geography of Latin America: Papers in Honor of Robert C. West**, edited by William V. Davidson and James J. Parsons, 1980, 163 pages. $14.00.

Vol. 22: **Traces of Prehistory: Papers in Honor of William G. Haag**, edited by Frederick H. West and Robert W. Neuman, 1981, 134 pages. $14.00.

Vol. 23: **Historical Archaeology of the Eastern United States: Papers from the R. J. Russell Symposium**, edited by Robert W. Neuman, 1983, 69 pages. $10.00.

Vol. 24: **Place: Experience and Symbol**, edited by Miles Richardson, 1984, 80 pages. $10.00.

Vol. 25: **The American South**, edited by Richard L. Nostrand and Sam B. Hilliard, 1988, 167 pages. $25.00.

Vol. 26: **Tojolabal Maya: Ethnographic and Linguistic Approaches**, edited by M. Jill Brody and John S. Thomas, 1988, 74 pages. $18.00.

Vol. 27: **Cultural Diffusion and Landscapes: Selections by Fred B. Kniffen**, edited by H. Jesse Walker and Randall A. Detro, 1990, 80 pages. $22.00.

Vol. 28: **Pioneers of Modern Geography: Translations Pertaining to German Geographers of the Late Nineteenth and Early Twentieth Centuries**, translated and edited by Robert C. West, 1990, 200 pages. $30.00.

Vol. 29: **Poverty Point Culture: Local Manifestations, Subsistence Practices, and Trade Networks**, edited by Kathleen M. Byrd, 1991, 203 pages. $35.00.

Vol. 30: **The Uneven Landscape: Geographical Studies in Post-Reform China**, edited by Gregory Veeck, 1991, 276 pages. $14.95.

Vol. 31: **Person, Place and Thing: Interpretative and Empirical Essays in Cultural Geography**, edited by Shue Tuck Wong, 1992, 440 pages. $25.00.

Vol. 32: **Culture, Form, and Place: Essays in Cultural and Historical Geography**, edited by Kent Mathewson, 1993, 356 pages. $24.95.

Vol. 33: **In Quest of Mineral Wealth: Aboriginal and Colonial Mining and Metallurgy in Spanish America**, edited by Alan K. Craig and Robert C. West, 1994, 360 pages. $27.00

THE FRED B. KNIFFEN CULTURAL LABORATORY MONOGRAPHS

No. 1: **Louisiana's Remarkable French Vernacular Architecture**, by Jay D. Edwards. 2nd edition, revised and enlarged. Forthcoming.

No. 2: **Historic Louisiana Nails: Aids to the Dating of Old Buildings**, by Jay D. Edwards and Tom Wells, 1994, 84 pages. $18.00.

No. 3: **Historic Maps of Louisiana: An Annotated Bibliography**, compiled by Joyce Nelson and Anne Stanton. Forthcoming.

No. 4. **Watercolor Paintings from St. Charles Parish, by Father Paret,** by Marcel Boyer. Forthcoming.

BOUNDARY FILES

In Atlas*Graphics

China County Boundary Files, by Nina S. Lam, also available in Arc/Info. $150.00, $5.00 postage and handling.

China Prefecture Map, by Nina S. Lam. $150.00, $5.00 postage and handling.

Historical United States County Boundary Files, 1850-1970, by Carville Earle and Changyong Cao. $250.00, $2.00 postage and handling.

Single disk files: all $50 plus $2 postage and handling

Kampuchea Provincial Boundary File.
Laos County Boundary File (2nd order sub-provincial units).
Laos Provincial Boundary File.
Pakistan County Boundary Files, by Russell Hires and Gregory Veeck.
Pakistan State Boundary File.
Philippines County Boundary Files, by Russell Hires and Gregory Veeck.
Philippines Provincial Boundary File.
Russia Oblast Boundary File, by Dmitry Mesyanzhinov and Greg Veeck.
Vietnam "County" Boundary Files, (2nd order sub-provincial units), by Russell Hires and Greg Veeck.
Vietnam Provincial Boundary File.

VIDEOTAPE

Bringing the Past Alive: Conversations between Bill Haag and George Quimby on WPA Archaeology in Louisiana. 2 hours. $20.00.

LINGUISTICS JOURNAL

The Journal of Mayan Linguistics, edited by M. Jill Brody. Vols. 8 and 9 in production, rate to be announced. Backcopies of nos. 1 and 2 of vols. 1 to 5, vols. 6 and 7: $12.00 each.

OTHER WORKS FROM GEOSCIENCE PUBLICATIONS

A Biographical Dictionary of American Geography in the Twentieth Century, compiled by Gary S. Dunbar, 1995, 2nd. ed. , revised and enlarged. $12.00.

Atlas of Louisiana Surnames of French and Spanish Origin, by Robert Cooper West, 1986, 212 pages. $37.50.

Atlas of Louisiana: A Guide for Students, by Milton B. Newton, Jr., 1972, 200 pages. $12.00.

Environment and Culture, compiled by H. J. Walker and M. B. Newton, Jr., 1978, 259 pages. $5.00.

Journal of John Landreth Surveyor: An Expedition to the Gulf Coast, November 15, 1818-May 19, 1819, edited by Milton B. Newton, Jr., 1985, 204 pages. $20.00. Also available in hard cover, $30.00.

Plantation Traits in the New World, edited by R. E. Chardon, 1983, 70 pages. $8.00.

Climatic Perspective of Louisiana Floods 1982-1983, edited by Robert A. Muller and Gregory E. Faiers, 1984, 48 pages. $4.00.

New Orleans Weather 1961-1980: A Climatology by Means of Synoptic Weather Types, by Robert A. Muller and James E. Willis, 1983, 70 pages. $8.00.

Louisiana Maps

$5 rolled, $3 folded

Each map about 22" x 33", except where noted

Energy Consumption 1940-1980.
Population Change, 1970-1980.
Louisiana Climate and Weather, 1951-1980.
Louisiana, An Historical Portrait, 1699-1812.
Respiratory Cancer Death Rates, 1950-1979.
Louisiana Commercial Fisheries Revenue, 1984.
Louisiana's Natural Heritage, 1984-1986.
Selected Agricultural Products, 1988.
The Course of the River Mississippi, 1765 (antique, 8" x 26").

mélanges

$3.00 each

No. 1: **Louisiana in North American Prehistory**, by William G. Haag, 1971, 45 pages.

No. 2: **Louisiana House Types: A Field Guide**, by Milton B. Newton, Jr., 1971, 45 pages.

No. 4: **Stewards of the Past**, by Charles R. McGimsey III, Hester A. Davis, and Carol Chapman, 1972, 21 pages.

No. 7: **Country and Small Town Stores of Louisiana: Legacy of the Greek Revival and the Frontier**, by Linda Pulliam and M. B. Newton, Jr., 1973, 11 pages.

No. 10: **Reconstructing the Forest Primeval, West Feliciana Parish**, by Hazel R. Delcourt, 1975, 14 pages.

No. 11: **An Archaeological Assessment of Coastal Louisiana**, by Robert W. Neuman, 1977, 44 pages.

No. 12: **The Spanish Moss Folk Industry of Louisiana**, by Fred B. Kniffen and Malcolm L. Comeaux, 1979, 19 pages.

No. 13: **The Lake St. Agnes Site: A Multi-Component Occupation of Avoyelles Parish, Louisiana**, by Alan Toth, 1979, 51 pages.

All books are paperbound unless otherwise indicated.

For orders or information, contact:
GEOSCIENCE PUBLICATIONS
Department of Geography and Anthropology
Louisiana State University
P.O. Box 16010
Baton Rouge, LA 70893

PHONE: (504) 388-6245 • FAX: 504-388-2912 • EMAIL: GAWILC@LSUVM.SNCC.LSU.EDU